Methods in Molecular Biology

Series Editor
John M. Walker
School of Life and Medical Sciences
University of Hertfordshire
Hatfield, Hertfordshire, AL10 9AB, UK

For further volumes:
http://www.springer.com/series/7651

Microfluidic Electrophoresis

Methods and Protocols

Edited by

Debashis Dutta

Department of Chemistry, University of Wyoming, Laramie, WY, USA

Humana Press

Editor
Debashis Dutta
Department of Chemistry
University of Wyoming
Laramie, WY, USA

ISSN 1064-3745 ISSN 1940-6029 (electronic)
Methods in Molecular Biology
ISBN 978-1-4939-8963-8 ISBN 978-1-4939-8964-5 (eBook)
https://doi.org/10.1007/978-1-4939-8964-5

Library of Congress Control Number: 2018962870

This Humana Press imprint is published by the registered company Springer Science+Business Media, LLC, part of Springer Nature.
The registered company address is: 233 Spring Street, New York, NY 10013, U.S.A.

Preface

Microfluidic electrophoresis has been gaining increased attention in academia and industry for sample analysis and purification applications since its first demonstration in the early 1990s due to the inherent advantages of miniaturizing charge-based separation techniques. The use of narrow analysis columns in electrophoretic assays, for example, is known to reduce Joule heating effects and band-broadening contributions and allow for narrow sample injections besides cutting down on the sample volume requirement. Moreover, microfluidic devices offer the opportunity to integrate a variety of sample preparation and analyte detection methods to the separation process on the same footprint permitting the realization of faster and high-efficiency assays at reduced costs. The ability to precisely manipulate fluid and analyte transport around micro-/nanoscale structures in these systems also makes it possible to exploit several unique physical processes prominent at shorter length scales leading to the development of novel functionalities often not realizable in capillary-based or macroscale instruments. Finally, microfluidic devices have been shown to improve portability and reduce instrumentation costs for electrophoretic assays besides presenting the potential for massively parallelizing and automating charge-based separations for high-throughput applications.

While the development of a vast number of microchip-based electrophoretic techniques over the past three decades has significantly advanced this separation field, limited effort has been invested in standardizing such assays. As a result, the ability to apply microfluidic electrophoresis to analyzing "real-world" samples still remains a challenge in most situations. With increased emphasis on translating academic research into practical technologies, several microfluidics-based separation platforms have emerged in the market recently. In spite of these developments, there is a strong need for literature that describes standardized approaches to implementing microfluidic electrophoresis both to broaden its utility in academia and industry and to allow for better comparison of such assays across different analytical laboratories. This book seeks to contribute to the noted standardization effort by providing a set of protocols necessary for the development of a variety of microchip-based electrophoretic assays. It compiles discussion on a range of such electrophoretic methods by some of the leading researchers in the field, covering subjects such as microfluidic device fabrication, on-chip sample preparation, theoretical/simulation protocols for assessing these separation methods, as well as common practices followed when applying them to important real-world applications. The contents of the book range from protocols for classical assays to those involving pioneering separation techniques recently developed by the scientific community for advancing our analytical capabilities. The discussion of these techniques has been carefully structured to be helpful for education and training purposes as well as for inspiring scientists in an effort to design the next generation of electrophoresis technologies. What you need to know and how to do it: each protocol offers step-by-step instructions, including an introductory overview of the technique, a list of materials and reagents required, as well as helpful tips and troubleshooting advice. Insightful reviews along with advice on how to successfully develop microchip-based electrophoretic assays make this volume indispensable reading for scientists entering the field as well as providing a reference text for those already established. Due to the multidisciplinary nature of the field, little background knowledge is assumed, providing an accessible text for academic researchers as well as practicing engineers, biochemists, and analytical laboratory professionals.

This book is organized into four sections of which the first one, titled "Microfluidic Device Fabrication, Sample Preparation, and Detection Protocols," contains established protocols for fabricating and operating microfluidic devices employed in electrophoresis applications. The second section, titled "Protocols for Classical and Nonclassical Microfluidic Electrophoresis Methods," discusses procedures for realizing traditional and cutting-edge electrophoretic separation methods developed on the microchip platform. Theoretical and simulation protocols often relied on for assessing and optimizing electrophoretic separations on microchips have been described in the third section, while the final section compiles protocols for specific applications of microfluidic electrophoresis to scientific problems of significant interest.

Laramie, WY, USA *Debashis Dutta*

Contents

Preface .... *v*
Contributors .... *ix*

1 Fabrication of Glass Microfluidic Devices 1
Christopher T. Culbertson, Jay Sibbitts, Kathleen Sellens, and Shu Jia

2 Soft Lithography, Molding, and Micromachining Techniques for Polymer Micro Devices 13
Ashis Kumar Sen, Abhishek Raj, Utsab Banerjee, and Sk Rameez Iqbal

3 Sample Injection Techniques 55
James M. Karlinsey

4 Sample Preconcentration Protocols in Microfluidic Electrophoresis 65
Fumihiko Kitagawa and Koji Otsuka

5 Microchip Electrophoresis Containing Electrodes for Integrated Electrochemical Detection 79
Lucas Paines Bressan, Dosil Pereira de Jesus, Dulan Bandara Gunasekara, Susan Marie Lunte, and José Alberto Fracassi da Silva

6 Micellar Electrokinetic Chromatography 87
Braden C. Giordano, Ronald Siefert, and Greg E. Collins

7 Microchip Isotachophoresis: Analysis of Pharmaceuticals 99
Marián Masár and Jasna Hradski

8 Microfluidic Free-Flow Isoelectric Focusing with Real-Time pI Determination 113
Stefan Nagl

9 Nanochannel Gradient Separations 125
Michael A. Startsev and David W. Inglis

10 Paper-Based Electrophoresis Microchip as a Powerful Tool for Bioanalytical Applications 133
Cyro L. S. Chagas, Thiago M.G. Cardoso, and Wendell K. T. Coltro

11 Band Broadening Theories in Capillary Electrophoresis 143
Sandip Ghosal

12 Estimating Stream Broadening in Free-Flow Electrophoretic Systems Based on the Method-of-Moments Formulation 167
Debashis Dutta

13 Microchip Electrophoresis Tools for the Analysis of Small Molecules 197
Federico J. V. Gomez and María Fernanda Silva

14 Integrated Microfluidic System for Rapid DNA Fingerprint Analysis: A Miniaturized Integrated DNA Analysis System (MiDAS)—Swab Sample-In to DNA Profile-Out 207
Jianing Yang, Cedric Hurth, Alan Nordquist, Stan Smith, and Frederic Zenhausern

15 Achieving Stable Electrospray Ionization Mass Spectrometry Detection from Microfluidic Chips 225
Iulia M. Lazar

16 Microchip-Based Electrophoretic Separations with a Pressure-Driven Backflow 239
Ling Xia and Debashis Dutta

Index *251*

Contributors

UTSAB BANERJEE • *Department of Mechanical Engineering, Indian Institute of Technology Madras, Chennai, India*

LUCAS PAINES BRESSAN • *Chemistry Institute, State University of Campinas, Campinas, SP, Brazil*

THIAGO M. G. CARDOSO • *Institute of Chemistry, Federal University of Goias, Goiania, GO, Brazil*

CYRO L. S. CHAGAS • *Institute of Chemistry, Federal University of Goias, Goiania, GO, Brazil*

GREG E. COLLINS • *Chemistry Division, U.S. Naval Research Laboratory, Washington, DC, USA*

WENDELL K. T. COLTRO • *Institute of Chemistry, Federal University of Goias, Goiania, GO, Brazil*

CHRISTOPHER T. CULBERTSON • *Department of Chemistry, Kansas State University, Manhattan, KS, USA*

JOSÉ ALBERTO FRACASSI DA SILVA • *Chemistry Institute, State University of Campinas, Campinas, SP, Brazil*

DOSIL PEREIRA DE JESUS • *Chemistry Institute, State University of Campinas, Campinas, SP, Brazil*

DEBASHIS DUTTA • *Department of Chemistry, University of Wyoming, Laramie, WY, USA*

SANDIP GHOSAL • *Department of Mechanical Engineering and Engineering Sciences and Applied Mathematics, Northwestern University, Evanston, IL, USA*

BRADEN C. GIORDANO • *Chemistry Division, U.S. Naval Research Laboratory, Washington, DC, USA*

FEDERICO J. V. GOMEZ • *Instituto de Biología Agrícola de Mendoza (IBAM-CONICET), Facultad de Ciencias Agrarias, Universidad Nacional de Cuyo, Mendoza, Argentina*

DULAN BANDARA GUNASEKARA • *Department of Chemistry, University of North Carolina at Chapel Hill, Chapel Hill, NC, USA*

JASNA HRADSKI • *Department of Analytical Chemistry, Faculty of Natural Sciences, Comenius University in Bratislava, Bratislava, Slovakia*

CEDRIC HURTH • *Center for Applied NanoBioscience and Medicine, University of Arizona College of Medicine—Phoenix, Phoenix, AZ, USA*

DAVID W. INGLIS • *School of Engineering, Macquarie University, Sydney, NSW, Australia*

SK RAMEEZ IQBAL • *Department of Mechanical Engineering, Indian Institute of Technology Madras, Chennai, India*

SHU JIA • *Department of Chemistry, Kansas State University, Manhattan, KS, USA*

JAMES M. KARLINSEY • *Department of Chemistry, Penn State Berks, Reading, PA, USA*

FUMIHIKO KITAGAWA • *Department of Frontier Materials Chemistry, Graduate School of Science and Technology, Hirosaki University, Hirosaki, Japan*

IULIA M. LAZAR • *Biological Sciences, Virginia Tech, Blacksburg, VA, USA*

SUSAN MARIE LUNTE • *Ralph N. Adams Institute for Bioanalytical Chemistry, University of Kansas, Lawrence, KS, USA*

MARIÁN MASÁR • *Department of Analytical Chemistry, Faculty of Natural Sciences, Comenius University in Bratislava, Bratislava, Slovakia*

Stefan Nagl • *Department of Chemistry, The Hong Kong University of Science and Technology, Kowloon, Hong Kong SAR, China*
Alan Nordquist • *Center for Applied NanoBioscience and Medicine, University of Arizona College of Medicine—Phoenix, Phoenix, AZ, USA*
Koji Otsuka • *Department of Material Chemistry, Graduate School of Engineering, Kyoto University, Kyoto, Japan*
Abhishek Raj • *Department of Mechanical Engineering, Indian Institute of Technology Madras, Chennai, India*
Kathleen Sellens • *Department of Chemistry, Kansas State University, Manhattan, KS, USA*
Ashis Kumar Sen • *Department of Mechanical Engineering, Indian Institute of Technology Madras, Chennai, India*
Jay Sibbitts • *Department of Chemistry, Kansas State University, Manhattan, KS, USA*
Ronald Siefert • *Chemistry Department, United States Naval Academy, Annapolis, MD, USA*
María Fernanda Silva • *Instituto de Biología Agrícola de Mendoza (IBAM-CONICET), Facultad de Ciencias Agrarias, Universidad Nacional de Cuyo, Mendoza, Argentina*
Stan Smith • *Center for Applied NanoBioscience and Medicine, University of Arizona College of Medicine—Phoenix, Phoenix, AZ, USA*
Michael A. Startsev • *School of Engineering, Macquarie University, Sydney, NSW, Australia*
Ling Xia • *Department of Applied Chemistry, Sun Yat-Sun University, Guangzhou, People's Republic of China*
Jianing Yang • *Center for Applied NanoBioscience and Medicine, University of Arizona College of Medicine—Phoenix, Phoenix, AZ, USA*
Frederic Zenhausern • *Center for Applied NanoBioscience and Medicine, University of Arizona College of Medicine—Phoenix, Phoenix, AZ, USA*

Chapter 1

Fabrication of Glass Microfluidic Devices

Christopher T. Culbertson, Jay Sibbitts, Kathleen Sellens, and Shu Jia

Abstract

This chapter provides step-by-step procedures for the fabrication of glass-based microfluidic devices. These procedures include device design, photomask generation, photolithography, channel etching, and high-temperature bonding.

Key words Microfluidics, Glass bonding, Etching, Photolithography, Channel manifold

1 Introduction

The first microfluidic devices were reported about 25 years ago and fabricated from glass [1–5]. Glass was chosen as the substrate material because of its well-characterized and advantageous physical properties, including high electrical resistivity, high thermal conductivity, and broad light transmissivity, and because of its well-known surface chemistry and chemical resistance. The initial operations performed on these devices also paralleled the electrophoretic separations in capillary electrophoresis, and so the use of glass substrates made the transfer of separation and coating methodologies easier.

Since these initial publications, the fabrication of microfluidic devices from a variety of other substrate materials has been reported including poly(dimethylsiloxane) (PDMS) [6], poly(methylmethacrylate) (PMMA), polycarbonate (PC), cyclic olefin copolymers (COC), polystyrene (PS), and silicon. While these other materials, especially PDMS, have become more common, many of them have lower thermal conductivities, poorer optical properties, and poorer chemical resistances than glass. In addition, the surface chemistries are less consistent, and the surfaces are generally more difficult to modify. As such, glass remains a common and important substrate material for fabrication; thus, several commercial firms offer design and fabrication services for glass devices.

Debashis Dutta (ed.), *Microfluidic Electrophoresis: Methods and Protocols*, Methods in Molecular Biology, vol. 1906, https://doi.org/10.1007/978-1-4939-8964-5_1, © Springer Science+Business Media, LLC, part of Springer Nature 2019

The channel manifold design layout, the transfer of the design to the glass substrate, and the etching of the glass substrate to form the actual channel manifold were all borrowed from the microelectronics industry, as indicated in the first reports [1–5]. The etching of the channel manifold, however, only results in a trough in the glass. In order to form an enclosed channel structure, the substrate with the etched pattern needs to be bonded to another flat piece of glass. Several bonding methods have been reported, but the most robust remains the use of high temperatures near the annealing point of glass, and so this approach is the focus of this chapter.

In general, the use of three major classes of glass for the fabrication of microfluidic devices has been reported: fused silica, borosilicate, and soda lime. Fused silica provides the best optical transparency and most consistent surface properties but, in addition to being expensive, requires an annealing (bonding) temperature of ~1140 °C, which is taxing on most ovens. As such, it is rarely used. Borosilicate (crown) glasses are much more economical, have good optical transparency down to ~320 nm, and only require annealing temperatures of ~565 °C. Even cheaper are soda-lime glasses that are transparent down to ~350 nm and have similar annealing temperatures compared to borosilicate glasses. Because the thermal expansion rates of the different types of glasses are different, different types of glasses cannot be annealed together. Etch rates also depend upon both the glass type and the composition of the etching solution. Fused silica glasses may etch up to ten times slower than borosilicate and soda-lime glasses.

The procedures below are specific to the fabrication of microfluidic devices from borosilicate and soda-lime glasses. They provide a detailed, step-by-step explanation of methods commonly reported in the literature, with improvements in terms of lowering costs and streamlining the fabrication. Two variations of the procedure are offered. In the first, a 4″ square borosilicate photomask blank (a glass plate pre-coated with chrome and photoresist) is used to etch eight chips simultaneously, while the second focuses on the use of standard microscope slides.

There are also previous MMB published protocols [7, 8] that are variations on the protocols reported here, and depending upon the glass and bonding method chosen, they may be more or less appropriate than the method described below.

Safety Precautions with HF. Hydrofluoric acid is particularly dangerous. It contains fluoride ions that can easily penetrate the skin and cause the destruction of deep tissue layers including the bone. Unlike other acids that are easily neutralized, the activity of the F^- ion can continue for days. As such, prevention of exposure should be paramount, and the appropriate personal protective gear should always be worn. Over 15 years of glass etching, we have never experienced an exposure incident; nonetheless, should one occur all personnel in the lab should be aware of the proper first aid response and how to clean up any spills.

A brief summary of first aid is given in the Appendix at the end of the chapter, but each investigator should have their own first aid and cleanup protocols approved by their local safety and hygiene officer. More detailed examples of safety protocols can be found online.

2 Materials

2.1 Major Lab Equipment Needed

1. UV flood exposure system (e.g., OAI (San Jose, CA) or ThermoOriel (Stratford, CT)) (*see* **Note 1**).
2. Radiometer/photometer with detector (e.g., Model IL1400A (radiometer/photometer), Model XRL340B (detector); International Light; Newburyport, MA).
3. Photoresist spinner (e.g., Model WS-400A-6NPP/LITE; Laurell; North Wales, PA) (*see* **Note 2**).
4. Custom chuck for coating glass slides (e.g., embedded vacuum chuck; Laurell; North Wales, PA) can be ordered special to fit various sizes of glass slide.
5. Surface profiler (KLA-Tencor, Milpitas, CA) (allows one to measure the channel depth).
6. Sonicator bath (e.g., 3510 Ultrasonic Cleaner; Branson; Danbury, CT).
7. Source for 18 MΩ·cm water (e.g., E-Pure System; Barnstead; Dubuque, IA).
8. Clean hood (bench) to provide a clean area for bonding (e.g., Purifier Horizontal Clean Bench; Labconco; Kansas City, MO) (*see* **Note 3**).
9. Dicing saw (if using photomask blanks as a substrate material) (e.g., Model EC-400; MTI Corp.; USA; inexpensive but more than adequate to the task) (*see* **Note 4**).
10. Air knife (optional) but makes drying the chips easier (e.g., Standard Air Knife; Exair; Cincinnati, OH).
11. Drying oven (e.g., Isotemp Vacuum Oven Model 280A; Fisher Scientific; Pittsburgh, PA).
12. Programmable high-temperature oven (e.g., Copper 120V small annealer; Evenheat; Caseville, MI).
13. Stirring plates (e.g., Isotemp Stirring Hotplate; Fisher Scientific; Pittsburgh, PA).

2.2 Consumable Materials

1. Substrate material (one can use 4″ square pre-coated glass photomask blanks or standard microscope slides):
   (a) 4″ in × 4″ (10.16 cm × 10.16 cm × 0.16 cm) soda-lime (or borosilicate or B-270) photomask blanks coated with chrome and AZ positive-tone photoresist (e.g., Telic Co.; Santa Monica, CA) (*see* **Note 5**).
   (b) 1″ × 3″ or 2″ × 3″ glass slides (e.g., Fisher Scientific).
2. Clean, inert air source (e.g., cylinder of N_2, He, or Ar).
3. 2-L polypropylene containers with lids to hold the etching and developing solutions (e.g., Nalgene Polypropylene Jar; Fisher Scientific; Pittsburgh, PA).
4. Cleaning detergent, e.g., Versa-Clean Liquid soap solution (Fisher Scientific; Pittsburgh, PA).
5. Cleanroom swabs (e.g., Texwipe Microdenier Swab; Fisher Scientific; Pittsburg, PA).
6. Positive-tone photoresist (for coating slides, e.g., AZ P4620; AZ Electronic Materials; Branchburg, NJ).
7. AZ 400 K Developer (Shipley Co.; Marlborough, MA).
8. Chrome Mask Etchant (Transene, Co.; Danvers, MA).
9. Buffered oxide etchant (Transene Co.; Danvers, MA).
10. Wax.
11. Concentrated HCl.
12. 5 M H_2SO_4.
13. Acetone.
14. Ethanol.
15. Ammonium hydroxide.
16. 30% hydrogen peroxide.
17. Binder clips.
18. Epoxy (e.g., Epo-Tek 353ND; Epoxy Technologies, Inc.; Billerica, MA).

3 Methods

3.1 Solution Preparation and Glass Slide Coating

3.1.1 Glass Etch Solution Preparation (See **Note 6**)

1. Add 1000 mL of 18 MΩ·cm water to 2 L polypropylene container.
2. Add 500 mL of concentrated HCl slowly to the same container to limit the temperature rise.
3. Wait for 5 min.

4. Add slowly 250 mL of buffered oxide etchant (NH_4F/HF, 10:1) to the same container.
5. Stir for 10 min before use.

3.1.2 Hydrolysis Solution Preparation (See Note 7)

1. Add 400 mL of water to 1 L beaker.
2. Add 200 mL concentrated ammonium hydroxide slowly to same beaker.
3. Add 200 mL 30% hydrogen peroxide to the basic solution and use immediately.

3.1.3 Coating Microscope Slides with AZ Photoresist (See Notes 8–10)

1. Pour ~3 mL of AZ photoresist into a light tight vial and ensure that bubbles are removed. Let it warm up to the room temperature for 1 h before using (*see* **Note 11**).
2. Turn on UV Source to warm up.
3. Preheat hot plate to 65 °C (*see* **Note 12**).
4. Check that the photoresist spin coater is clean and clean with acetone if necessary.
5. Clean glass slide thoroughly using cleanroom swab and glassware detergent.
6. Rinse slide with 18 MΩ·cm water and dry with compressed inert air.
7. Place slide in oven at 100 °C for 20 min and then let it cool in clean hood.
8. Turn on vacuum to the spin coater.
9. Place slide on spin coater in a custom slide chuck and turn on vacuum.
10. Pour ~1–2 mL of AZ onto the center of slide and close lid.
11. Spin at 1000 rpm for 17 s, with an acceleration of 300 rpm/s, which will yield a ~17 μm thick film (*see* **Note 12**).
12. Open the lid slowly to prevent dripping.
13. Soft bake the slide (covered to prevent exposure to light) on the hot plate for 2 min at 65 °C and then ramp to 95 °C, and hold for 2 min then ramp to 120 °C, and hold for 4 min and let it cool covered in clean hood for 20 min (*see* **Note 12**).
14. Begin procedure for patterning immediately (refer to Subheading 3.3).

3.2 Microchip Design

Chip designs can be drawn in-house using a CAD drawing program (e.g., AutoCAD LT; Thompson Learning; Albany, NY) and electronically sent to photomask fabricators for translation and fabrication (e.g., Fineline Imaging; Colorado Springs, CO). Laser photoplotted photomasks at a variety of resolutions (10,000–50,000 DPI) can be obtained from several sources and

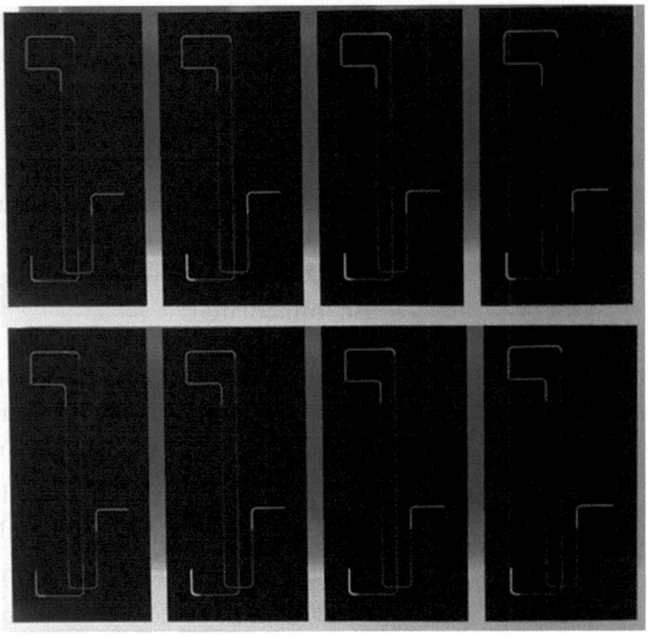

Fig. 1 Example of a 4″ × 4″ photoplotted mask with eight 2″ × 1″ microchip designs that can be transferred to a 4″ × 4″ glass photomask blank

are much more economical than trying to create such masks in-house. An example of such a mash has been included in Fig. 1. For mask features below 10 μm, a more expensive glass photomask needs to be ordered (e.g., Advance Reproductions Corporation, Andover, MA).

3.3 Patterning Microfluidic Substrate Material (See Note 13)

1. Place photomask PRINTED SIDE DOWN and quartz block (to ensure flatness of photomask) onto photoresist-covered substrate (i.e., photomask blank or coated glass slide).
2. Place assembled substrate, photomask, and quartz block into UV flood exposure enclosure, and expose for the calculated exposure time (e.g., 25.2 s for 17 μm-thick film of AZ P4620 at a measured output power of 25 mW/cm^2) (*see* **Note 14**).
3. Submerge the exposed plate (or slide) in a stirred solution of Microposit Developer (Shipley Co.; Marlborough, MA) for ~90 s, followed by a thorough rinse with 18 MΩ·cm water.
4. (Plate only) Submerge the plate in a stirred solution of Chrome Mask Etchant (Transene, Co.; Danvers, MA) for 3 min, rinse with 18 MΩ·cm water, and dry with inert gas.
5. Place the developed plate/slide into a stirred, dilute buffered oxide etchant made up in Subheading 3.1.1 for chemical etching (*see* Fig. 2).

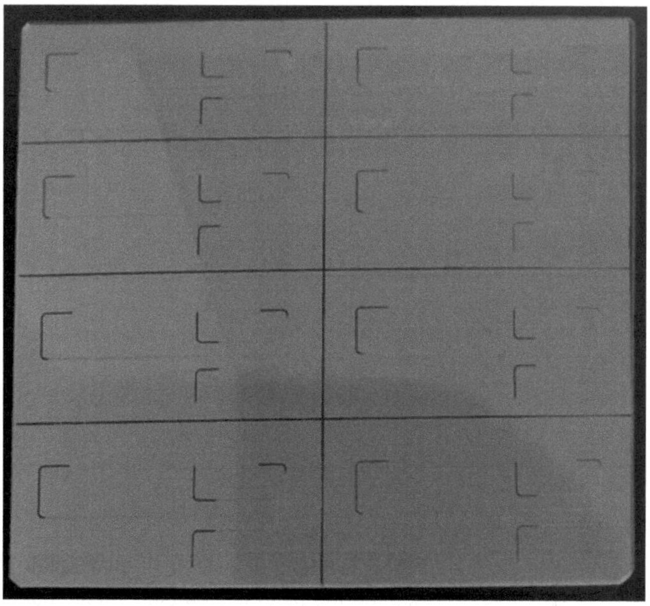

Fig. 2 Image of an etched 4″ × 4″ plate prior to removal of photoresist and chrome

6. Periodically measure the channel dimensions during the etching process using a stylus-based surface profiler. The average etch rate for this solution will be ~1.03 μm/min for borosilicate glass. Every 5 min the chips should be turned 90° in order to ensure even etching (*see* **Note 15**).
7. Once the channels are the desired depth, remove the remaining photoresist by rinsing the plate with acetone followed by an18 MΩ·cm water rinse.
8. Remove the remaining chrome (for glass plate only) by immersing the plate in a stirred solution of Chrome Mask Etchant for 10 min followed by rinses in 1 M sulfuric acid and 18 MΩ·cm water.
9. (Plate only) Dry the cleaned plate with inert gas, and dice into eight individual 2″ × 1″ (2.54 cm × 5.08 cm) slides using a dicing and cutting saw (Model EC-400; MTI Corp.; USA) (*see* **Notes 16** and **17**).
10. Mechanically drill access holes into 10.16 cm × 10.16 cm × 0.15 cm cover plates using diamond-tipped drill bits or an ultrasonic mill, or outsource the drilling to a specialty glass shop (e.g., Technical Glass Inc.; Aurora, CO) (*see* Fig. 3) and then cut into eight individual slides in a similar manner to that in **step 9** directly above prior to bonding (*see* **Note 18**).

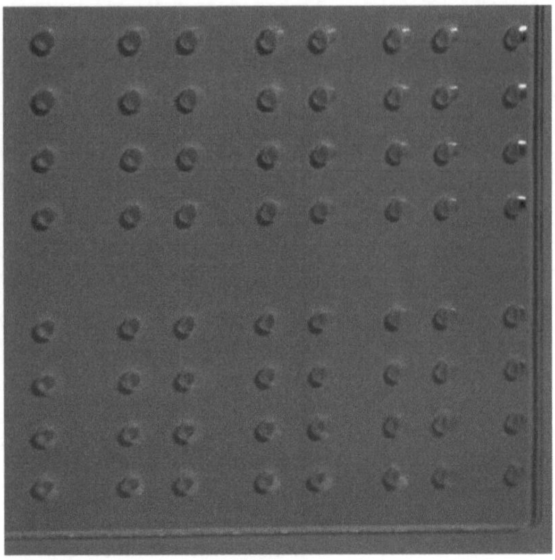

Fig. 3 Example of a 4″ × 4″ glass plate with holes drilled that can serve as a generic top plate for the etched channel plate

3.4 Microchip Bonding

1. Thoroughly clean the etched and cover slides by swabbing with acetone followed by rinses with ethanol and 18 MΩ·cm water.
2. Dry the slides under inert gas.
3. Submerge the slides in a stirred 5 M sulfuric acid solution for 5 min, rinse with 18 MΩ·cm water, and then clean by swabbing with acetone followed by rinses with ethanol and 18 MΩ·cm water.
4. Dry the slides under a flow of inert gas.
5. In a laminar flow hood, immerse the slides in a Versa-Clean Liquid soap solution (Fisher Scientific; Pittsburgh, PA), sonicate (3510 ultrasonic cleaner; Branson; Danbury, CT) for 15 min, rinse with 18 MΩ·cm water, and dry under inert gas.
6. Sonicate the slides in acetone for 10 min, dry, and place in the previously described dilute buffered oxide etch solution for 10 s.
7. Immediately rinse the slides with 18 MΩ·cm water and place in a dilute hydrolysis solution (1:1:2 parts NH_4OH, H_2O_2, and H_2O, respectively) for 12 min at 60 °C [3].
8. Rinse the slides with 18 MΩ·cm water and sonicate in flowing 18 MΩ·cm water for 60 s before joining.
9. Remove the etched slides one at a time from the flowing stream of 18 MΩ·cm water and place on a Cleanroom Wiper (DURX 670; Great Barrington, MA).
10. Remove the drilled cover slides from the flowing water and place on top of the respective etched slide.
11. Fasten binder clips on the perimeter of the chip to ensure contact between the two surfaces.

12. Remove the water from the channels with a vacuum hose.
13. Place the joined chips in the oven at 95 °C for 15 min to drive out any remaining water (*see* **Note 19**).
14. Place the chips in the high-temperature oven, and anneal using the following temperature program.
15. Ramp the temperature from room temperature to 530 °C at 400 °C/min.
16. Once 530 °C is reached, ramp to 560 °C at 100 °C/min until the annealing temperature is reached (*see* **Note 20**).
17. Hold at the annealing temperature for 25 min and then allow to cool to room temperature overnight before removing (*see* **Notes 21** and **22**).
18. Attach cylindrical glass reservoirs (~140 μL capacity) using Epo-tek 353ND Epoxy (Epoxy Technologies, Inc.; Billerica, MA) where the access holes are located.

4 Notes

1. Alternatively some researchers have used "blacklights," but the output is variable, and exposure timing is harder to optimize.
2. The photoresist spinner is needed if coating microscope slides with photoresist. Alternatively, one could purchase glass plates pre-coated with the photoresist from a commercial vendor.
3. A clean room could be used, but it is significantly more expensive to maintain and operate.
4. Alternatively, the glass plate can be scored and split using a glass scoring tool.
5. This plate can be used to make eight, 2″ × 1″ chips in parallel.
6. The buffered oxide etchant solution is a stable glass etch solution that can be used for several months depending upon the number of slides etched.
7. The hydrolysis solution should be prepared immediately prior to use and disposed of properly afterward.
8. New, clean microscope slides should be coated with photoresist for transferring the pattern from the photomask to the substrate material.
9. Chrome is not needed underneath the photoresist. The photoresist will hold up in the glass etching solution.
10. All steps in this process should be carried out under lighting conditions that do not expose the photoresist to excessive amounts of UV light.
11. Do not let the photoresist solution stand at room temperature for more than 3 h.

12. These processing parameters are specific for AZ P4620. Refer to datasheet associated with photoresist for these parameters if using a different photoresist.
13. Measure flood exposure system output using photometer/radiometer with detector to calculate exposure time to achieve proper UV dosage for photoresist used and film thickness, i.e., exposure time (s) = dosage (mJ/cm^2) ÷ output power (mW/cm^2).
14. From this point on in the protocol, make sure that you have donned the proper personal protective gear to ensure safety including lab coat, safety glasses, and the appropriate gloves.
15. Over time the etching solution will get cloudy and the etch rate will decrease. Once the etch rate decreases below 75% of the initial rate, replace it.
16. The best way to dice the plate is to attach it via low melting point wax to another unetched plate and place the combined plates onto the vacuum chuck on the saw. In this way you can completely dice through the etched glass plate without risking damage to the vacuum chuck.
17. Alternatively, you may score and break the glass plate. However, the saw gives a more reproducible chip size.
18. We have had much greater success cutting the 4″ × 4″ plate into the individual chips and bonding them individually rather than trying to bond the uncut plates and then dicing the individual chips out of the plate.
19. At this point if you see Newton's rings (*see* Fig. 4) near or over a channel, place the chip in a water bath, separate the two pieces, and repeat **steps 1–13** in Subheading 3.4. The rings indicate

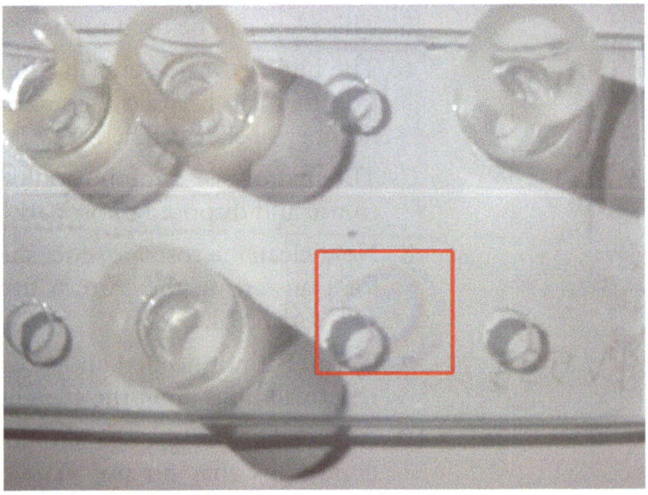

Fig. 4 Image of chip with a Newton ring in a non-critical area (i.e., an area not over or near channel)

something is interfering with the bonding and that the chip will be unusable if you continue.

20. The substrate annealing temperature can vary slightly depending upon the glass composition. We have found generally that 560 °C works well.
21. We have successfully performed the bonding by simply placing the chips in the oven, by placing the chips in the oven with the binder clips still on, or by placing weights on the chips.
22. In the past, we have successfully bonded 112 of 112 microchips using this method for a bonding success rate of 100%.

Appendix

First Aid Measures for Hydrofluoric Acid Exposure

Skin Contact

1. Remove victim from contaminated area and immediately (within seconds) shower and flush with plenty of water for 5 min.
2. Remove all clothing while in the shower. (Remove goggles last and double-bag-contaminated clothes.)
3. Take a tube of calcium gluconate gel from the first aid kit. Use gloves while applying to prevent contact with uncontaminated skin. Massage the gel promptly and repeatedly into burned area until pain is relieved. Even if pain subsides within 20–30 min, get medical help.

Breathing Vapor

1. Immediately get to fresh air.
2. Keep the victim lying down, quiet, and warm.
3. Call or have someone call for medical help.

Ingestion

1. Drink large amount of water. Do not induce vomiting.
2. Drink several glasses of milk or several ounces of magnesia may be given for a soothing effect.
3. Get medical help for the victim.

Eye Contact

1. Irrigate eyes for at least 15 min with large amount of gently flowing water. Keep the eyelids open and gently pulled away from the eyeballs during irrigation.
2. Seek medical attention immediately after flushing the eyes.
3. Apply ice water compresses or continue irrigating the eyes until medical aid arrives.

Advice to Physician

For large skin area burns (totaling greater than 25 square inches), for ingestion, and for significant inhalation exposure, severe systemic effects may occur. Monitor and correct for hypocalcemia, cardiac arrhythmias, hypomagnesemia, and hyperkalemia. In some cases, renal dialysis may be indicated. For certain burns, especially of the digits, use of intra-arterial calcium gluconate may be indicated. For inhalation exposures, treat as chemical pneumonia. Monitor for hypocalcemia. 2.5% calcium gluconate in normal saline by nebulizer or by IPPB with 100% oxygen may decrease pulmonary damage. Bronchodilators may also be administered.

Neutralizing Buffered HF Oxide Etch Spill

1. Use an acid spill emergency cleanup kit.
2. Protect yourself. Put on facial covers, gloves, and shoe coverings.
3. Contain the spill by encircling it with the absorbent/neutralizer mixture and then fill the circle with the remaining mixture.
4. After the mixture turns blue, put it into a waste bag using a dustpan and brush. Seal the bag tightly and place it in the kit along with kit contents. Close kit securely and dispose of in the appropriate manner.
5. Call the Department of Environmental Health and Safety for waste pickup.

References

1. Harrison DJ, Manz A, Fan Z, Luedi H, Widmer HM (1992) Capillary electrophoresis and sample injection systems integrated on a planar glass chip. Anal Chem 64(17):1926–1932. https://doi.org/10.1021/ac00041a030
2. Seiler K, Harrison DJ, Manz A (1993) Planar glass chips for capillary electrophoresis: repetitive sample injection, quantitation, and separation efficiency. Anal Chem 65(10):1481–1488. https://doi.org/10.1021/ac00058a029
3. Jacobson SC, Hergenroder R, Koutny LB, Warmack RJ, Ramsey JM (1994) Effects of injection schemes and column geometry on the performance of microchip electrophoresis devices. Anal Chem 66(7):1107–1113. https://doi.org/10.1021/ac00079a028
4. Jacobson SC, Hergenroder R, Koutny LB, Ramsey JM (1994) High-speed separations on a microchip. Anal Chem 66(7):1114–1118. https://doi.org/10.1021/ac00079a029
5. Jacobson SC, Hergenroeder R, Koutny LB, Ramsey JM (1994) Open channel electrochromatography on a microchip. Anal Chem 66 (14):2369–2373. https://doi.org/10.1021/ac00086a024
6. Duffy DC, McDonald JC, Schueller OJA, Whitesides GM (1998) Rapid prototyping of microfluidic systems in poly(dimethylsiloxane). Anal Chem 70(23):4974–4984. https://doi.org/10.1021/AC980656Z
7. Crain MM, Keynton RS, Walsh KM, Roussel TJ Jr, Baldwin RP, Naber JF, Jackson DJ (2006) Fabrication of a glass capillary electrophoresis microchip with integrated electrodes. Methods Mol Biol 339:13–26
8. Mazurczyk R, Mansfield CD, Lygan M (2013) Glass microstructure capping and bonding techniques. Methods Mol Biol 949:141–151. https://doi.org/10.1007/978-1-62703-134-9_10

Chapter 2

Soft Lithography, Molding, and Micromachining Techniques for Polymer Micro Devices

Ashis Kumar Sen, Abhishek Raj, Utsab Banerjee, and Sk Rameez Iqbal

Abstract

This chapter enumerates the methods, protocol, and safety procedures of various fabrication techniques for polymer-based microfluidic devices. The polymer materials can be a solid or a liquid, and the fabrication protocol needs to be executed accordingly. Various techniques demonstrating the fabrication of microfluidic devices using solid and liquid polymers are described. Procedure for each fabrication process is delineated with detailed images. Further, dos and don'ts for all the fabrication techniques are explained in the notes of each section. This chapter will benefit those interested in the microfluidic device fabrication using polymers and guide them to avoid mistakes so as to obtain an elegant device.

The techniques are listed as follows:

1. Replica molding
2. Microcontact printing
3. Micro-transfer molding
4. Solvent-assisted molding
5. Hot embossing
6. Injection molding
7. CNC micromachining
8. Laser photo ablation
9. X-ray lithography
10. UV patterning
11. Plasma etching
12. Ion beam etching
13. Capillary molding
14. Micro-stereolithography

Key words Etching, Micromachining, Molding, Polymer micro devices, Soft lithography

Debashis Dutta (ed.), *Microfluidic Electrophoresis: Methods and Protocols*, Methods in Molecular Biology, vol. 1906, https://doi.org/10.1007/978-1-4939-8964-5_2, © Springer Science+Business Media, LLC, part of Springer Nature 2019

1 Introduction

1.1 Replica Molding

Replica molding (REM) is one of the well-established soft lithography techniques, which is used for replicating the information present on an original master [1] onto a substrate. The original master can be fabricated by various processes such as photolithography and micromachining. This method of soft lithography accommodates a wider range of materials and provides a high frequency of replication [2] as compared to photolithography. Various types of materials such as PDMS, polyurethane, organic polymers, and epoxy resin can be used for replica molding. Replication against the elastomeric masters enhances the ease of separating the replica and the master, which protects the fragile structure and also minimizes the cost and damage to the master [3]. The replica molding procedure is detailed in Subheading 2.1, and the schematic for the same is shown in Fig. 1. Examples of some molds fabricated by the replica molding process have been included in Fig. 2.

1.2 Microcontact Imprinting

Microcontact printing is the only soft lithographic technique that is capable of generating chemical patterns on a surface [4]. It allows engineering of a surface with molecular-level detail. The process can provide sub-100 nm self-assembled monolayer (SAM) patterns of biomolecules over macroscopic areas [5]. This method is similar to transfer ink from an ink pad to a paper using a stamp. The mold is inked with the material (to be transferred) to the substrate (metal coated) by making contact between the substrate and the

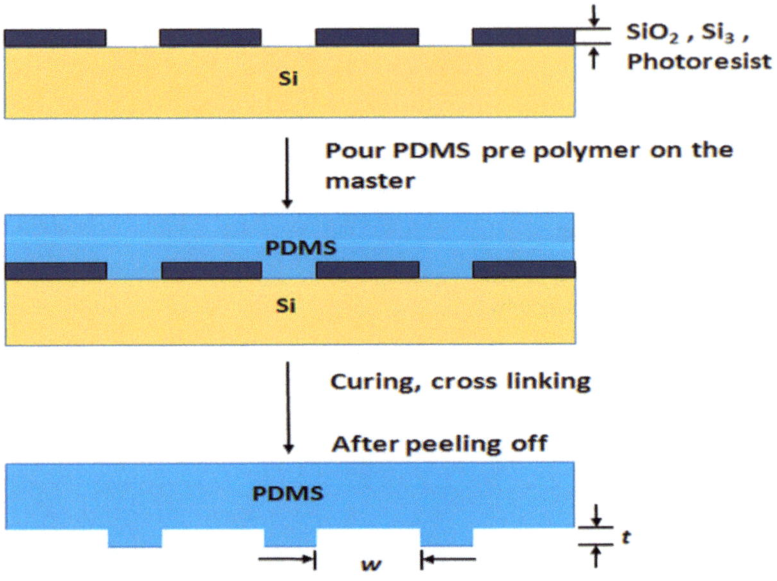

Fig. 1 Process outline of the replica molding technique

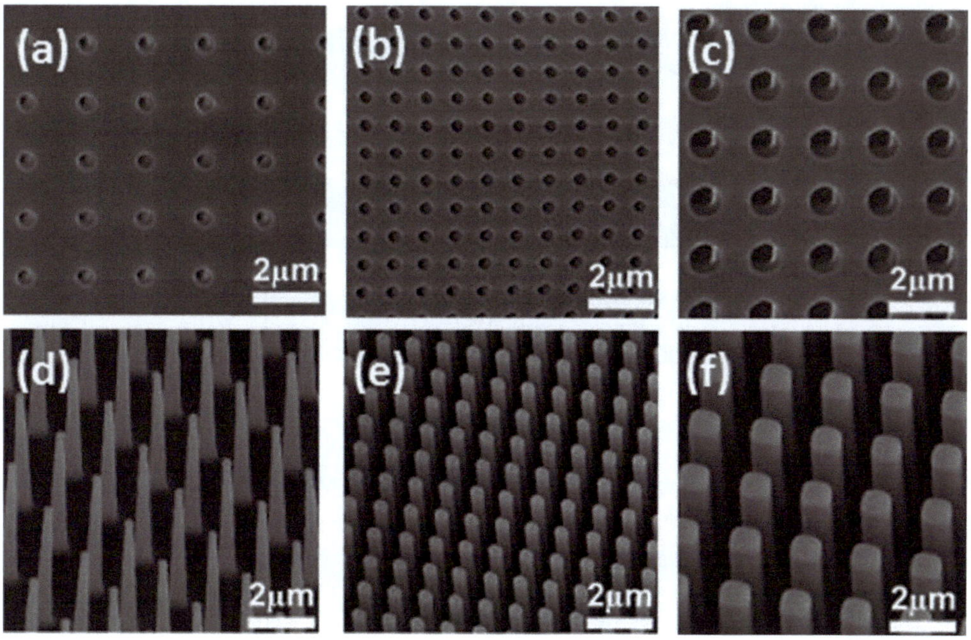

Fig. 2 (**a–c**) PDMS molds replicated from corresponding silicon masters, (**d–f**) epoxy nanopillars molded from corresponding PDMS molds (Reprinted with permission from [5]). (Copyright (2006) American Chemical Society)

protruding features of the mold [3]. The substrates are prepared by vapor deposition methods, e.g., thermal beam evaporation and electron beam evaporation. The inks can be small biomolecules, proteins, or suspensions of cells [6]. The schematic of the process described in Subheading 2.2 is shown in Fig. 3. Representative SEM images of the master and mold fabricated by microcontact imprinting have been included in Fig. 4 for reference.

1.3 Micro-transfer Molding

In this technique, a liquid precursor is filled inside a PDMS microchannel; then the mold is brought into contact with a planar or non-planar substrate [1]. The liquid precursor is solidified in situ either thermally or photochemically [7]. Then the elastomer can be peeled off, leaving the desired microstructure. The technique can generate isolated and interconnected microstructures. It can generate microstructures over larger areas (~3 cm^2) within a short period of time [8]. The PDMS stamps can be utilized further [9]. The schematic of the process outline is shown in Fig. 5. A representative SEM image of a micro-transfer molded sample has been included in Fig. 6 for reference.

1.4 Solvent-Assisted Molding

Solvent-assisted micromolding (SAMIM) technique makes use of a solvent to wet the PDMS stamp and restructure a polymer film by swelling or dissolving the polymer [1]. This room-temperature processing is suitable for polymers whose T_g (glass transition

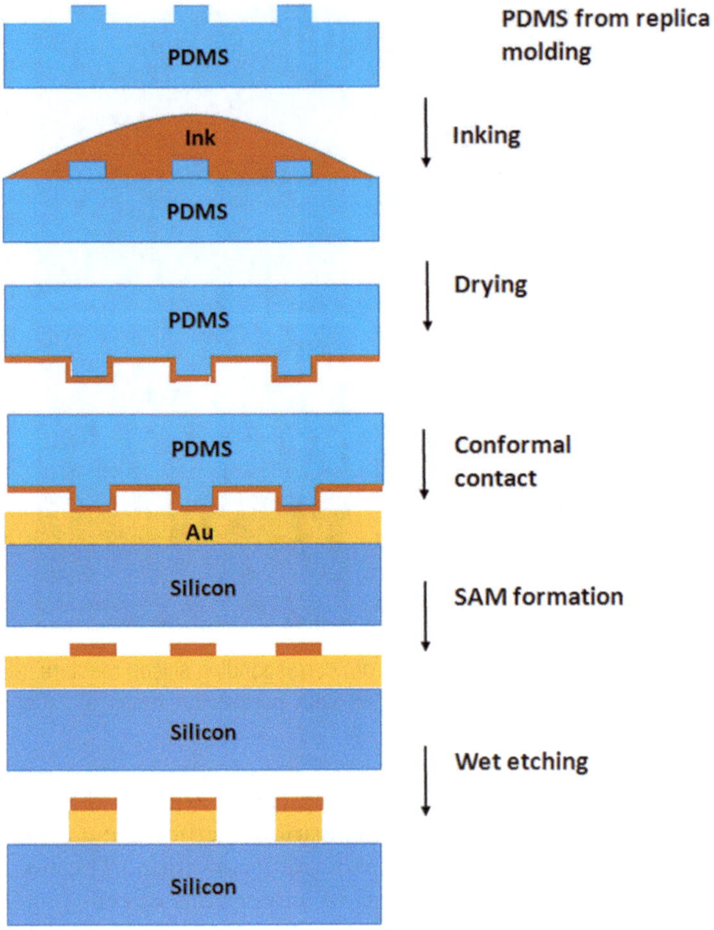

Fig. 3 Process outline of microcontact imprinting technique

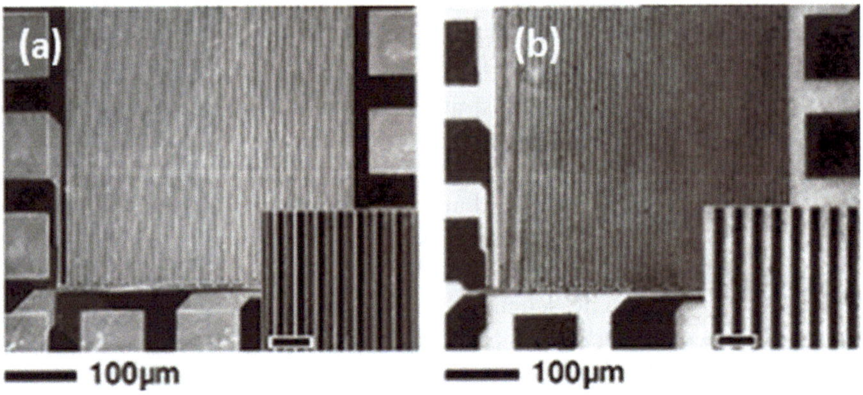

Fig. 4 (**a**–**b**) SEM images (at two different magnifications) of a master and the pattern of a SAM of HDT on gold formed by μCP (contact time of 10 s) with a PDMS stamp cast from this master [4] (Copyright Wiley-VCH Verlag GmbH & Co. KGaA. Reproduced with permission)

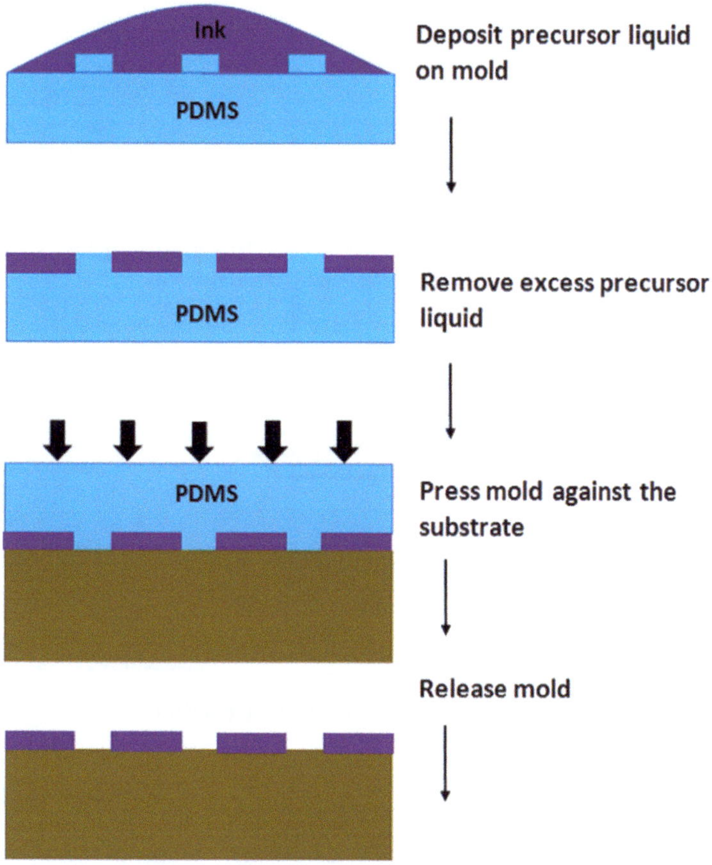

Fig. 5 Process outline of the microcontact imprinting technique

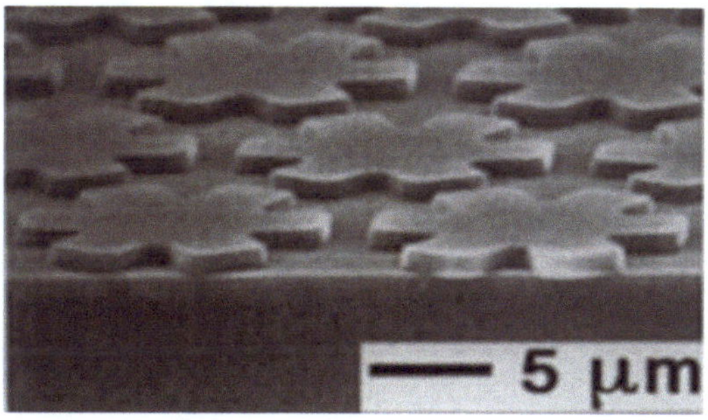

Fig. 6 An SEM image of a fractured sample showing a pattern of isolated stars of UV-cured polyurethane (NOA 73) on Ag, fabricated by μTM [8] (Reproduced with permission from the author and publisher)

temperature) is close to the degradation temperature and hence cannot be molded by imprinting techniques [10]. This technique avoids thermal cycling of the substrate, which can be time intensive and lead to oxidation of the substrates [3]. The solvent-assisted molding is suitable for many polymers and their precursors due to

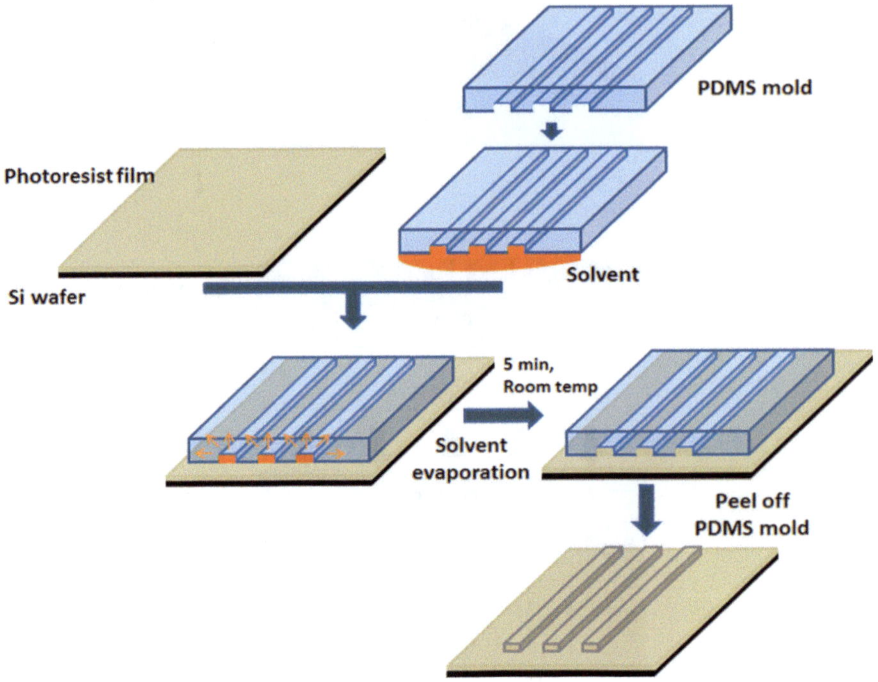

Fig. 7 Schematic of the solvent-assisted molding technique

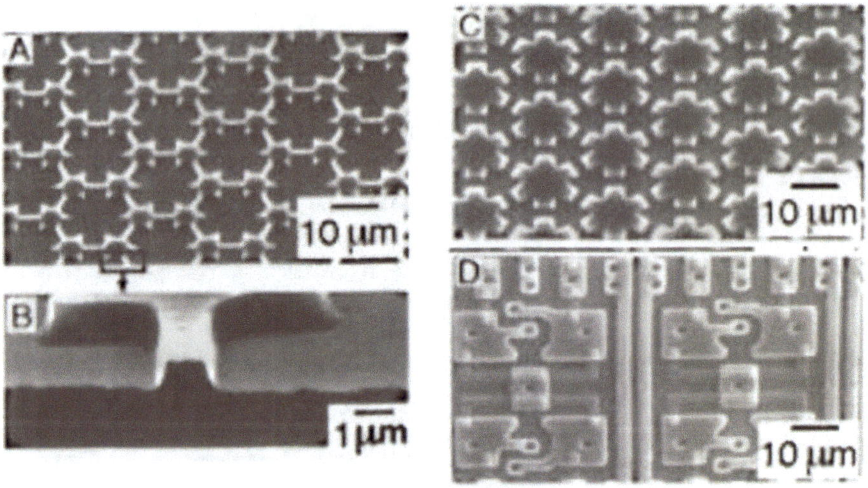

Fig. 8 SEM images of test structures formed in thin films of novolac photoresist using solvent-assisted molding, with ethanol as the solvent [12] (Reproduced with permission from the author and publisher)

its simple procedures and high production efficiency and since it does not require special molding equipment and system [10]. The general procedure of the technique is given below, and the schematic is shown in Fig. 7. Representative SEM images of the 3D structures formed in thin films of novolac photoresist using this technique have been included in Fig. 8 [11].

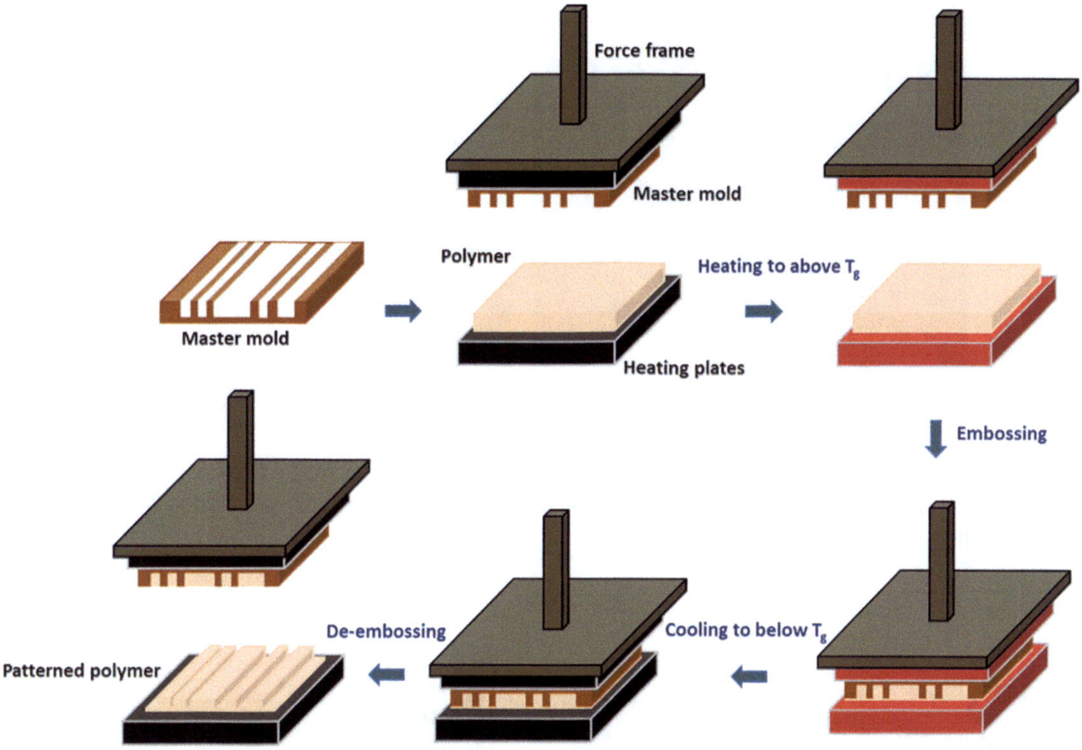

Fig. 9 Schematic of the hot embossing technique

1.5 Hot Embossing

Hot embossing is used for fabrication of simple microchannels. It is essentially the stamping of a mold pattern onto a polymer softened by increase in the temperature of the polymer just above its glass transition temperature, which ranges from 50 °C to 150 °C. Embossing force (0.5–2 kN/cm^2) is then applied on the substrate under vacuum conditions [12]. The polymers experience two stages of deformation during embossing: stress concentration and strain hardening during heating and embossing steps and stress relaxation and deformation recovery stage during cooling and demolding steps [13]. The schematic of the process flow is shown in Fig. 9. Some representative images of hot embossed in polycarbonate (a) and PMMA (b) substrate have been included in Fig. 10 [14].

1.6 Injection Molding

Injection molding is a polymer molding technique in which molten polymer is injected under high pressure into a mold cavity through an opening (sprue) [15]. For amorphous thermoplastics such as polymethyl methacrylate (PMMA), polycarbonate (PC), and polysulfone (PSU), the temperature should be higher than the glass transition temperature T_g. For semicrystalline thermoplastics such as polyoxymethylene (POM) and polyamide (PA), the temperature should be higher than the crystallite melting point. The pressure

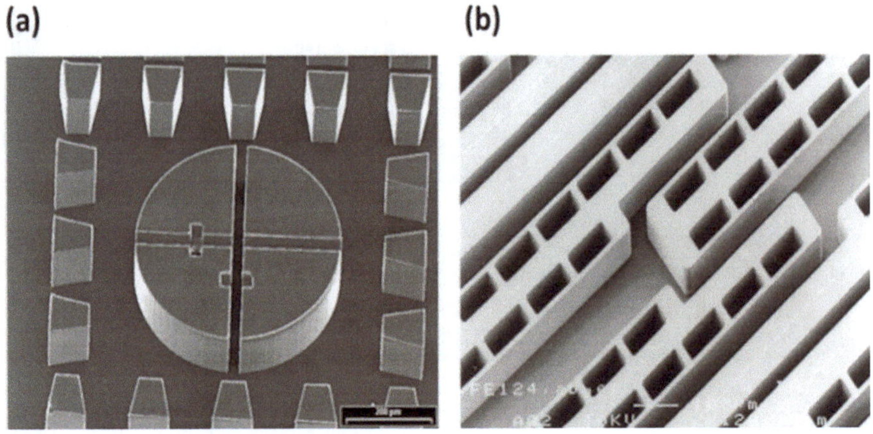

Fig. 10 (**a**) Test structures in polycarbonate (PC), (**b**) PMMA structure made from a LIGA mold for the fabrication of a three-dimensional acceleration sensor [15] (Reproduced with permission from the author and publisher)

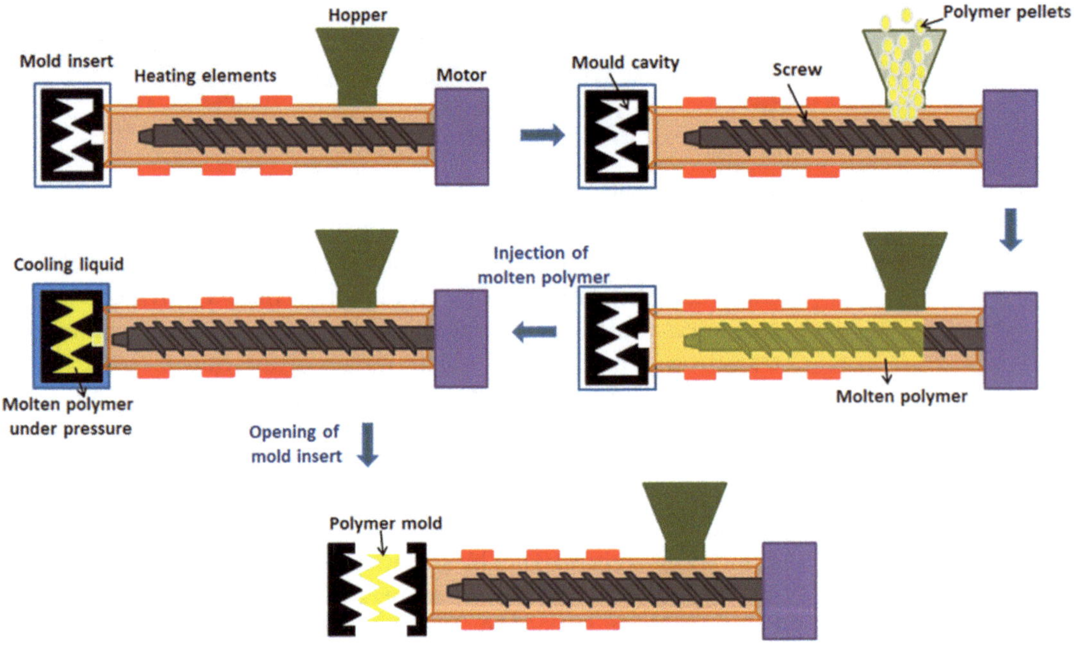

Fig. 11 Schematic of the injection molding technique

applied during the process is of the order of 500–2000 bar [1]. The advantages of this technique include high production levels, high tolerance, low labor cost, and scrap cost [16–18]. The step-by-step procedure for this technique is shown in Fig. 11. An example of an injection-molded microfluidic chip has been included in Fig. 12 [19].

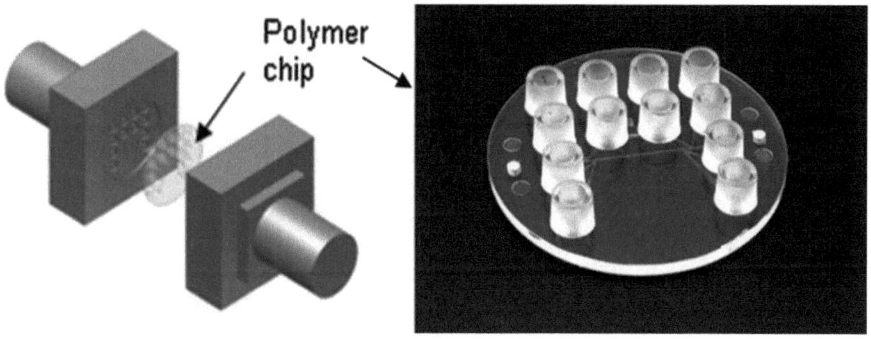

Fig. 12 (**a**) Image showing the molds and the injection-molded polymer chip, (**b**) molded part of the cell capture device [20] (Reproduced with permission from the author and publisher)

1.7 CNC Micromachining

Micromachining is a subtractive manufacturing technique in which rotating cutting tools remove material from a starting workpiece. Modern milling machines use computer numerical control (CNC) for automating XYZ positioning which provide improved repeatability and precision (*see* Fig. 13) [20]. The CNC machine has mini-/microcomputer that acts as the controller unit of the machine to which a set of instructions are fed using a small board. All the cutting processes and dimensions of the final device are programmed into the computer. In addition to its advanced capabilities, it is also flexible since it is easy to setup and run a different program for a different device [21]. The following procedure is generally used for micromachining. An example of a microfluidic device fabricated by CNC micromachining has been included in Fig. 14 [22].

1.8 Laser Photoablation

Laser photoablation is a localized, noncontact machining technique which can be used to cut, engrave, drill, mark, and texture substrates such as metal, ceramics, plastics, and wood [1]. Photoablation is the spontaneous etching of material from a polymer surface that occurs upon the absorption of a pulse of laser, whose energy is greater than the ablation threshold value (*see* Fig. 15) [23]. Micromachining with controlled accuracy is achievable since it is possible to remove materials in small amounts with a small heat-affected zone [1]. This helps in low-cost rapid prototyping. However, controlling the quality of the machined surface is difficult due to the redeposition of substrate material [1]. Though UV lasers are expensive than CO_2 laser, the ablation is strictly confined to the area that absorbs the energy [24, 25]. As a result of UV absorption, bond breaking within the long-chain polymer molecules occurs. A shock wave and ejection of decomposed polymer fragments ensues forming of a photo-ablated cavity. The ejected material contains gas, polymer molecules, and small particulate matter [24]. Some representative structures formed by UV laser photoablation in triazene polymer and Kapton have been included in Fig. 16a, b [24].

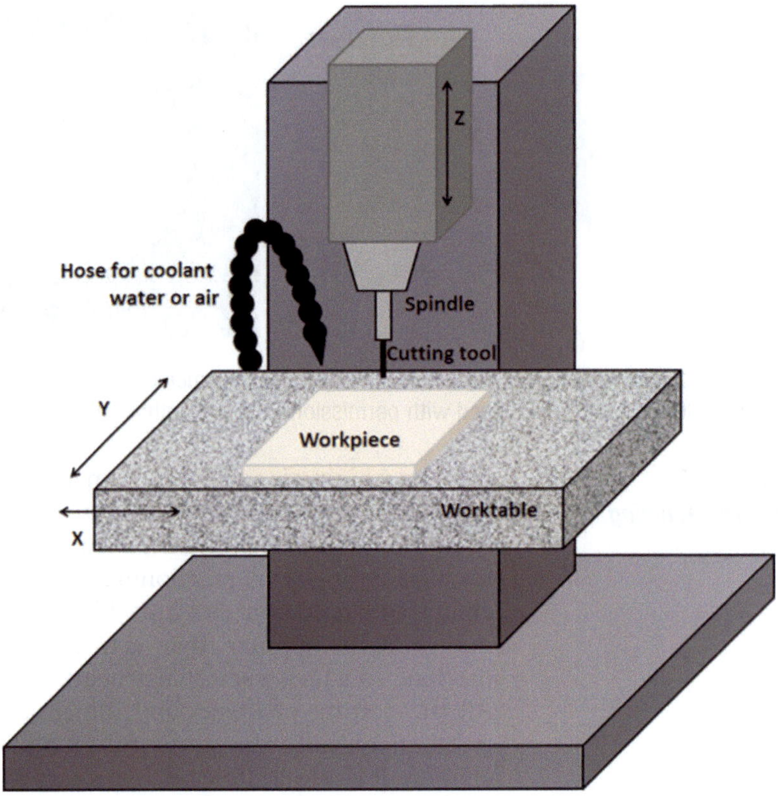

Fig. 13 Schematic of the CNC micromilling machine

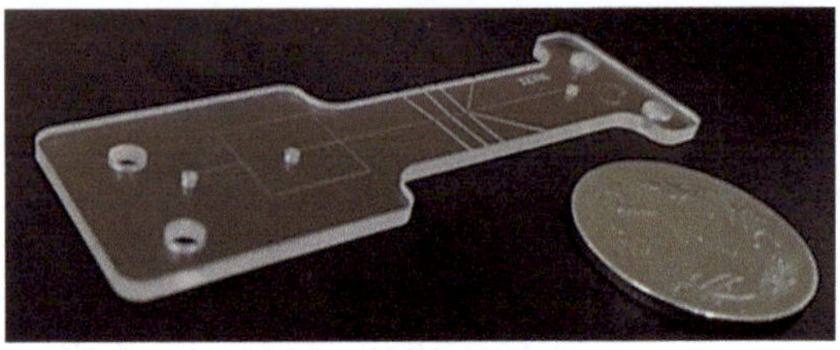

Fig. 14 Image of a CNC micromilled PMMA device

1.9 X-Ray Lithography

X-ray lithography (XRL) is a shadow printing-based technique which utilizes the patterns coated on a mask to create certain three-dimensional features in the resist material (*see* Fig. 17) [26]. Normally, polymethyl methacrylate (PMMA) is utilized as a resist material. Further, chemical process is incorporated to dissolve away the material volume, affected by X-ray exposure (*see* **Note 52**). X-rays having shorter wavelength than UV light allow this technique to increase the lateral resolution of the fabricated features by reducing the diffraction effect of the light compared to optical lithography technique [27]. Also, the higher penetrating ability of

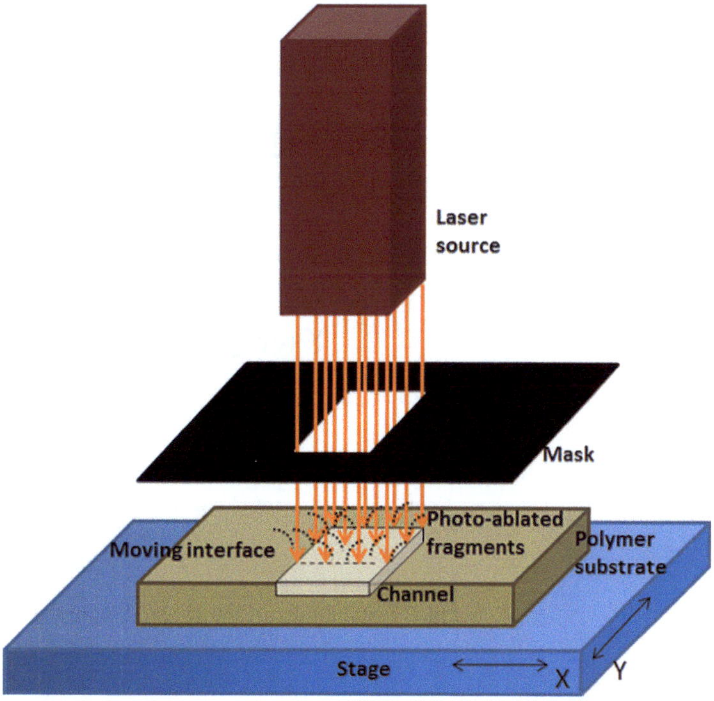

Fig. 15 Schematic of the laser ablation technique

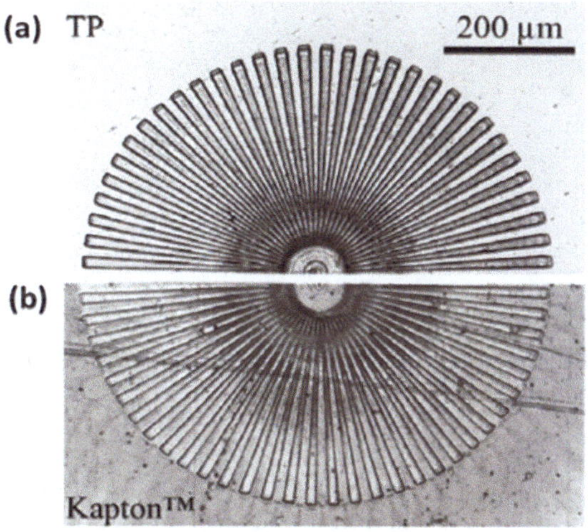

Fig. 16 SEM images of Siemens stars in (**a**) TP and (**b**) Kapton, both produced with five laser pulses at 308 nm [25] (Copyright Wiley-VCH Verlag GmbH & Co. KGaA. Reproduced with permission)

X-rays makes this technique capable of fabricating the microstructures with high aspect ratios of vertical dimensions of the order of hundreds of microns to millimeters and horizontal dimensions of few microns. Further, smooth and near 90° vertical side walls can

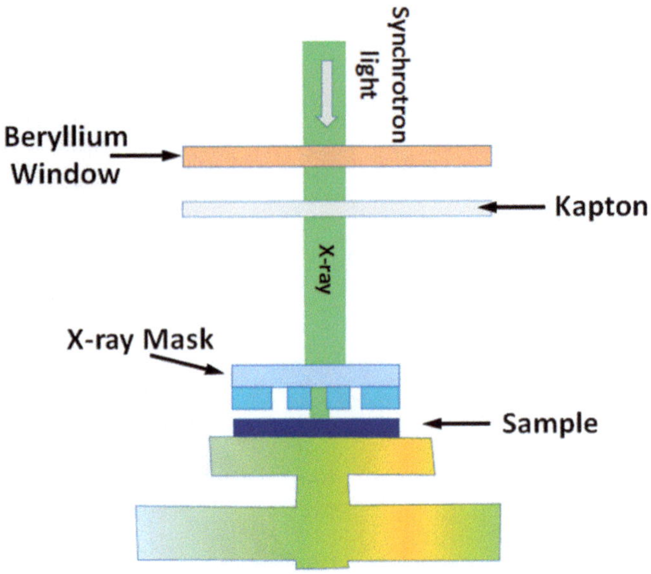

Fig. 17 Schematic of the X-ray lithography setup

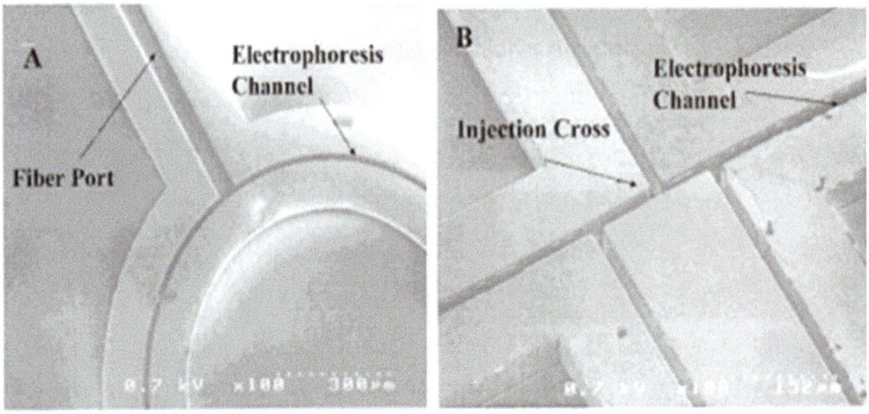

Fig. 18 Microfluidic PMMA device fabricated using X-ray lithography (Reproduced with the permission from the author and publisher [30])

be achieved by the technique [28]. Some representative microscopic images of device fabricated using X-ray lithography can be found in Fig. 18 [29].

1.10 UV Patterning

UV-patterning technique is a relatively low-cost machining process. The technique can be scaled to perform large area exposures, which makes it a good candidate for fabricating low-cost microfluidic components [30]. Moreover, this process can be accomplished using commercial grade PMMA, further lowering cost. UV light with intensity 4 mW/cm^2 is utilized to expose the PMMA surface (which acts as a positive-type resist). This high-energy radiation

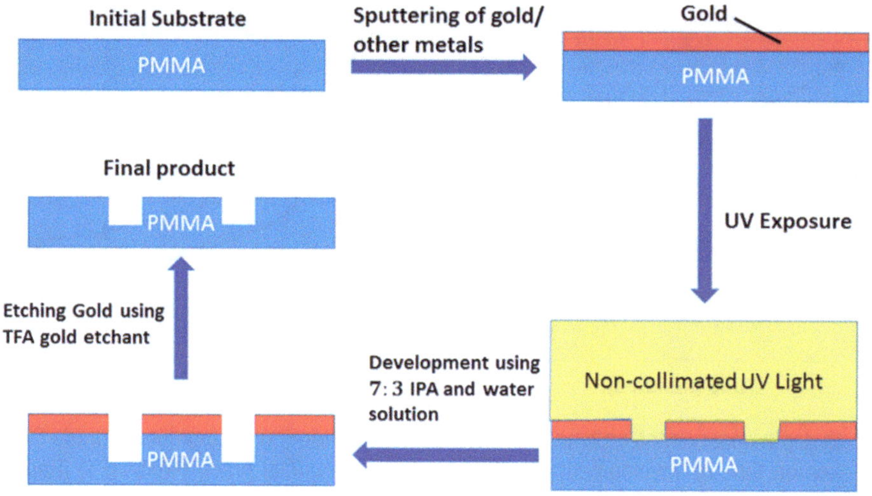

Fig. 19 Schematic of the UV-patterning technique

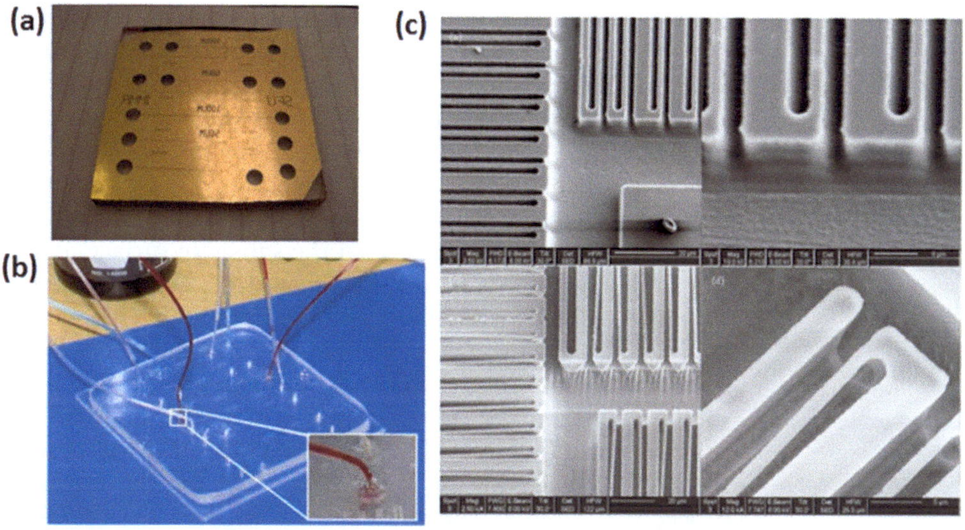

Fig. 20 (**a**) PMMA substrate coated with gold film, (**b**) fabricated microfluidics device using UV patterning, (**c**) SEM image of the fabricated PMMA device using UV patterning [31] (Reproduced with permission from the author and publisher)

results into the cleavage of the molecular bonds in PMMA, which is further developed using 7:3 mixture of IPA and water. In this technique, PMMA is mostly used as the substrate material. Further, the procedure of the UV-patterning technique is described below. Please refer to Fig. 19 for the flowchart of the fabrication procedure. Figure 20 shows some representative images of PMMA devices fabricated using the UV-patterning technique. Note that the exposure time should be kept optimal to achieve the desired depth of the microchannel as this depth varies with the exposure and development time as shown in Fig. 21.

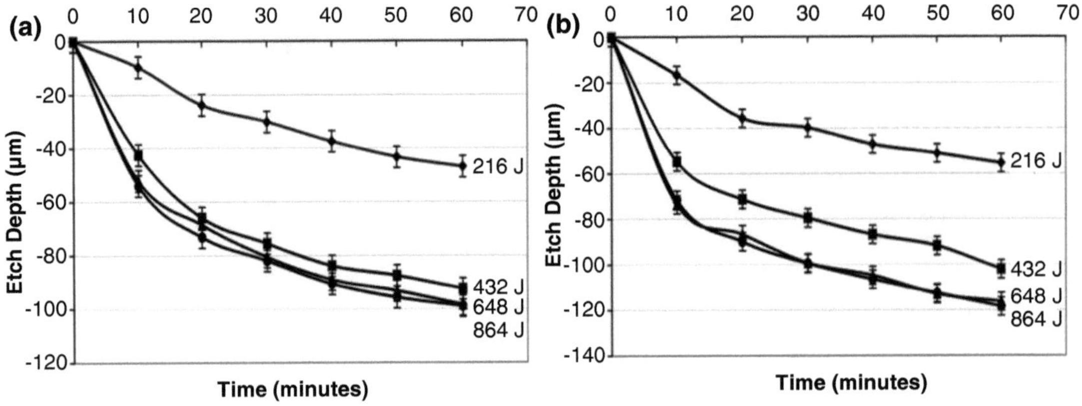

Fig. 21 (**a**) Etch depth dependence on time of exposure for UV patterning of commercial grade PMMA, (**b**) etch rate dependence on time of development UV patterning of commercial grade PMMA [31] (Reproduced with permission from the author)

1.11 Plasma Etching

Plasma etching is a dry etching technique in which certain types of chemical etchants in gas phase are utilized to etch substrate materials. Applying energy to the gaseous state of such etchants beyond a limit causes the existing shell of the atom to break up. Electrically charged, excited particles and molecule fragments are formed (negatively charged electrons and positively charged ions and radicals), which is called plasma or the fourth state of matter (aggregate state). The etch species in the plasma (charged or neutral particles) are utilized to bombard and etch the substrate surface. A continuous supply of the gas is maintained, and the existing gas enriched with the etched material is expelled. Although etch rate in this technique (1–10 μm/h) is much lower than the wet etching case, it provides much better control over thickness [31]. Figure 22 shows the schematic of the experimental setup. This method is suitable for polymers like POM, PTFE, PMMA, PEEK, and PDMS [32, 33]. Please refer to Fig. 23 for the layout of the procedure required for the fabrication. The device fabrication involves two stages: in the first stage, a mask is created over the polymer substrate, and in the second stage, the sample is exposed to plasma. The dependence of the etch rate and nano-roughness observed in this method on the various processing parameters has been shown in Figs. 24 and 25 [33]. Representative SEM images of a microfluidic channel fabricated using the plasma etching method have been included in Fig. 26. Typical plasma etching chemicals used for different film materials and the corresponding gaseous products are shown in Table 1 [34].

1.12 Ion Beam Etching

Ion beam etching is a physical process in which the ionized inert gas ions (with energy greater than 100 eV) are used to etch the material from the substrate by sputtering process [35]. Noble gases such as argon, neon, krypton, and xenon can be used as the inert gas. For

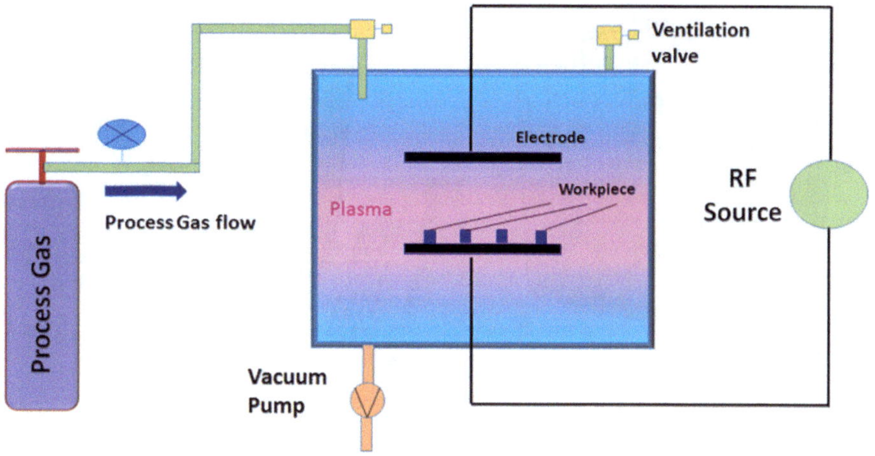

Fig. 22 Schematic of experimental setup used in plasma etching technique

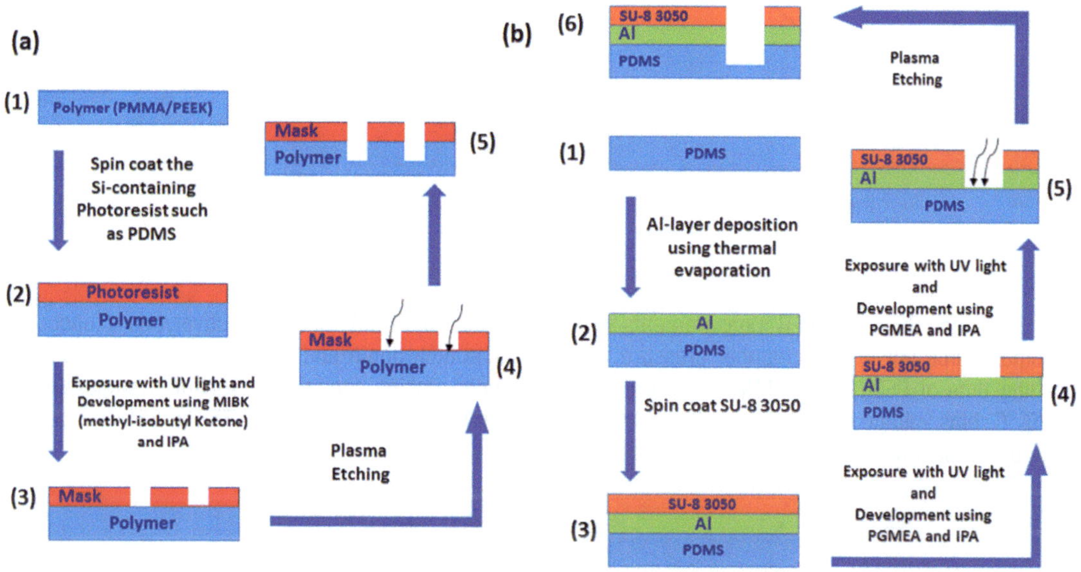

Fig. 23 Flowchart of the process protocol for plasma etching of (**a**) PMMA and PEEK and (**b**) PDMS

the etching of light materials, neon is best suited, whereas krypton and xenon having higher masses are more suitable for removal of heavy elements [36]. This process is highly directional rather than selective. Etch rates are dependent on the sputter yield, which again is a function of the sputter ion and the material to be sputtered [35]. Further, the sputter yield is also affected by the energy of the ion and the incident angle. Initial procedure to create the mask material is similar to that used in the plasma etching technique. Once the mask is created, the material can be removed from the desired locations by applying the beam of ions over it [37]. Please

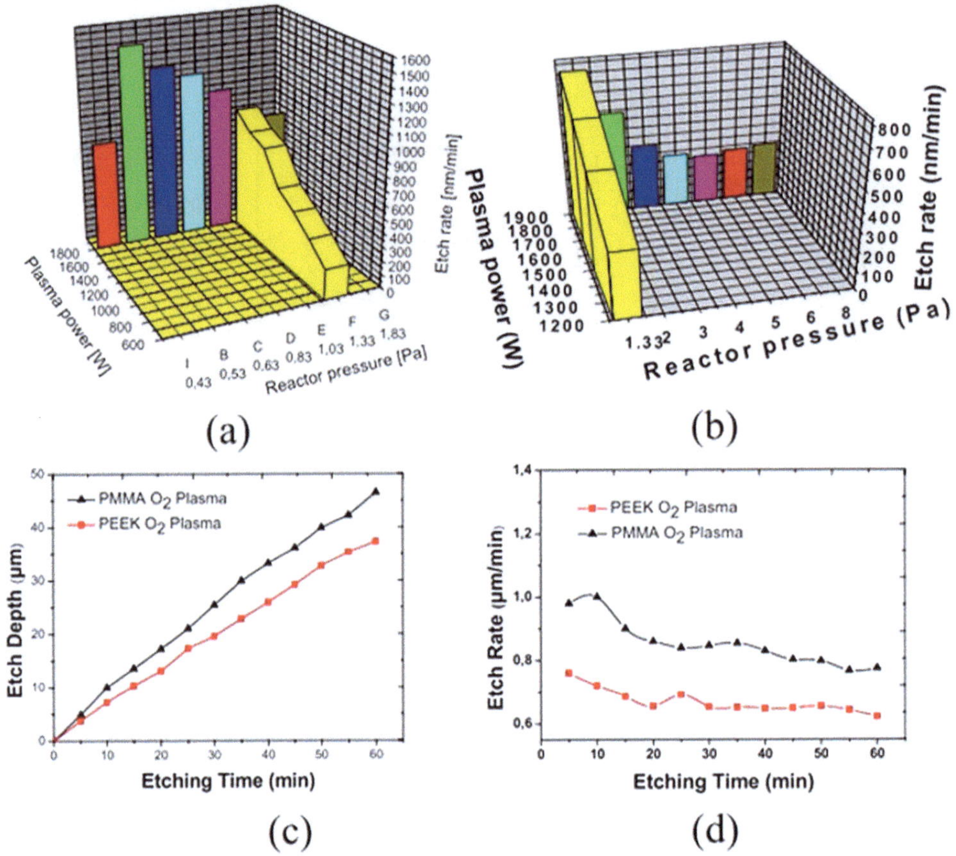

Fig. 24 O_2 and SF_6 plasma ER measurements of (**a**) PMMA and (**b**) PDMS films, respectively, as a function of plasma source power and chamber pressure. (**c**) O_2 plasma etch depth and (**d**) etch rate as a function of etching time of PEEK and PMMA plates are presented under etching conditions: electrode temperature of −20 °C, bias voltage of −120 V, O_2 flow of 100 sccm, and chamber pressure of 0.75 Pa [33] (Reproduced with permission from the author and publisher)

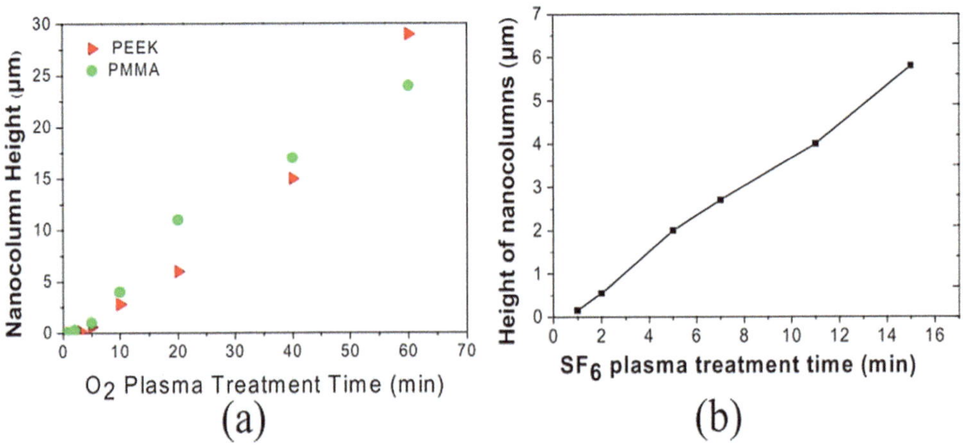

Fig. 25 Plasma etching: dependence of the nano-column height of (**a**) PEEK, PMMA, and (**b**) PDMS on O_2 and SF_6 plasma treatment time, respectively [33] (Reproduced with permission from the author)

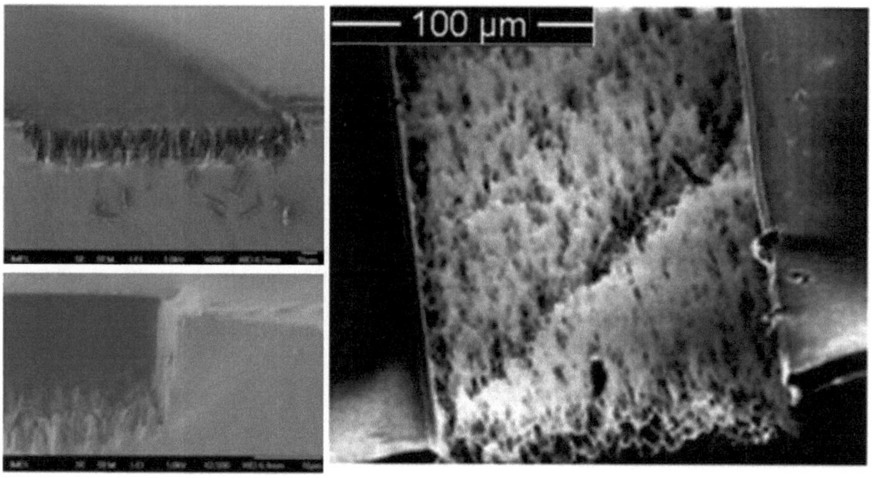

Fig. 26 Plasma etching for device fabrication: An SF_6 etched PDMS microchannel before and after sealing (**a**) an O_2 etched PMMA microchannel and (**b**) O_2 etched PMMA and PEEK microchannels, before and after sealing [33] (Reproduced with permission from the author and publisher)

refer to Fig. 27a for the schematic of the experimental setup used in this technique. An SEM image of a nanochannel fabricated using the ion beam etching method has been shown in Fig. 28.

1.13 Capillary Molding

Miniaturization in chemical, analytical, and diagnostic applications has recently gained a formidable interest [38, 39]. Capillary molding is one of the techniques proven to be simple, cost-effective, and easy to fabricate for microfabrication. The technique is composed of three components—soft elastomeric mold, solid support, and fluid prepolymer—where the mold can be used to impart various patterns. This technique has several advantages over photolithography which requires two steps—forming (usually by spin coating) and patterning photoresist films by single exposure per structure. In capillary molding, forming and patterning of polymeric film can be performed simultaneously, and the master can be used several times [40]. Photolithography is limited to a special class of polymers, whereas capillary molding is applicable to most of the polymers including polymers with low viscosity. It has found various applications, such as fabrication of compact disks, encapsulation of electronic devices [41], in the semiconductor industries. The capillary molding procedure is shown as a schematic in Fig. 29 and explained below. In spite of few limitations, the simplicity, flexibility, and ease of use of this technique proved to be advantageous over other microfabrication methods [42–44]. Figure 30 shows some representative patterns of polymeric structures formed by capillary molding [45, 46].

Table 1
Choices of plasma etching chemicals for various film materials

Film	Etchant	Typical gas compounds
Al	Chlorine	BCl_3,CCl_4,Cl_2,$SiCl_4$
Mo	Fluorine	CF_4,SF_4,SF_8
Polymers	Oxygen	DF_4,SF_4,SF_8
Si	Chlorine Fluorine	BCl_3,CCl_4,Cl_2,$SiCl_4$ C SF_4,SF_6
SiO_2	Fluorine	CF_4,CHF_3,C_2F_6,C_3F_8
Ta	Fluorine	CF_4,CHF_3,C_2F_6,C_3F_8
Ti	Fluorine	CF_4,CHF_3,C_2F_6,C_3F_8
W	Fluorine	CF_4,CHF_3,C_2F_6,C_3F_8

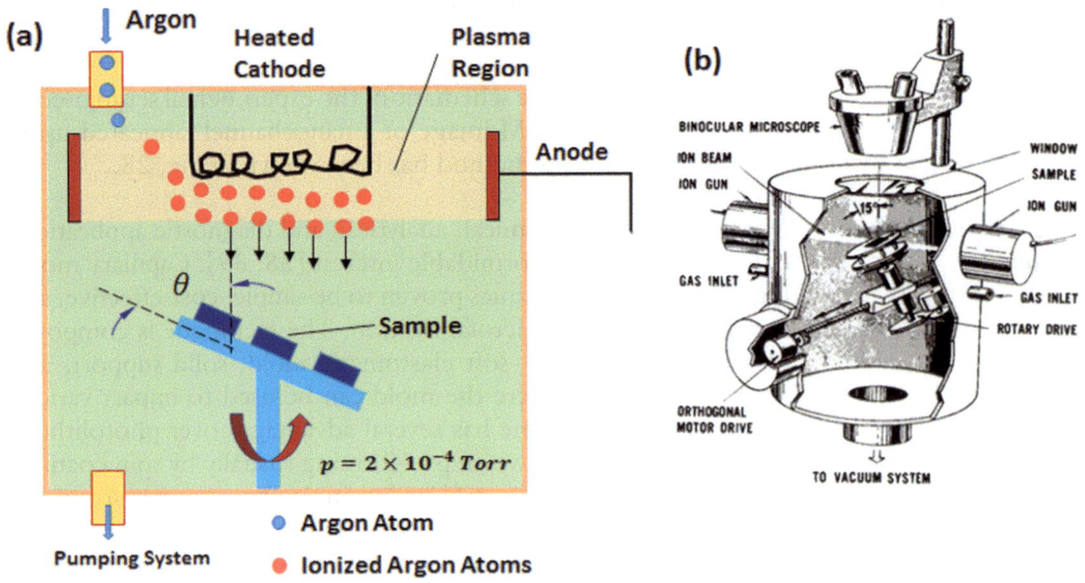

Fig. 27 (**a**) Schematic of ion beam etching system and (**b**) cross-sectional view of an actual setup [36] (Reproduced with permission from the author and publisher)

1.14 Micro-stereolithography

The micro-stereolithography technique evolved from the rapid prototyping industry provides an ideal solution for the fabrication of complex 3D shapes with high aspect ratio in a wide variety of materials, due to its unique characteristics of high resolution, high liability, and lower cost. It uses UV laser beam or light source which is scanned on a photo-polymerizable resin, followed by the curing of the resin, layer-by-layer, and then stacking the layers. The beam scanning techniques can generally be classified into two methods: (a) scanning micro-stereolithography (or vector-by-vector micro-

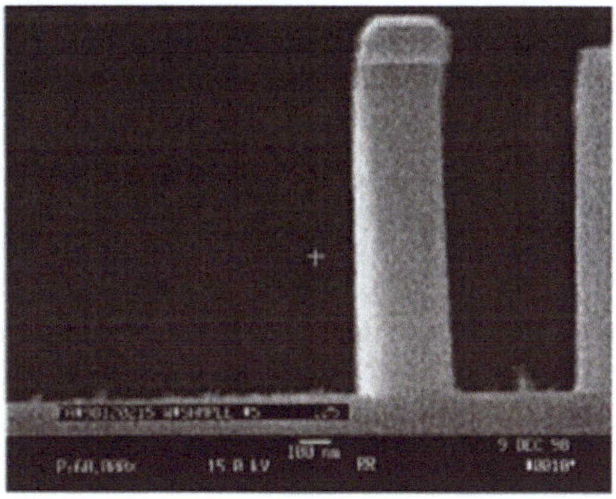

Fig. 28 An SEM image of fabricated nanochannel [38] (Reproduced with permission from the author and publisher)

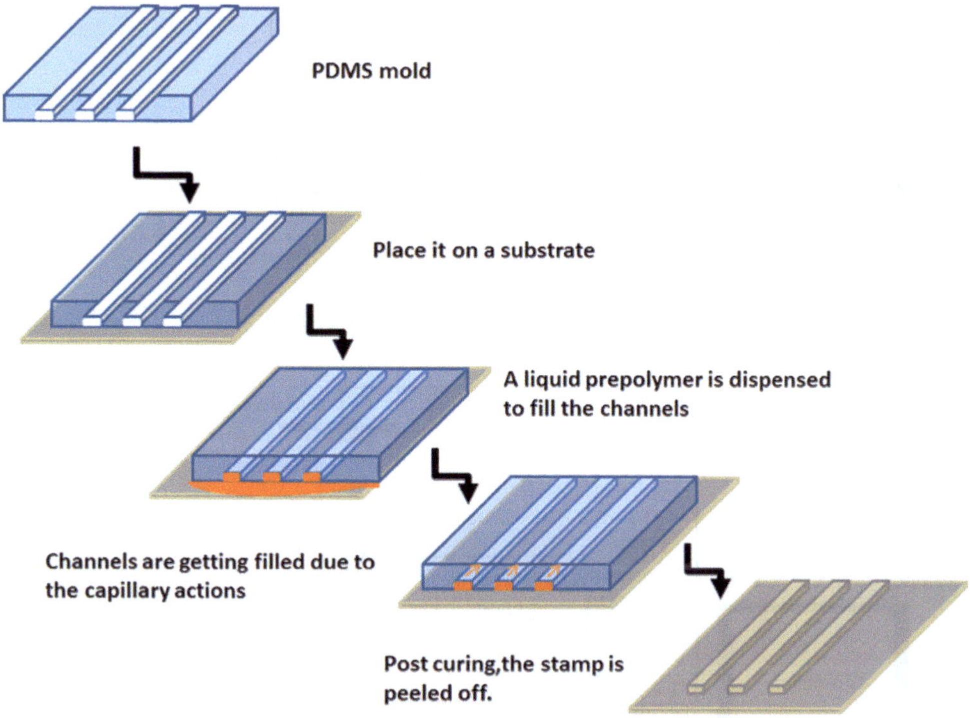

Fig. 29 Schematic of the capillary molding technique

stereolithography) (Fig. 31) and (b) projection micro-stereolithography (Fig. 32). Micro-stereolithography being an assembly-free process can be used to fabricate components having complex structures in a single step. True 3D devices from the μm to mm scale including curvilinear and reentrant microstructures that are difficult to make using conventional micromachining can easily be fabricated. The micro-stereolithography, originally generated

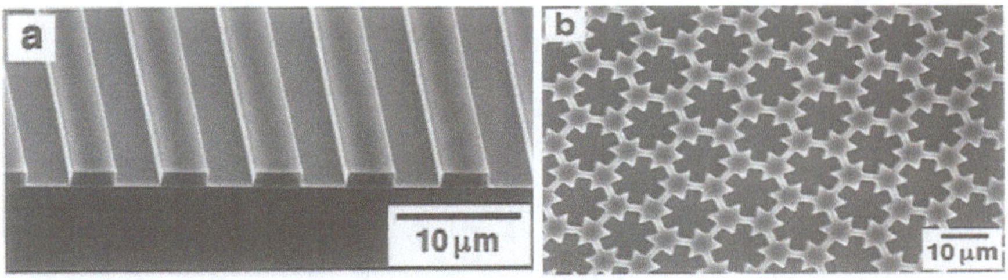

Fig. 30 Scanning electron micrographs (SEM) of patterned polymeric structures formed using liquid precursors to the polymers. (**a**) Polyurethane on Si/SiO_2 using an elastomeric mold with rectangular recessed pattern. (**b**) Polyurethane pattern on Si/SiO_2 made using a mold containing a more complex pattern (Reproduced with permission from the author and publisher)

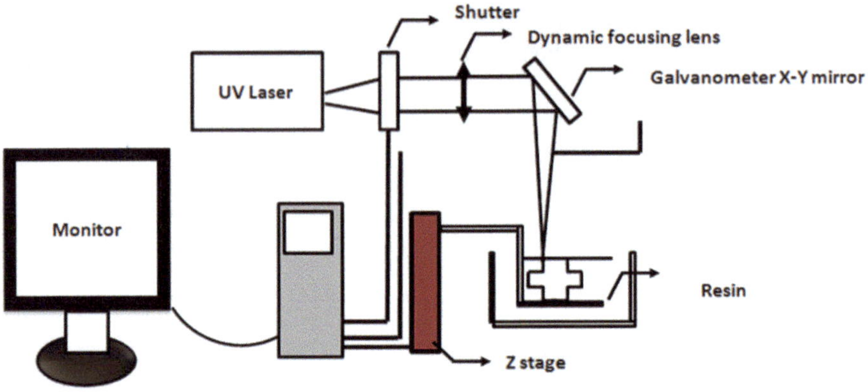

Fig. 31 Schematic of the system used in micro-stereolithography technique

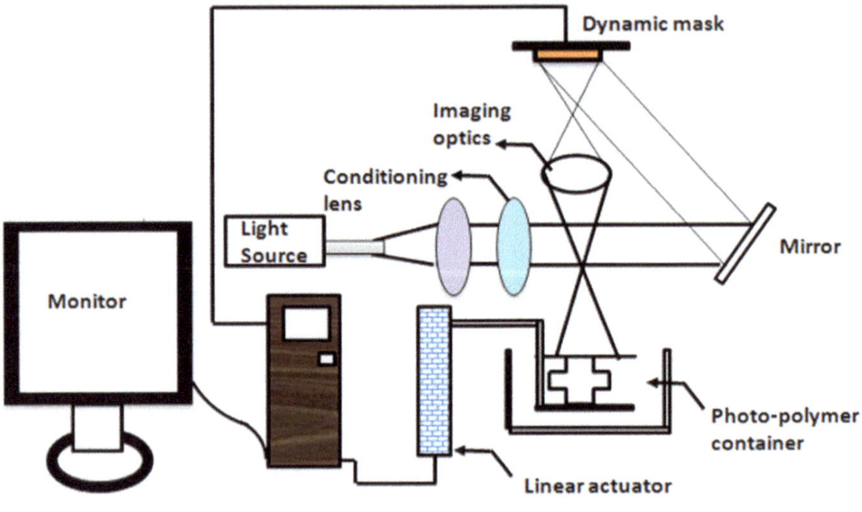

Fig. 32 Schematic of the projection micro-stereolithography system

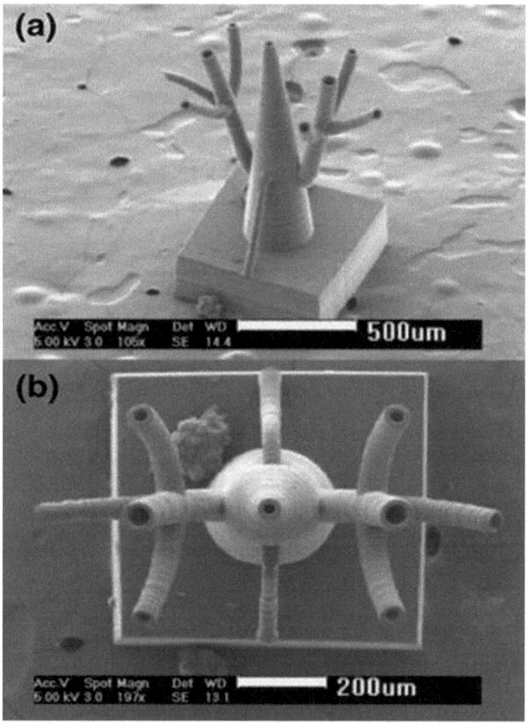

Fig. 33 (**a**, **b**) High-resolution 3D microstructures fabricated by micro-stereolithography (Reproduced with permission from the author and publisher)

from conventional stereolithography process patented by Chuck Hull in 1986 [47], was first proposed by Ikuta et al. 1993 [48]. Some representative structure made using micro-stereolithography technique have been included in Fig. 33 [49].

2 Materials

2.1 Replica Molding

1. A SiO_2, Si_3N_4, metals, or PMMA original master patterned with the desired features.
2. Weighing balance.
3. Sylgard 184 kit (has both the silicone elastomer base and curing agent).
4. Disposable syringe.
5. Stir rod.
6. Tissue wipe.
7. Desiccator.
8. Vacuum oven.

2.2 Microcontact Imprinting

1. A patterned cross-linked PDMS stamp.
2. Ink (e.g., a thiol solution).
3. Wet etching reagent (buffered hydrofluoric acid).

2.3 Micro-transfer Molding

1. A patterned PDMS stamp.
2. Polymer or precursor liquid (polydimethylsiloxane).
3. Heating oven.

2.4 Solvent-Assisted Molding

1. Photoresist (Microposit 1813, Shipley, a positive-tone novolac resin).
2. Si wafer.
3. Spin coater.
4. A PDMS stamp.
5. Ethanol.
6. Q-tips.

2.5 Hot Embossing

1. A master stamp (silicone, SiO_2, nickel).
2. Embossing machine.
3. Polymer substrate (PMMA).
4. A heating plate with active temperature control.
5. Vacuum chamber.

2.6 Injection Molding

1. Mold inserts (hardened steel, pre-hardened steel, aluminum, and/or beryllium-copper alloy).
2. Injection molding machine.
3. Polymer pellets (e.g., made of PMMA, PC, PSU, POM, PA).

2.7 CNC Micromachining

1. AutoCAD software (AutoCAD 2016).
2. CNC machine.
3. CNC program created with softwares such as LazyCAM and SprutCAM.
4. Workpiece (PMMA).
5. A sacrificial material (e.g., PC, PS, PMMA).
6. Ethanol.
7. Deionized water.
8. Compressed air.

2.8 Laser Photoablation

1. Deionized water.
2. 0.1 M NaOH.
3. Compressed air.

4. Polymer substrate (PMMA, polyvinyl chloride, polyethylene terephthalate).
5. Photomask.
6. Laser.
7. Focusing lens.
8. Motorized XY stage.
9. Compressed air.

2.9 X-Ray Lithography

1. Substrate (e.g., silicon, alumina, glass).
2. Photoresist (e.g., PMMA).
3. Photomask.
4. Titanium pellet.
5. PMMA sheet.
6. Vacuum oven.
7. Methyl methacrylate.
8. Compressed air.
9. Mask holder.
10. Scanner.
11. Shims.
12. Substrate holder.
13. Synchrotron X-ray source.
14. Solvent for developing substrate (e.g., 7:3 mixture of IPA and water for PMMA substrates).

2.10 UV Patterning

1. PMMA sheet.
2. Deionized water.
3. Dishwasher gel.
4. Methanol.
5. Nitrogen gas.
6. Gold wire for sputtering gold.
7. TFA gold etchant.
8. UV light source (~254 nm).
9. Solvent for developing PMMA sheet (e.g., 7:3 mixtures of IPA and water).

2.11 Plasma Etching

2.11.1 For Creating a Mask over a PMMA/PEEK Substrate

1. PMMA/PEEK sheet.
2. Si-containing photoresist (e.g., PDMS).
3. Photomask.

4. UV light source (~365 nm).
5. Methyl isobutyl ketone.
6. Isopropyl alcohol.

2.11.2 For Creating a Mask over a PDMS Substrate

1. PDMS sheet.
2. Aluminum wire for thermal deposition.
3. SU-8 3050 photoresist.
4. Spin coater.
5. UV light source.
6. Propylene glycol methyl ether acetate.
7. Isopropyl alcohol.

2.11.3 For Plasma Etching

1. PMMA/PEEK/PDMS workpiece.
2. Plasma etching system.
3. Vacuum pumps.
4. Oxygen gas.

2.12 Ion Beam Etching

1. Polymer substrate (PMMA/PEEK/PDMS).
2. Substrate holder.
3. Ion beam etching system.
4. Inert gas (e.g., argon).

2.13 Capillary Molding

1. A master stamp.
2. Polymer precursor (e.g., Sylgard 184 and its cross-linking agent for PDMS devices).
3. Heating oven.

2.14 Micro-stereolithography

1. CAD software.
2. UV curable photopolymer (epoxy-based photopolymer (3D systems, Valencia, CA).
3. Tank.
4. Programmable UV beam scanner.
5. Washing solvent (DBE, a mixture of dibasic esters).

3 Methods

3.1 Replica Molding

1. Obtain an original master patterned with the desired nanometer-sized features and prepared using advanced lithographic techniques.
2. Measure the mass of a plastic cup (empty) on a weighing balance and tare it.

3. Pour the required amount of polymer (Sylgard 184 Silicone elastomer base) into the cup and measure the mass.
4. Use a disposable syringe to inject the curing agent into the silicone elastomer base. Usually a mass ratio (curing agent/ elastomer base) of 1:10 is recommended. After injecting, dispose the syringe into the bin.
5. Stir the mixture vigorously for 2 min using a stir rod and wipe the rod with tissue wipes after stirring (*see* **Note 1**).
6. Degas the mixture by placing it inside the desiccator for 50–60 min till the bubbles disappear (*see* **Note 2**).
7. Prepare a boat structure by wrapping a foil around the master which will restrict the PDMS to spill out of the master (*see* **Note 3**).
8. Now, pour the mixture (elastomer base and curing agent) gently onto the master within 10–15 min of degassing and wait for the additional bubbles (generated during pouring) to disappear.
9. Place the master into a vacuum oven at a temperature of 70 °C for 1 h for curing (*see* **Notes 4** and **5**).
10. After curing, peel off the PDMS stamp carefully from the master to obtain the required microstructure (*see* **Note 6**).

3.2 Microcontact Imprinting

1. Fabricate a patterned cross-linked PDMS stamp by replica molding (REM) (*see* **Note 7**).
2. Next, immerse the stamp into the ink to cover the stamp. The inking step can also be accomplished by vapor deposition and by placing the stamp above a beaker containing the ink (*see* **Note 8**).
3. The inked stamp should be subsequently placed in contact with the substrate to be patterned. Due to the patterned structure of the stamp, only protruded sections can make contact with the substrate (*see* **Note 9**).
4. Remove the PDMS stamp after the SAM formation.
5. Finally, wet etching can be performed to remove the unprotected parts.

3.3 Micro-transfer Molding

1. Prepare a PDMS mold, known as stamp, which contains the microchannels.
2. Fill the PDMS microchannels with prepolymer or precursor liquid (*see* **Note 10**).
3. Invert the stamp filled with the liquid polymer (*see* **Note 11**).
4. Make a conformal contact between the inverted stamp and the substrate to be patterned.
5. Cure the polymer at the required temperature and time, and then peel off the PDMS stamp (*see* **Note 12**).

3.4 Solvent-Assisted Molding

1. Prepare a thin film of photoresist on a Si wafer by spin coating at 5500 rpm for 30 s and baking at 105 °C for 3.5 min.
2. Select a suitable elastomeric mold, usually PDMS, and fabricate the desired patterns to make the master mold (*see* **Note 13**).
3. Wet the PDMS mold with ethanol with the help of Q-tips (*see* **Notes 14** and **15**). Allow the solvent to fill the recessed regions on the surface of PDMS.
4. Place the mold on top of the photoresist film such that the elastomer makes conformal contact with the resist forming microscale channels between them (*see* **Note 16**).
5. The compliant PDMS will adhere spontaneously with the surface and squeeze out the extra ethanol from the regions of contact. The permeability of solvent and gas through PDMS enables the uniform evaporation of solvent and escape of trapped air bubbles (*see* **Note 17**).
6. The remaining solvent dissolves or swells the photoresist, and a negative pattern forms in the resulting polymeric fluid (*see* **Note 18**).
7. Allow the PDMS mold to remain on the resist for ~5 min at room temperature until most of the solvent has dissipated [11].
8. Peel away the PDMS mold to obtain the microstructured photoresist.

3.5 Hot Embossing

1. Firstly, pattern the master stamp by silicon micromachining, LIGA process, or CNC micromachining (*see* **Note 19**).
2. Mount the master onto the embossing machine, which consists of a force frame that delivers the embossing force via a spindle and a T-bar to the boss.
3. Mount the master and the flat polymer substrate (*see* **Note 20**) on the heating plates. Circulate oil with high heat capacity through the cooling channels, which allows active cooling and isothermal heating of the plates.
4. Before operation, heat the plates in a vacuum chamber at 10^{-1} mbar to a temperature just above the glass transition temperature T_g of the polymer material (*see* **Note 21**). Vacuum is required to remove gas bubbles in the microstructures and also to prevent corrosion of the master [14].
5. Bring the master mold on the frame into contact with the substrate (*see* **Notes 22** and **23**), and emboss with a sensor feedback-controlled force, typically of the order of 20–30 kN [14].
6. While still applying the embossing force, cool the tool-substrate sandwich to just below T_g to stabilize the patterned microstructure on the polymer.

7. The embossing master can then be mechanically driven apart from the substrate (*see* **Note 24**). The polymer on the bottom plate will now have the desired pattern.

3.6 Injection Molding

1. Fabricate the mold inserts by bulk micromachining or LIGA.
2. The injection molding machine consists of a screw, an injection nozzle, a heater, and a mold insert.
3. Feed the polymer pellets through an open-bottomed container (hopper) into the screw (*see* **Note 25**).
4. An electric or hydraulic motor rotates the screw inside a barrel which is surrounded by heating elements as shown in Fig. 11. As the temperature is raised to a desired temperature, the polymer softens and melts (*see* **Note 26**).
5. The screw pushes the molten polymer into the mold cavity through its grooves at high pressure (*see* **Note 27**).
6. A gate before the mold cavity restricts the flow of the melt into the mold and limits backflow (*see* **Note 28**).
7. The screw injects the molten polymer into the mold, holds it under pressure, and adds more molten polymer to avoid formation of air gaps in the final product as a result of cooling and solidification (*see* **Note 29**).
8. The gate solidifies and isolates the mold from the injection cylinder (*see* **Note 30**).
9. Circulate a cooling liquid such as water through the small holes in the mold to cool the molten polymer inside the mold cavity. This step consumes 85% of the cycle time [16] (*see* **Note 31**).
10. After cooling and solidification of polymer, remove the patterned polymer mold by opening the two halves of the mold cavity holding the solidified polymer.
11. Clean any extra polymer from the mold.

3.7 CNC Micromachining

1. Design the microstructure with the help of AutoCAD/ SolidWorks.
2. Choose appropriate cutting tools, speed (RPM), feed rates, and depth of cuts (*see* **Notes 32–36**). Prepare the CNC program with the help of softwares such as LazyCAM, SprutCAM, etc. Load the CNC program (*see* **Note 37**).
3. Check the levelness of the worktable with a spirit level.
4. Fix the workpiece onto the worktable (granite) with the help of clamps or adhesive tape. The schematic of the machine is shown in Fig. 13. Add a sacrificial layer (PC), polystyrene (PS), and polymethyl methacrylate (PMMA) to avoid accidental impact on the worktable [20].

5. Properly align the cutting tool to ensure accuracy. For Z alignment, lower (step) the tool toward the workpiece while the spindle is running. As soon as the tool touches the surface, a chip forms. Stop the Z movement and set it as "zero" (*see* **Note 38**).
6. For X and Y alignment, move the tool sideways and lower it while the spindle is running till a chip forms at the desired location. Set it as "zero" for X and Y. Check the levelness of the workpiece.
7. Start the spindle coolant and run the program (*see* **Notes 39** and **40**).
8. During machining, milling debris are cleared from the workpiece by blowing air and vacuuming or with a flood coolant of DI water mixed 20:1 with a synthetic coolant [20] (*see* **Note 41**).
9. After machining, clean the patterned polymer workpiece with 70% ethanol in DI water, and dry with compressed air [20].

3.8 Laser Photoablation

1. Clean the substrate initially with DI water, and then immerse it to a 0.1 M NaOH solution for 30 min. Rinse it with DI water, and dry with pressurized air [24] (*see* **Notes 42** and **43**).
2. Mount the cleaned substrate onto the stage.
3. A photomask with the desired pattern should also be mounted as shown in Fig. 15 (*see* **Note 44**).
4. Fire UV laser pulses (193 nm) onto the substrate through the mask and a 10:1 lens with a frequency of 50 Hz at 200 mJ/pulse [25] (*see* **Notes 45–47**).
5. Move the XY stage with the polymer substrate horizontally at a speed of 0.15–0.2 mm/s to desired channels of desired lengths (*see* **Notes 48–51**).
6. Pattern reservoirs by firing sufficient pulses to penetrate the whole polymer sheet.
7. As a result of "ejection by pressure," each successive pulse of the laser cleans away any debris that might have accumulated in the ablated region (*see* **Note 52**).
8. Clean the patterned polymer using a jet of compressed air.

3.9 X-Ray Lithography

1. Select a suitable substrate for the fabrication, such as silicon, alumina, and glass.
2. Select a suitable photoresist, either positive or negative, for the technique.
3. Fabricate a mask with a material having low atomic number, and keep it ready for the exposure step (*see* **Note 53**).
4. Sputter coat a titanium layer onto the substrate.

5. Oxidize the titanium layer.
6. Cut PMMA sheets (with thickness ~1.5 mm) to either rectangular or circular shape depending upon the type of mask and design layout.
7. Anneal the cut PMMA sheet at 110 °C inside a vacuum oven.
8. Bond the annealed PMMA sheet over the titanium oxide side of the substrate by applying methyl methacrylate monomer on the substrate as well as substrate and pressing them using a compressed air source for overnight.
9. Normally, most of the applications utilize PMMA resist layer of thickness >100 μm. So, thin down the glued PMMA sheet by fly-cutting process (utilizing a diamond cutting tool) in two steps. In first step, remove the material from the PMMA bulk at a rate of 100 μm thickness per cycle and, in the second step, smoothen out the PMMA surface to transparency.
10. Mount the mask onto the customized holder and load into the scanner such that it is close to the water-cooled copper ring.
11. In order to maintain an optimal gap between mask and the polymer substrate, utilize shims of precise thickness. Typically, a gap of 100 μm and 1 mm is utilized for micrometer size features and millimeter size features, respectively.
12. Mount the substrate over the sample holder and fix in alignment with the mask and synchrotron X-ray source.
13. Expose the substrate to X-rays with an optimal level of intensity depending upon the feature size and thickness of the microstructure with the mask in between the substrate and X-ray synchrotron radiation source.
14. Develop the exposed substrate further using a suitable solvent. For PMMA, popular solvent used is 7:3 mixture of IPA and water.

3.10 UV Patterning

1. Cut the PMMA sheet of thickness 5 mm into desired size (mostly 3×3 in. square size) in order to proceed further with the fabrication procedure.
2. Clean the substrate with deionized water and mild dishwasher gel (*see* **Note 54**).
3. Immerse the sample in methanol bath for 10 min just to remove any oil residual on the surface, and dry it with N_2 gas.
4. Apply a 100 nm thick layer of gold over the substrate by sputtering at 80 W (see Fig. 20a for the image of the gold-coated substrate).
5. Then pattern the gold film using optical photolithography technique. TFA gold etchant can be used for etching gold.

6. Expose the PMMA sample with non-collimated UV light having a nominal power of 4 mW/cm^2 and a spectrum whose strongest peak is at 254 nm. The exposure time depends upon the depth of the microstructure required (*see* **Notes 55** and **56**).
7. After finishing the exposure process, transfer the sample to the development bath at 28 °C. A 7:3 mixture of IPA and water is used as the development solution. Development time of sample is dependent on the depth of the microchannel.
8. The development process is monitored continuously at a period of 10 min. Once the development is completed, the sample is dried by blowing N_2 gas (*see* **Note 55**).
9. Once the sample is developed, the gold layer is dissolved using TFA gold etchant.

3.11 Plasma Etching

3.11.1 Procedure for Creating a Mask over a PMMA/PEEK Substrate [32]

1. Cut the PMMA/PEEK sheet to optimal size.
2. Coat a thin layer (~2 μm) of Si-containing photoresist such as PDMS or inorganic-organic hybrid polymer (ORMOCER) over the PMMA/PEEK sheet.
3. Expose the substrate with UV light broadband at 365 nm through the desired mask with the pattern which has to be replicated.
4. Remove the soluble part of the Si-containing polymer in MIBK (methyl isobutyl ketone) and IPA. This makes the sample ready for creating the masking layer plasma etching.

3.11.2 Procedure for Creating a Mask over PDMS Substrate [32]

1. Deposit a thin layer of Al over the substrate in thermal evaporator by maintaining the largest possible gap between substrate and Al target to minimize formation of wrinkles at Al-PDMS interface.
2. Spin coat photoresist SU-8 3050 over the Al layer side of the substrate.
3. Expose the SU-8 3050 layer with UV light of optimum power intensity depending upon the thickness of the photoresist layer.
4. Develop the nonexposed part of photoresist using PGMEA and IPA.

3.11.3 Procedure for Plasma Etching [33]

1. Load the sample containing the masking layer into the reaction chamber of plasma etching system.
2. Create a low pressure of approximately 0.1 mbar inside the reaction chamber with the help of vacuum pump.
3. Feed the process gas (oxygen) into the chamber, and create a working pressure of 0.1 mbar to 1 mbar.

4. Once the working pressure is achieved, switch on the generator, which ionizes the process gas converting it into the plasma state.
5. Expose the substrate to plasma for certain time duration (1 min to few hours) in order to achieve the desired etching depth (*see* **Notes 57–59**).
6. Switch off the generator, and evacuate the chamber to bring the chamber pressure to atmospheric condition.
7. Remove the workpiece.

3.12 Ion Beam Etching

1. Create a mask over the polymer substrate using similar procedure as explained in the plasma etching technique.
2. Load the sample onto the substrate holder.
3. Create a vacuum level of pressure ~2×10^{-4} Torr inside the chamber [35].
4. Introduce the continuous supply of inert gas (often argon) into the plasma chamber made of ceramics with a coil wound around.
5. Start the neutralizer, which emits few electrons to the plasma chamber.
6. Switch on the RF power to the coil, which sets the electron to vibration, induces the generation of more electrons and ions, and ultimately generates the plasma inside the chamber.
7. Direct the ions toward the substrate to remove the atoms from the substrate.
8. Keep the cooling systems switched on to flush away the heat generated because of machining.
9. Keep the etching system on till the etch depth is achieved (*see* **Notes 60–62**).
10. Switch off the RF power source, which essentially stops the generation of plasma inside the chamber.
11. Evacuate the chamber to atmospheric pressure and unload the sample carefully.

3.13 Capillary Molding

1. Firstly, fabricate a master (having a network of recessed channels of desired dimensions) using lithographic or other non-lithographic techniques [42, 43] which holds the complementary structures of the elastomeric mold.
2. Dispense the polymer on the master to fabricate the elastomeric mold. Usually, PDMS (Sylgard 184, Dow Corning, USA) with its cross-linking agent is used as the stamp material due to its high elasticity and low surface energy. The transparency of PDMS to UV light, down to 300 nm, has made it compatible with the photochemical polymerization employed

for the capillary molding. Alternative materials such as polyimide [44], polyurethane, and novolac (phenol-formaldehyde) resin have also been used to prepare these stamps (*see* **Notes 63** and **64**).

3. Then, bring the elastomeric stamp which contains the inverse features of the master to an intimate contact with the solid substrate. Thus, the recessed microchannels on the stamp form a network of empty capillaries (*see* **Notes 65–69**).
4. Place a low-viscosity fluid precursor in close contact at one end, which spontaneously fills the channels by capillary action (*see* **Notes 63** and **70**).
5. Then, cross-link the prepolymers thermally (at 65 °C, 1–2 h for thermally cured epoxies) or photochemically (for thin PDMS films, which are optically transparent).
6. Once the polymer has cross-linked completely, peel off the elastomeric stamp from the substrate, thus leaving the patterned microstructures on the substrate surface (*see* **Notes 71** and **72**).

3.14 Micro-stereolithography

1. Design a 3D solid model of the product to be fabricated using CAD software.
2. After designing the 3D solid model with CAD software, slice it into a series of 2D layers of uniform thickness and convert it into a bitmap file.
3. Then, take the UV laser curable photopolymer in the tank.
4. Dip the elevator into the tank which is initially above the liquid level, and then bring it up again to ensure that a suitable thickness of liquid remains on top of it.
5. Then execute the NC codes generated from each sliced 2D file to control the UV beam scanning (*see* **Notes 73–80**).
6. Finally, wash the product with suitable solvents.

4 Notes

1. The mixture of base elastomer and curing agent should be stirred properly, to avoid any nonuniformity [1].
2. Degassing vacuum gauge level should go beyond 20 in./Hg for proper removal of bubbles [2].
3. Surface tension will restrict the spillage of PDMS from the master; still a proper boat structure should be prepared to avoid wastage and nonuniformity in the final microstructure [3].

4. Place the master on a flat surface; otherwise it will lead to an uneven thickness of the PDMS microstructure [3].
5. Curing temperature and time is a very crucial parameter for REM, which will decide flexibility of the final PDMS structure. Curing above the recommended time and temperature can lead to hard and brittle microstructure. Curing below the recommended time and temperature can make the structure sticky and floppy [6].
6. After peeling off the PDMS stamp, rinse it with isopropyl alcohol (IPA) to wash off the PDMS debris during peeling and cutting. Do not use any polar solvents (e.g., hexane, toluene, and methylene chloride) as these can swell up the cross-linked PDMS [50].
7. Before casting the stamp, it is very important to functionalize the master with the silane vapor so that the cast polymer does not adhere to the master.
8. During inking of the stamp, it is very important to maintain an appropriate concentration of the ink solution.
9. A gentle pressure should be applied to make a conformal contact between the inked stamp and the substrate.
10. Remove the extra liquid by scrubbing or using nitrogen stream.
11. During inversion, care has to be taken to avoid spillage of prepolymer liquid.
12. The appropriate curing temperature and time needs to be used to avoid distorted patterns.
13. Choosing appropriate solvent and mold material plays an important role in the effectiveness of this molding technique [50].
14. Many nonpolar solvents (e.g., hexane, toluene, and methylene chloride) are capable of swelling the cross-linked PDMS and therefore should not be used for the process [11].
15. The solvent should have a relatively high vapor pressure and a moderately high surface tension (e.g., methanol, ethanol, and acetone) to ensure rapid evaporation of the excess solvent and minimal swelling of the PDMS mold. Solvents with low vapor pressures (e.g., ethylene glycol and dimethyl sulfoxide) are not suitable. Solvents with high surface tension (e.g., water) wet the PDMS surface only partially and therefore cannot be used [11].
16. If the height of the pattern structure is much higher than the thickness of the resist, a hard mold can be used since the solvent can be expelled from the gap between the resist and the mold [10].

Table 2
Embossing parameters for PMMA and PC [15]

Material	T_g (°C)	Embossing temperature (°C)	De-embossing temperature (°C)	Hold time (s)
PMMA	106	120–130	95	30–60
PC	150	160–175	135	30–60

17. The distortion of PDMS mold as a result of solvent adsorption can be avoided by changing PDMS composition and surface modification and reducing surface energy [10]. Deng et al. [8] have used a coating of amorphous fluorinated polymer, a low surface energy material to avoid swelling and deformation of the PDMS. The fluoropolymer molds are easier to release from the resists and avoid swelling by solvent. However, its chemical inertness and low surface energy make functionalization and inking difficult during the microcontact printing.
18. Moderate pressure can be applied for improved engraving of pattern into the mold [10].
19. To avoid frictional forces between the mold and the polymer microstructures during de-embossing leading to defects, the master mold needs to have minimum sidewall roughness. LIGA is best suited (in place of hot embossing) for high aspect ratio channels [14].
20. Polymethyl methacrylate (PMMA) and polycarbonate (PC) are the most commonly used polymers for hot embossing. Table 2 shows the various embossing parameters for these polymers [14].
21. Glass transition temperature T_g, forming pressure, and holding time are the most important parameters for the micro-hot embossing process [1].
22. The surfaces of the mold and the substrate should have minimal chemical surface bonding sites as this will offer additional stiction force during de-embossing [14].
23. Si mold can be coated with amorphous silicon carbide, Teflon-like fluoropolymer, or self-assembled n-octadecyltrichlorosilane for reduction of adhesion between the mold and patterned polymers [14].
24. Three different configurations can be used for hot embossing: plate to plate, roll to plate, and roll to roll [1].
25. Cracking and crazing can occur due to improper reinforcement content, loading, under-curing, and resin richness. This can be solved by increasing glass and filler content, extending molding

cycle time, and making sure the reinforcement is not displaced during mold closing [17].

26. Blisters on the mold can indicate faulty cooling or heating facility [18].
27. Flash (excess material on mold) may be due to over-packing of mold, high injection speed, low clamping force, or contamination of polymer. Too slow injection can lead to formation of flow lines on the mold [18].
28. Poor tool design, gate position, and high injection speed can lead to jetting by turbulent flow of material [18].
29. Voids in the mold can be caused by low pack or holding pressure, out of registration of mold halves, or improper melting of polymer [17, 18].
30. Weld lines can appear on the mold when the material or the mold temperatures are set too low or when the time between injection and holding is too low [18].
31. Warping can occur when the cooling is too short and when the cooling water is of incorrect temperature.
32. Take sufficient care while choosing the tool size, spindle speed, feed rate, and depth of cut. Incorrect input may lead to tool failure or premature tool wear and poor quality, accuracy, and surface finish.
33. Longer tools deflect more easily. Therefore, use shortest possible tool to maximize rigidity [22].
34. Ensure that the tips of the tools do not come into contact with each other in storage [22].
35. Always check for the sharpness of the tools. Replace when it loses its edge.
36. Too high spindle speed and too low feed rate can lead to friction and can generate heat. Excess heat can cause material ductility resulting in burrs, polymer melting, or tool breakage [20].
37. Writing the code manually will help in expanding the utility of the mill [20].
38. To avoid material removal during XYZ alignment, place a paper below the tool, and move it side to side while lowering the tool. At the position, where the paper gets caught between the tool and the workpiece, offset the tool with the thickness of the paper to reach the zero position [20].
39. It is advantageous to start the coolant well before starting the spindle in order to provide the coolant enough time to reach the tool [22].

40. Avoid driving the tools into deep slots and pockets except with absolutely minimal cut depth and stepover per pass [22].
41. For smaller tools, use mist instead of coolant flood to avoid deflection and breakage [22].
42. Polymers that show a photochemical ablation behavior at the irradiation wavelength would be most preferable for structuring. In these polymers, the damage of the surrounding material is minimized, and less carbonization is observed [24].
43. If the polymers do not have intrinsic absorption at the irradiation wavelength, a dopant can be added to induce the necessary absorption. The dopant can be added at the molecular level into the polymer backbone and side chains or as absorber particles at nanometer or micrometer range [24].
44. Laser machining can be done in two ways: direct writing and using a mask. In the direct writing mode, the laser beam is focused and scanned on the substrate surface to make pattern. Here, the smallest structure depends on the accuracy of the scanning system and is of the order of 25–50 mm. In the masking mode, the mask determines the detailed shape of the structure. Therefore, the minimum structure size can be brought down to twice that of the laser wavelength [1].
45. The three types of lasers that can be used for laser photoablation are [1]:
    (a) Excimer laser—UV (351, 308, 248, 193 nm)
    (b) Nd:YAG laser—Near infrared (1067 nm), visible (533 nm), and UV (355, 266 nm)
    (c) CO_2 laser—Deep infrared (10.6 μm)
46. The choice of laser wavelength depends on the minimum structure size and the optical properties (absorption and reflection characteristics) of the substrate material. The minimum achievable focal spot diameter and the smallest structure size are about twice the laser wavelength [1].
47. The process depends on the thermal energy of the laser beam. The important parameters of photoablation are ablation rate *d* (*F*), ablation threshold fluence F_{th}, and effective absorption coefficient $\propto_{eff}$. The ablation process is defined as [24]

$$d(F) = \frac{1}{\alpha_{eff}} \ln\left(\frac{F}{F_{th}}\right)$$

48. The cross section of the microchannel depends on the energy distribution of the laser beam, its speed, the laser power, and the thermal diffusivity of the substrate material [1].

Table 3
Ablation depth per laser pulse for different materials [1]

Material	Depth per pulse (μm)
Polymers	0.3–0.7
Ceramics and glass	0.1–0.2
Diamond	0.05–0.1
Metals	0.1–1

Table 4
Choices of the mask for X-ray lithography technique

Mask material	Quality
Beryllium	High quality and cost (including submicron sizes)
Graphite	Intermediate quality and cost: 2 μm feature size and above
Graphite	Intermediate quality and cost: 10 μm feature size and above

49. The energy of the laser beam has a Gaussian distribution; thus the cross section of the channel also has a Gaussian shape [1].
50. The choice of laser power depends on the desired structure size and ablation rate. The typical ablation depths per pulse for different materials are given in Table 3 [1].
51. Short-pulsed lasers are advantageous because they create clean and accurate structures since they avoid heat flow to surrounding materials [1].
52. Depending on the chemical nature of the resist material, the X-ray exposed areas may cause cross-linking (for negative resists) or bond breaking (for positive resists). After exposure, the resist pattern on the substrate can be developed utilizing the proper solvent. The exposed areas in a positive resist will dissolve, and the unexposed areas will remain. Alternatively, the exposed areas in a negative resist will not be affected, while the unexposed areas will dissolve.
53. For the choice of the mask, the following materials as shown in Table 4 can be utilized. Please refer elsewhere for the fabrication procedure of the mask [26].
54. Instead of dishwater gel, neutral detergent solution can also be used for the cleaning purpose.
55. The depth of the microstructure varies with the exposure time and the development time as shown in Fig. 21a, b, respectively.

Hence, the exposure time should be kept optimal to achieve the desired depth of the microchannel.

56. After every successive exposure time (say 10 min), the sample should be taken out of the bath and should be quenched in an ultrasonic IPA bath at room temperature (18 °C) for 10 s.
57. The main process parameters of this technique are plasma power (in W), total process pressure (in Pa), gas flow (in sccm), substrate bias voltage (in V), and substrate temperature (in °C) [33]. The etching rates of PMMA, PEEK, and PDMS materials increase with decreasing chamber pressure, increasing plasma power, and increasing bias voltage as shown in Fig. 24a, b.
58. Please refer to Fig. 24c, d for relation between etch depth, etch rate, and etching time, respectively, for PMMA and PEEK materials [33].
59. One disadvantage with this technique is the nano-roughness created at the microchannel wall surface. Figure 25a, b shows the variation of the nano-roughness level for PMMA, PEEK, and PDMS materials [33].
60. The parameters governing the ion beam etching process are the type of gas used, kinetic energy of the ions, ion flux, and the angle of incidence of the ion beam with respect to the sample surface [35].
61. Angle of incidence θ plays an important role in the yield of the process. Yield increases to 3.5 times for $\theta \approx 70°$ and falls off rapidly at $\theta \approx 80°$ [35].
62. Yield of the experiment increases with the increase in the incident ion energy [38].
63. The prepolymer has to be nonreactive to the soft elastomeric substrate. It should have low viscosity (<400 cP) as well. Upon curing, it is expected that the volume should change (<3%), and most importantly it should wet the support or the substrate partially or completely. UV-curable polymethyl methacrylate (SK-9, Summers Optical), ultraviolet-curable epoxies (UV 15, UV 15-7, Master Bond), heat-curable epoxies (F113, F114, TRACON), ultraviolet-curable polyurethanes (J-91 Summers Optical, NOA 60, 71, 72, 73, Norland), and heat-curable polyurethanes (NOA 133) are normally being used as liquid prepolymer for the capillary micro-molding process [41].
64. Once the master mold is made, many elastomeric stamps can be cast and used repeatedly. Clean (with compressed nitrogen gas or argon gas) the master mold every time before use, to avoid any contamination.

65. The softness of the PDMS enables the mold to make a conformal contact with the substrate which limits the lateral spreading of the prepolymer.
66. The mold and the substrates adhere without any external adhesive forces enabling reproduction of channels with high accuracy.
67. Si, SiO_2 (both native and thermally grown), NiO, Ni, TiO, TiO_2, platinum, glass, and gold can be used as supports in capillary molding [40].
68. The surface energy of the support is often preferred to be higher than that of PDMS to ensure selective adhesion of the polymer to the support rather than to the master.
69. The support need not be necessarily flat. For curved surfaces, a PDMS master is fabricated as a thin film (~100 μm) that can adhere to the curved substrate without losing the fidelity of the stamp [40].
70. When brought in contact with the organic solvents including chloroform, methylene chloride, toluene tetrahydrofuran, etc., PDMS stamps swell and lose the conformal contact with support [40].
71. A PDMS thin layer can be spin casted before the stamp is placed on the substrate to avoid the partial adhesion of the fabricated pattern or the channels with the support of the elastomeric stamps [46].
72. Some access holes through the PDMS stamp may be provided to improve the capillary filling into the network of capillary channels.
73. The scanning micro-stereolithography techniques work on the principle where the solid micro parts are built in a point-by-point and line-by-line approach as shown in Fig. 31. A laser beam is focused (through a dynamic lens and followed by two low-inertia galvanometric mirrors) on the surface of a resin system containing UV photo initiator, monomer, and other additives.
74. Generally, the scanning is done by line-by-line or point-by-point approach, so there can be some error associated with this. Generally, components of dimensions above 30 μm are built using this technique.
75. With the rotation of the mirrors, the focal point of the laser beam may remain slightly above or below the liquid surface, instead of remaining on top of the liquid surface. To avoid such problems, the optical setup can be fixed, while the X-Y tank can be assigned with movements.

76. To get a better resolution, the beam is focused more precisely in order to reduce the spot size to a few micrometers in diameter.

77. Projection micro-stereolithography (shown in Fig. 32) builds 3D microstructure in a layer-by-layer fashion. Here the UV laser source is replaced by the UV light source and a dynamic mask or dynamic pattern generator which is used to digitally pattern and project the incident light as an image and expose an entire cross-sectional layer at once. Feature sizes smaller than 10 μm can be fabricated as it is not limited by laser beam radius or its scan speed. After passing through a series of optical lens, the pattern light is focused on the surface of liquid photopolymer resin. Polymerization of the resin starts as soon as it comes in contact with light, and once the first layer is cured, it is repositioned such that additional resin is recoated onto the previously cured photopolymer. An image of the next cross-sectional layer is projected on top of the previous layer for curing. After the final cross-section is projected, the completed part is removed and post-processed.

78. The factors affecting the resolution of components are (a) laser intensity, (b) motion and quality of the beam, (c) photopolymer/monomer used, and (d) focusing and exposure.

79. The desired properties of the UV curable resins are (a) photosensitivity at the operating wavelength, (b) low viscosity to produce a highly smooth surface, (c) high curing speed, (d) low shrinkage during polymerization, and (e) high absorption for the low penetration of light.

80. Typically epoxy and acrylate resins are normally used, e.g., HDDA (1,6-hexanediol diacrylate) with 4% by weight of photo-initiator.

References

1. Nguyen NT (2008) Micromixers: fundamentals, design, and fabrication. William Andrew, New York
2. Mcdonald JC, Duffy DC, Anderson JR, Chiu DT, Wu H, Schueller OJ, Whitesides GM (2000) Fabrication of microfluidic systems in poly (dimethyl siloxane). Electrophoresis 21:27–40
3. Gates BD (2005) Nanofabrication with molds & stamps. Mater Today 8(2):44–49
4. Andras P, Reinhoudt DN, Jurriaan H (2009) Microcontact printing: limitations and achievements. Adv Mater 21:2257–2268
5. Ruizab SA, Chen CS (2006) Micro contact printing: a tool to pattern. Soft Matter 3:168–177
6. Xia Y, Whitesides GM (1998) Soft lithography. Angew Chem Int Ed 37:550–575
7. Zhao XM, Xia Y, Whitesides GM (1996) Fabrication of three-dimensional micro-structures: microtransfer molding. Adv Mater 8 (10):837–840
8. Deng T, Ha YH, Cheng JY, Ross CA, Thomas EL (2002) Micropatterning of block copolymer solutions. Langmuir 18:6719–6722
9. Lee JH, Kim CH, Ho KM, Constant K (2005) Two-polymer microtransfer molding for highly

layered microstructures. Adv Mater 17 (20):2481–2485

10. Han LL, Zhou J, Xiao G, Gao CY (2009) Solvent-assisted polymer micro-molding. Chin Sci Bull 54:2193–2204
11. King E, Xia Y, Zhao XM, Whitesides GM (1997) Solvent-assisted microcontact molding: a convenient method for fabricating three-dimensional structures on surfaces of polymers. Adv Mater 9(8):651–654
12. Peng L, Deng Y, Yi P, Lai X (2014) Micro hot embossing of thermoplastic polymers: a review. Micromech Microeng 24:013001
13. Liu C, Li J, Liu J, Wang L (2010) Deformation behavior of solid polymer during hot embossing process. Microelectron Eng 87:200–207
14. Becker H, Heim U (2000) Hot embossing as a method for the fabrication of polymer high aspect ratio structures. Sensors Actuators 83:130–135
15. Kopeliovich D (2014) Injection molding of polymers. In: Methods of polymers fabrication. Available via SubsTech. http://www.substech.com/dokuwiki/doku.php?id=injection_molding_of_polymers. Accessed 11 July 2017
16. The plastic injection moulding process. In: All about plastic moulding. Available via D&M. Plastics Inc. http://www.plasticmoulding.ca/techniques/injection_moulding.htm. Accessed 11 July 2017
17. Rosato DV, Rosato MG (2000) Injection molding handbook. Springer, New York
18. Malloy RA (1994) Plastic part design for injection molding. Hanser, Munich Vienna
19. Tanzi S, Ostergaard PF, Matteucci M, Christiansen TL, Cech J, Marie R, Taboryski R (2012) Fabrication of combined-scale nano- and microfluidic polymer systems using a multilevel dry etching, electroplating and molding process. J Micromech Microeng 22:115008
20. Guckenberger DJ, Groot TE, Wan AMD, Beebe DJ, Young EWK (2015) Micromilling: a method for ultra-rapid prototyping of plastic microfluidic devices. Lab Chip 15 (11):2364–2378
21. Khemani H (2009) What is the CNC machine? How CNC machine works? In: Manufacturing technology. Available via Bright Hub Engineering. http://www.brighthubengineering.com/manufacturing-technology/55787-what-is-the-cnc-machine-how-cnc-machine-works/. Accessed 12 July 2017
22. Introduction to micromachining. In: Enjoy CNC micromachining & micro-cutter success: easy guide. Available via CNC Cookbook. http://www.cnccookbook.com/CCCNCMillFeedsSpeedsMicroMachining.htm. Accessed 12 July 2017
23. Lazare S, Granier V (1988) Ultraviolet laser photoablation of polymers: a review and recent results. Laser Chem 10:25–40
24. Urech L, Lippert T (2010) Photoablation of polymer materials. In: Allen NS (ed) Photochemistry and photophysics of polymer materials. Wiley, New York
25. Roberts MA, Rossier JS, Bercier P, Girault H (1997) UV laser machined polymer substrates for the development of microdiagnostic systems. Anal Chem 69:2035–2042
26. Jackson MJ (2007) Micro and nanomanufacturing. Springer, New York, pp 55–98
27. Maldonado JR, Peckerar M (2016) X-ray lithography: some history, current status and future prospects. Microelectron Eng 161:87–93. https://doi.org/10.1016/j.mee.2016.03.052
28. Sylmand X-ray lithography protocol. https://www.youtube.com/watch?v=ZEwRy2Zv0JA. Accessed 6 Aug
29. Ford SM, Kar B, McWhorter S, Davies J, Soper SA, Klopf M, Calderon G, Saile V (1998) Microcapillary electrophoresis devices fabricated using polymeric substrates and x-ray lithography. J Microcolumn Sep 10 (5):413–422
30. Haiducu M, Rahbar M, Foulds IG, Johnstone RW, Sameoto D (2008) Deep UV patterning of the commercial grade PMMA for low-cost, large-scale microfluidics. J Micromech Microeng 18:115029 (7pp)
31. Etching and deposition. Available Via NPTEL lecture series. http://nptel.ac.in/courses/113106062/Lec26.pdf. Accessed 31 July 2017
32. Vlachopoulou ME, Tsougeni K, Kontakis K, Vourdas N, Tserepi A, Gogolides E. Plasma etching technology for fabrication and surface modification of plastic microfluidic devices. http://www.ispc-conference.org/ispcproc/papers/678.pdf. Accessed 7 Aug 2017
33. Plasma technology. Available via Diener plasma-surface technology. http://www.plasmasurfacetechnology.eu/media/Plasmatechnik_en_web.pdf. Accessed on 10 Aug 2017
34. Van Zant P (2004) Microchip fabrication, 5th edn. McGraw-Hill, New York
35. Spencer EG, Schmidt PH (1971) Ion beam technique for device fabrication. J Vac Sci Technol 8(5):52–70
36. Rasmussen KH (2014) Advanced dry etching studies for micro- and nano-systems. Dissertation, Technical University of Denmark

37. Panda S, Economou DJ (2000) Anisotropic etching of polymer films by high energy (~100 s of eV) oxygen atom neutral beams. J Vac Sci Technol A 19(2):398–404
38. Yeo LY, Chang H-C, Chan PPY, Friend JR (2011) Microfluidic devices for bioapplications. Small 7:12–48
39. Feng D, Nguyen AV (2017) Effect of contact angle and contact angle hysteresis on the floatability of spheres at the air-water interface. Adv Colloid Interf Sci. https://doi.org/10.1016/j.cis.2017.07.031
40. Kim E, Xia YN, Whitesides GM (1995) Polymer microstructures formed by molding in capillaries. Nature 376:581–584. https://doi.org/10.1038/376581a0
41. Shaw JEA (1993) Capillary fill encapsulation of ISFETs. Sensors Actuators A Phys 37–38:74–76. https://doi.org/10.1016/0924-4247(93)80015-9
42. Kumar A, Biebuyck HA, Whitesides GM (1994) Patterning self-assembled monolayers: applications in materials science. Langmuir 10:1498–1511. https://doi.org/10.1021/la00017a030
43. Kumar A, Whitesides GM (1993) Features of gold having micrometer to centimeter dimensions can be formed through a combination of stamping with an elastomeric stamp and an alkanethiol “ink” followed by chemical etching. Appl Phys Lett 63:2002–2004. https://doi.org/10.1063/1.110628
44. Bride JA, Baskaran S, Taylor N, Halloran JW, Juan WH, Pang SW, O’Donnell M (1993) Photolithographic micromolding of ceramics using plasma etched polyimide patterns. Appl Phys Lett 63:3379–3381. https://doi.org/10.1063/1.110151
45. Kim E, Xia Y, Whitesides GM (1996) Micromolding in capillaries: applications in materials science. J Am Chem Soc 118:5722–5731. https://doi.org/10.1021/ja960151v
46. Filaments IP (2002) Defense technical information center compilation part notice title: micromolding in capillaries for the generation of molecularly imprinted polymer filaments and microstructures
47. Hull CW (1984) Apparatus for the production of three-dimensional objects by stereolithography
48. Ikuta K, Hirowatari K (1993) Real three dimensional micro fabrication using stereo lithography and metal molding. Proc IEEE Micro Electro Mech Syst IEEE:42–47
49. Xia C, Fang NX (2009) 3D microfabricated bioreactor with capillaries. Biomed Microdevices 11:1309–1315. https://doi.org/10.1007/s10544-009-9350-4
50. Ying Z, Chi-Wei L, Ashley TJ, Shu Y (2006) Replica molding of high-aspect-ratio polymeric nanopillar arrays with high fidelity. Langmuir 22(20):8595–8601

Chapter 3

Sample Injection Techniques

James M. Karlinsey

Abstract

Sample introduction is an important consideration in any microchip electrophoresis (ME) separation. The number of applications and level of complexity of ME devices continue to increase, but the introduction of sample into the separation channel has remained relatively constant. This work describes the most common electrophoretic methods of performing sample injection for ME applications using electroosmotic flow (EOF), providing both a transition from capillary electrophoresis (CE) separations and a straightforward entry point into ME.

Key words Microchip electrophoresis, Electrokinetic injection, Microfluidics, Electroosmotic flow

1 Introduction

Starting with the initial reports of voltage-driven separations on microchip [1, 2], sample has been injected using voltage as well. This is not surprising, but it does introduce an additional consideration when ensuring that a representative sample is introduced into the separation channel when the same electrokinetic means are used for both injection and separation. The procedures described here deal exclusively with voltage-based injections for microchip electrophoresis (ME), but there have been an increasing number of pressure-based injections reported for ME [3–5]. It is worth noting that researchers are consistently developing novel ways of introducing sample for reasons ranging from sample composition to the integrated nature of the device, especially in micro total analysis systems (μTAS) [6–10].

Early ME applications, initially referred to as "capillary electrophoresis on a chip" [1], featured intersecting channels to define the volume of sample (i.e., sample plug) introduced into the separation channel. In most capillary electrophoresis (CE) applications, a single continuous silica capillary is employed. Sample is introduced at one end, and analytes are separated as they migrate through the capillary to the opposite end (Fig. 1a). Sample can be introduced

Debashis Dutta (ed.), *Microfluidic Electrophoresis: Methods and Protocols*, Methods in Molecular Biology, vol. 1906, https://doi.org/10.1007/978-1-4939-8964-5_3, © Springer Science+Business Media, LLC, part of Springer Nature 2019

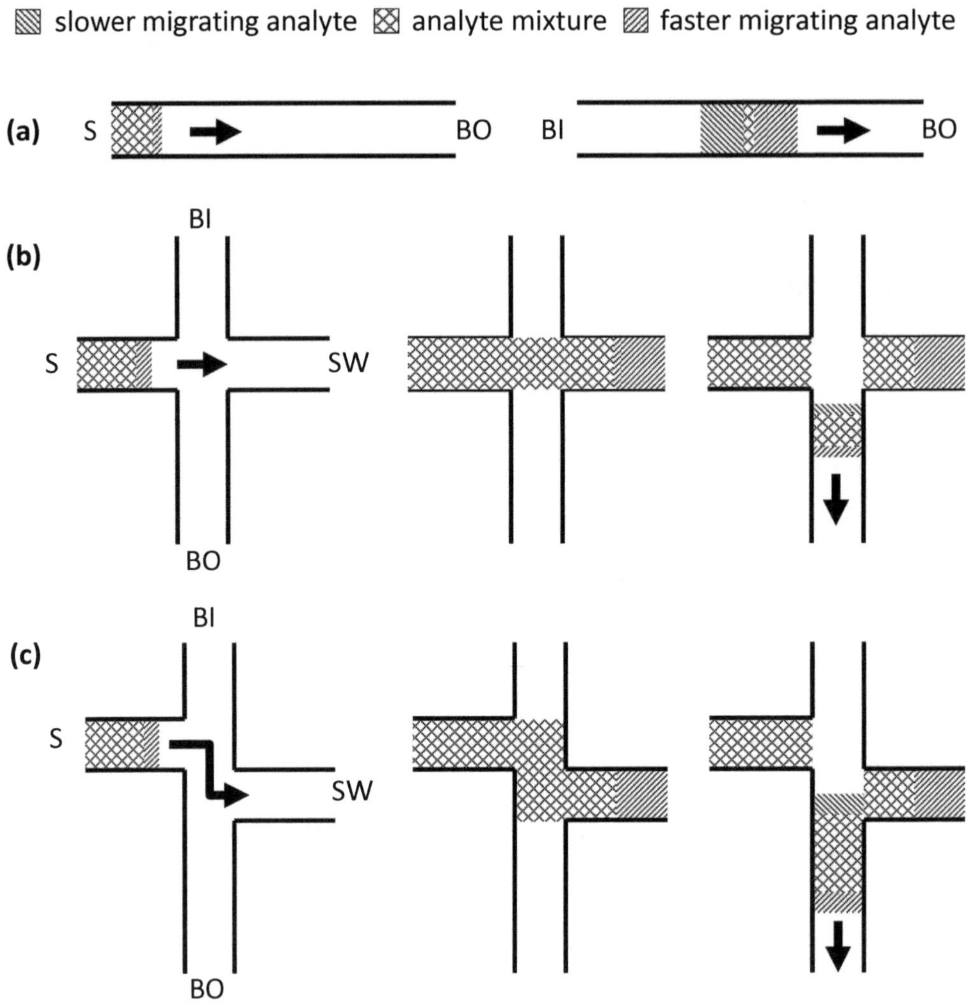

Fig. 1 Injection designs. (**a**) A conventional CE separation features a single microchannel with a single inlet and outlet. (**b**) A cross tee ME design features sample and sample waste arms that cross the separation channel. (**c**) An injection offset design features sample and sample waste arms that intersect the separation channel with some distance, or offset, between them. *S* sample reservoir, *SW* sample waste, *BI* buffer inlet, *BO* buffer outlet

using electrokinetic means, applying voltage across the capillary, or by hydrodynamic means, pressurizing the sample solution. By taking into account such parameters as capillary dimensions and solution viscosity, it is possible to calculate the amount of sample introduced in the capillary [11], regardless of which technique is used. Commercial instrumentation even allows the user to seamlessly alternate between voltage- and pressure-based injections.

In ME, the injection process is not as straightforward as in capillary, but the procedures that follow will describe the most common electrokinetic injection schemes in cross tee and offset tee microfluidic devices (Fig. 1b, c, respectively) for a separation using electroosmotic flow (EOF) and fluorescence detection. While the cross tee is traditionally employed to introduce a plug of sample into the separation channel, an offset tee allows for a larger sample plug. In either case, there are two steps that need to be performed: injection and dispensing. During the injection step, voltage applied across the sample (S) and sample waste (SW) arms will mobilize analyte(s) across the separation channel using EOF. During the dispensing step, voltage applied across the buffer inlet (BI) and buffer outlet (BO) arms will mobilize injected analyte(s) into the separation channel.

Different types of electrokinetic sample injections can be performed using the tee designs. In a floating injection, voltage is only applied to the S and SW arms. Though this simplifies the experiment, requiring only one applied voltage while the opposite reservoir is electrically grounded, it does suffer from the plug broadening that will occur at the tee intersection due to diffusion. This effect is significant and variable, especially when a longer injection time is employed to allow slower migrating species to cross the separation channel. The preferential injection of faster migrating species is a phenomenon that has previously been reported as electrokinetic bias in CE [12], when sample is introduced directly into the separation channel. One of the ways around the plug broadening that results from a long injection time is to employ additional voltage at the BI and BO reservoirs to "pinch" the injection. The pinched injection was first described by Jacobson and co-workers [13], along with a sample dispensing step in which a "pullback" voltage was used to prevent sample from leaking into the separation channel after the injection. When predicting how flow will be directed in a network of fluidic channels, it is worth noting that the behavior can be modeled as current flowing through a circuit. With channels having known resistance, owing to physical dimensions and buffer characteristics, and voltage control at each of the reservoirs, EOF can be determined in each of the flow channels. This makes it possible to choose voltages that will produce the pinched and pullback effects (Fig. 2). A similar approach has also been demonstrated for the offset tee design [14].

2 Materials

All buffer solutions for ME should be prepared with ultrapure water (18 MΩ.cm at 25 °C) using analytical grade reagents. Solutions should be prepared and used at room temperature. It is assumed that the glass chip presented here has already been fabricated and that the fluorescence detection system (i.e., excitation source, detector) for this work is already in place.

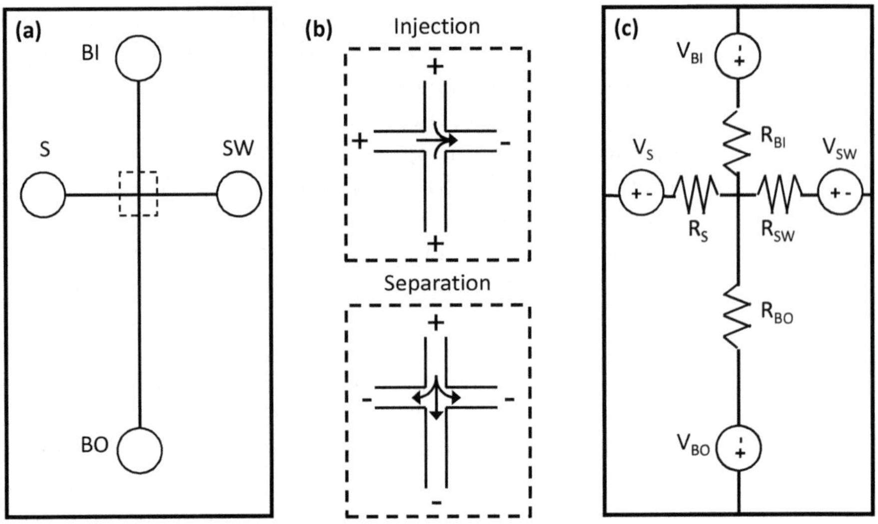

Fig. 2 (**a**) Cross tee schematic for ME device. (**b**) Cross tee close-up of pinched injection (top) and dispensing with pullback (bottom). Voltage polarity is indicated with ±, and arrows indicate the flow of EOF. (**c**) Circuit representation of cross tee design, with channel resistance (R) and applied voltage (V) indicated. *S* sample, *SW* sample waste, *BI* buffer inlet, *BO* buffer outlet

1. *Stock borate buffer*: 20 mm boric acid, pH 8.1 (*see* **Note 1**). Add about 100 mL water to a 500 mL volumetric flask. Weigh 3.1 g boric acid and transfer to flask. Add about 300 mL water to flask and dissolve boric acid with vigorous stirring using a magnetic stir bar. Adjust pH with 1 M NaOH. Make up to 500 mL with water. Degas and filter with 0.45 μm syringe filter. Store at 4 °C for up to 2 weeks.
2. *Run buffer*: Dilute stock buffer to 20 mm daily.
3. *Methanol/HCl*: 1:1 ratio using concentrated HCl.
4. Concentrated sulfuric acid.
5. 0.5 M sodium hydroxide
6. *Stock fluorescein (100 μm)*: Add about 1 mL ethanol to a 100 mL amber volumetric flask. Weigh 3.3 mg fluorescein and transfer to flask. Add about 80 mL water to flask and dissolve with vigorous stirring using a magnetic stir bar. Make up to 100 mL with water. Store at 4 °C.
7. *Fluorescein sample*: Dilute to 1 μm in run buffer daily (*see* **Note 2**).
8. *Glass microchip*: Photolithography, wet etching, and high temperature bonding used to fabricate glass microchips [15] (*see* **Note 3**).
9. *Fluid reservoirs*: Upchurch Scientific® NanoPort™ assemblies placed around access holes drilled in the glass according to manufacturer's instructions.

10. Laminar flow hood with pressure (N_2) and vacuum lines (*see* **Note 4**).
11. *Power supply*: Can be obtained commercially for ME applications or built in-house using high-voltage supplies and relays.
12. Pt electrodes.
13. *Excitation source*: 488 nm argon ion laser (*see* **Note 5**).
14. *Photodetector*: photomultiplier tube or CCD camera.
15. Computer with data acquisition capability (*see* **Note 6**).

3 Methods

All procedures are carried out at room temperature.

3.1 Microchannel Conditioning

1. Check glass microchip for channel integrity by observing channel wetting under a microscope (*see* **Note 7**).
2. Condition channels with acidic solution to increase the number of exposed silanol groups (Si-OH) at the start of the day (*see* **Note 8**). Add 5 μL of 1:1 methanol/HCl to BI, S, and SW reservoirs, and apply vacuum to BW reservoir to ensure channel wetting. Add 50–100 μL of ddH_2O (double-distilled water) to reservoirs (depending on reservoir size) and agitate with pipette to rinse acidic residue from reservoir walls.
3. Incubate for 15–30 min and flush ddH_2O through channels by pulling vacuum on BW reservoir.
4. Repeat **steps 2** and **3** using concentrated H_2SO_4 in place of methanol/HCl.
5. Condition channels with basic solution to reduce exposed silanol groups (Si-O^-) in order to promote EOF. Add 50–100 μL of 1 M NaOH to BI, S, and SW reservoirs, and apply vacuum to BW reservoir to ensure channel wetting. Add 50 μL of 1 M NaOH to BO reservoir and incubate for 15–30 min.
6. Replace NaOH solution in each reservoir with 50–100 μL run buffer, and flush by pulling vacuum on BO reservoir. Add 50 μL of run buffer to BO reservoir and incubate for 15–30 min.

3.2 Microchip Electrophoresis

1. Replace run buffer in reservoirs, and transfer microchip to the experimental setup for fluorescence detection.
2. Align microchip in the fluorescence detection system so that signal is maximized when a fluorescent species passes through the channel during the ME separation. Special attention should be paid to focal depth (*see* **Note 9**).
3. Perform voltage conditioning with run buffer using voltages that simulate a "blank" injection and separation.

4. Remove buffer solution from the reservoirs in the following order: S, SW, BI, BO.
5. Add fresh run buffer to the reservoirs in the following order: BO, BI, SW.
6. Add fluorescein sample to S reservoir.
7. Perform injection and separation protocol (*see* **Note 10**).
8. Remove microchip from experimental setup and repeat channel conditioning starting with 1 M NaOH (**step 5**).
9. Repeat ME protocol to produce fluorescein peak with desirable Gaussian profile (using plug shaping described above). For reproducibility, it is recommended that multiple sequences of injection, dispensing, and separation steps are programmed into the ME experiment (*see* **Note 11**).
10. At the end of the day, the channels should be rinsed with NaOH followed by ddH_2O and stored dry, ideally, in the laminar hood or in an enclosed space to prevent particles from settling in the reservoirs.

4 Notes

1. For the purpose of establishing conditions for EOF, it is important that the pH is greater than the pK_a of the silanol groups on the surface of the glass in order to establish a negatively charged wall for EOF mobilization. A borate buffer with pH 8.1 works well for this, but it should be noted that many different buffer compositions have been reported in the literature.
2. Fluorescein is used here because of its high quantum yield when excited with the 488 nm line of an argon ion laser. Depending on the nature of the excitation source and the available optics, a different dye may be used to evaluate the sample injection. With the conditions outlined here, fluorescein would be negatively charged in solution, resulting in an electrophoretic mobility opposite that of EOF. The ability to observe fluorescein therefore suggests that the faster migrating positive and neutral species in solution would also be injected using EOF.
3. Glass is listed here as a starting point for the microfluidic device, but poly(dimethyl siloxane) (PDMS) and glass/PDMS hybrid devices continue to be popular choices, owing primarily to the ease with which PDMS can be patterned [16].
4. The laminar flow hood ensures that there is a steady flow of "clean" air over the microchip during the conditioning steps to prevent particles in the environment from entering the channels. The vacuum line in the hood should be equipped with a trap. Despite the small volumes associated with ME, the

conditioning steps generate waste that will need to be collected and neutralized prior to being poured down the drain.

5. The excitation source can easily be changed to take advantage of the optics on hand, but the dye should then be paired with the source in a way that maximizes signal. Because the nature of EOF will mobilize neutral species within the channel, it is not necessary to ensure the dye will be charged.
6. Low-cost data acquisition (DAQ) devices such as National Instruments™ USB-6008 multifunction I/O device can be programmed in a user-friendly way with LabVIEW software to interface with the power supply as well as the input signal from the photodetector to ensure precise control over the ME experiment.
7. High magnification is not necessary when examining flow in channels with dimensions ranging from 10 to 100 μm. A stereo zoom (or dissecting) microscope with relatively low magnification allows visualization of the channel and easy manipulation of the microchip. When removing a channel blockage, a combination of vacuum and pressure can be utilized with 1 M NaOH solution to clear the channel. Care must be taken when pressurizing the NaOH.
8. Microchannel conditioning steps follow a method reported by Cras and co-workers for the preparation of glass for silanization [17]. Their work showed that exposing the glass surface to a 1:1 methanol/HCl solution followed by concentrated H_2SO_4 to clean the surface increased the number of exposed silanol groups on glass slides. For this reason, the methanol/HCl and concentrated H_2SO_4 conditioning is recommended at the start of each day.
9. Focusing the excitation source into the microfluidic channel can be optimized for depth by incorporating an alignment channel into the design prior to etching [18]. A channel having the same cross section and etch depth as the separation channel is recommended, and it should have access holes drilled to allow fluorescent dye to be added to the channel for the depth alignment.
10. An injection and separation protocol is going to be highly dependent on the design of the microchip. For the following example, a cross tee design is used with 6-mm-long S, SW, and BI channels and a 30-mm-long BW channel. The initial line width w_x for each channel is 50 μm, and the etch depth d_x is 50 μm. It is also possible to account for differences in channel dimensions using the concept of equivalent channel length [19]. Before selecting voltages, it is helpful to use the circuit analogy mentioned in the Introduction as a starting point. The resistance R_x of each channel can be approximated using the resistivity ρ of the buffer solution, the length l_x of the channel,

and the cross-sectional area A_x of the channel (where x = S, SW, BI, BO).

$$A_x = \left(1/2 \cdot \pi \cdot {d_x}^2\right) + (w_x \cdot d_x) \quad (1)$$

$$R_x = \rho \cdot l_x \cdot A_x \quad (2)$$

The total resistance R_T can be found by combining the individual R_x values in parallel. Using the calculated resistances and the voltage V_x applied to each channel, it is possible to calculate the voltage V_{node} at the injection cross using Kirchhoff's rules. Once V_{node} is found, it is possible to determine the electric field strength E_x in each channel.

$$R_T^{-1} = R_S^{-1} + R_{SW}^{-1} + R_{BI}^{-1} + R_{BO}^{-1} \quad (3)$$

$$V_{node} = \left(\left(V_S \cdot R_S^{-1}\right) \cdot \left(V_{SW} \cdot R_{SW}^{-1}\right) \cdot \left(V_{BI} \cdot R_{BI}^{-1}\right) \cdot \left(V_{BO} \cdot R_{BO}^{-1}\right)\right) \cdot R_T^{-1} \quad (4)$$

$$E_x = (V_x - V_{node}) \cdot R_x^{-1} \quad (5)$$

These calculations can readily be incorporated into an Excel® spreadsheet or a LabVIEW program to quickly determine the electric field strength in each channel. A more useful way of reporting and comparing the electric fields is to use the relative field strength ε_x [20]. By setting each field strength relative to the one with the largest magnitude, it is possible to quickly compare flow into and out of the node. For example, performing a separation with pullback using the chip described above, and applying 415 V at the BI and −1460 V at the BO while the S and SW reservoirs are grounded, results in a node voltage of 38 V. The relative field in the BI channel is 1.00 as

Table 1
Table of channel and voltage parameters for a cross tee microchip

	Channel	S	SW	BI	BO
Channel parameters	Length (mm)	6	6	6	30
	Width (μm)	50	50	50	50
	Depth (μm)	50	50	50	50
Injection	Voltage (V)	0	−600	–	–
	Electric field ($V \cdot cm^{-1}$)	−500	−500	–	–
	Relative field strength	–	–	–	–
Dispensing	Voltage (V)	0	0	415	−1460
	Electric field ($V\ cm^{-1}$)	−64	−64	628	−499
	Relative field strength	−0.1	−0.1	1.0	−0.8

ME conditions include a floating injection followed by sample dispensing with pullback. *S* sample, *SW* sample waste, *BI* buffer inlet, *BO* buffer outlet

EOF is directed toward the node, $\varepsilon_x = -0.80$ as the majority of the field is directed toward the BO reservoir, and $\varepsilon_x = -0.10$ in the S and SW arms for pullback. Table 1 provides a starting point for using a floating injection followed by dispensing with pullback using EOF to mobilize analyte(s) with the chip parameters presented above. A floating injection represents a good starting point to ensure sample is introduced into the separation channel. Additional separations can then focus on optimizing the sample injection.

11. When performing multiple sequences of ME, equilibrium effects (i.e., diffusion) dominate the flow when no voltage is applied [21]. For that reason, it may be useful to program consecutive runs with no time interval in between.

References

1. Harrison DJ, Manz A, Fan ZH, Ludi H, Widmer HM (1992) Capillary electrophoresis and sample injection systems integrated on a planar glass chip. Anal Chem 64(17):1926–1932
2. Manz A, Harrison DJ, Verpoorte EMJ, Fettinger JC, Paulus A, Ludi H, Widmer HM (1992) Planar chips technology for miniaturization and integration of separation techniques into monitoring systems: capillary electrophoresis on a chip. J Chrom A 593(1–2):253–258
3. Castro ER, Manz A (2015) Present state of microchip electrophoresis: state of the art and routine applications. J Chromatogr A 1382:66–85
4. Ou GZ, Feng XJ, Du W, Liu X, Liu BF (2013) Recent advances in microchip electrophoresis for amino acid analysis. Anal Bioanal Chem 405(25):7907–7918
5. Karlinsey JM (2012) Sample introduction techniques for microchip electrophoresis: a review. Anal Chim Acta 725:1–13
6. Arora A, Simone G, Salieb-Beugelaar GB, Kim JT, Manz A (2010) Latest developments in micro total analysis systems. Anal Chem 82(12):4830–4847
7. West J, Becker M, Tombrink S, Manz A (2008) Micro total analysis systems: latest achievements. Anal Chem 80(12):4403–4419
8. Dittrich PS, Tachikawa K, Manz A (2006) Micro total analysis systems. Latest advancements and trends. Anal Chem 78(12):3887–3907
9. Vilkner T, Janasek D, Manz A (2004) Micro total analysis systems. Recent developments. Anal Chem 76(12):3373–3385
10. Auroux PA, Iossifidis D, Reyes DR, Manz A (2002) Micro total analysis systems. 2. Analytical standard operations and applications. Anal Chem 74(12):2637–2652
11. Ermakov SV, Zhukov M, Capelli L, Righetti PG (1994) Quantitative studies of different injection systems in capillary electrophoresis. Electrophoresis 15(8–9):1158–1166
12. Huang X, Gordon MJ, Zare RN (1988) Bias in quantitative capillary zone electrophoresis caused by electrokinetic sample injection. Anal Chem 60(4):375–377
13. Jacobson SC, Hergenroder R, Koutny LB, Warmack RJ, Ramsey JM (1994) Effects of injection schemes and column geometry on the performance of microchip electrophoresis devices. Anal Chem 66(7):1107–1113
14. Shultz-Lockyear LL, Colyer CL, Fan ZH, Roy KI, Harrison DJ (1999) Effects of injector geometry and sample matrix on injection and sample loading in integrated capillary electrophoresis devices. Electrophoresis 20(3):529–538
15. Manz A, Fettinger JC, Verpoorte E, Ludi H, Widmer HM, Harrison DJ (1991) Micromachining of Monocrystalline silicon and glass for chemical-analysis systems—a look into next century technology or just a fashionable craze. Trac-Trend Anal Chem 10(5):144–149
16. Duffy DC, McDonald JC, Schueller OJ, Whitesides GM (1998) Rapid prototyping of microfluidic systems in Poly(dimethylsiloxane). Anal Chem 70(23):4974–4984
17. Cras JJ, Rowe-Taitt CA, Nivens DA, Ligler FS (1999) Comparison of chemical cleaning methods of glass in preparation for silanization. Biosens Bioelectron 14(8–9):683–688
18. Karlinsey JM, Landers JP (2006) Multicolor fluorescence detection on an electrophoretic

microdevice using an acoustooptic tunable filter. Anal Chem 78(15):5590–5596

19. Seiler K, Fan ZHH, Fluri K, Harrison DJ (1994) Electroosmotic pumping and valveless control of fluid-flow within a manifold of capillaries on a glass chip. Anal Chem 66 (20):3485–3491
20. Ermakov SV, Jacobson SC, Ramsey JM (2000) Computer simulations of electrokinetic injection techniques in microfluidic devices. Anal Chem 72(15):3512–3517
21. Crabtree HJ, Cheong EC, Tilroe DA, Backhouse CJ (2001) Microchip injection and separation anomalies due to pressure effects. Anal Chem 73(17):4079–4086

Chapter 4

Sample Preconcentration Protocols in Microfluidic Electrophoresis

Fumihiko Kitagawa and Koji Otsuka

Abstract

Electrophoretic on-line sample preconcentration techniques in microfluidic channels improve the sensitivity prior to the separation. Among various techniques, the most important field-amplified sample stacking and sweeping on cross-channel microchips are demonstrated. As a novel microfluidic preconcentration approach, a large-volume sample stacking with electroosmotic flow pump (LVSEP) on straight-channel chips is also presented, which can omit a complicated voltage program for sample injection processes. In this chapter, we describe how to prepare and how to run these on-line sample preconcentration methods in microchip electrophoresis.

Key words Microchip electrophoresis, On-line sample preconcentration, Field-amplified sample stacking, Sweeping, LVSEP

1 Introduction

In this chapter, protocols for several on-line sample preconcentration techniques in microchip electrophoresis (MCE) are presented. We focus on two fundamental sample preconcentration techniques, field-amplified sample stacking (FASS) and sweeping, since they can provide the sensitivity enhancements with simple experimental procedures and short analysis times. Additionally, protocols for a large-volume sample stacking with electroosmotic flow (EOF) pump (LVSEP) are demonstrated.

1.1 Sample Injection in MCE

In the gated injection technique [1] on a cross-channel microchip, a sample solution is loaded with a variable injection volume by controlling a voltage program and an injection time. As shown in Fig. 1, three reservoirs connected to the loading channels are used as BGS (B), sample (S), and sample waste (SW), while that to the separation channel as BGS waste (BW). Under a faster EOF condition, the sample solution flows from the S reservoir to the SW reservoir waste in the first loading step. Simultaneously, a

Debashis Dutta (ed.), *Microfluidic Electrophoresis: Methods and Protocols*, Methods in Molecular Biology, vol. 1906, https://doi.org/10.1007/978-1-4939-8964-5_4,

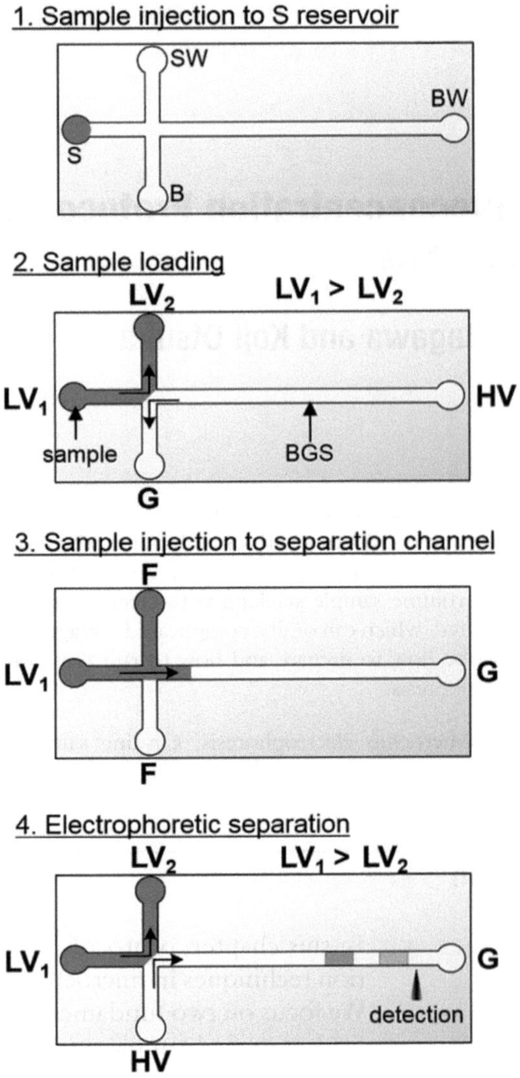

Fig. 1 Schematics of the gated injection on the cross-channel microchip

background buffer solution (BGS) flows from the BW reservoir to the B reservoir to prevent a sample leakage into the separation channel. In the injection step, the voltage in the BW reservoir is set to be grounded to inject the sample plug into the separation channel. The selection of the duration for the injection step (injection time) allows a variable volume introduction of the sample. In the subsequent separation step, a high voltage is applied to the B reservoir to deliver the BGS from the B to the BW reservoir.

1.2 FASS

Among various on-line sample preconcentration techniques, FASS is the simplest and most fundamental approach in capillary electrophoresis (CE) [2]. Under the FASS condition in CE, a large

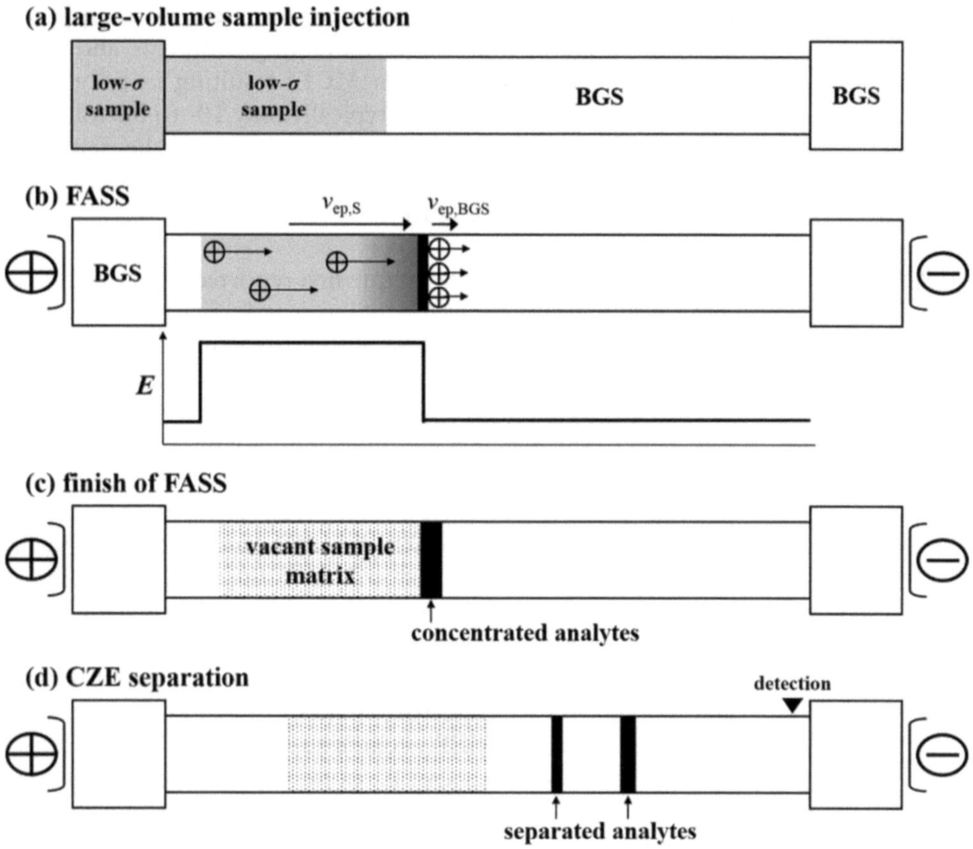

Fig. 2 Principle of FASS. $v_{ep,S}$ and $v_{ep,BGS}$ are the electrophoretic velocities in the sample and BGS zones, respectively

volume of an ionic sample solution with a low-electrolyte concentration is introduced into the separation channel filled with a BGS containing a large amount of electrolytes (Fig. 2a). In MCE, a large-volume injection of the sample is carried out by the gated injection technique. Since the electric current flowing throughout a capillary is constant, the electric field strength is inversely proportional to the electrical conductivity (σ). A higher field is applied to the low-σ sample zone relative to the high-σ BGS zone as shown in Fig. 2b. Because the electrophoretic mobility is directly proportional to the field strength, the migration velocity of ionic analytes in the sample zone is faster than that in the BGS. After the faster migration in the sample zone, the ionic analytes slow down at the sample/BGS boundary. Consequently, the analytes are stacked into a narrower zone than the original sample plug (Fig. 2c). In FASS, a large difference in the conductivity between the two solutions gives a higher preconcentration efficiency. Finally, the concentrated analytes are separated by the capillary zone electrophoresis (CZE) as shown in Fig. 2d. To improve the enrichment efficiency, a short plug of water is often injected in front of the sample zone in CE. However, the injection of water plug is difficult, and the sample injection volume is limited in MCE, resulting in poor sensitivity enhancement factors (SEFs), typically only 10- to 50-fold improvements. FASS is commonly used to improve the sensitivity for various ionic analytes such as amino acids, peptides, proteins, DNA, amines, and metal ions.

1.3 LVSEP

In the original LVSEP technique first reported by He and Lee [3], a large volume (up to entire column volume) of a low-σ sample solution, typically containing anionic analytes in deionized water, is fully injected into a bare fused-silica capillary, and then a separation voltage is applied to both reservoirs filled with an acidic BGS in CE. Due to the difference in the conductivities between the sample and the BGS, the anionic analytes are stacked at the sample/BGS boundary. Since most of the capillary is filled with the almost neutral-pH sample, the stacked analytes move toward the cathode by a faster EOF. After the stacking, the removal of the vacant sample matrix (SM) and the introduction of the acidic BGS into the capillary suppress the EOF. When the electrophoretic velocity of the anionic analyte becomes higher than the EOF rate, the stacked analytes migrate to the anode. Finally, they are separated by CZE. In the LVSEP technique, both higher separation and enrichment efficiencies can be obtained because the total column length is utilized both for the preconcentration and the effective separation length.

In the recent LVSEP applications, the polymer-coated capillaries and microchannels are employed to suppress both the EOF and the surface adsorption of analytes [4], which is effective for the on-line preconcentration of biogenic compounds. As shown in Fig. 3a, a neutral polymer-coated straight channel is filled with a low ionic strength (I) solution containing anionic analytes. After applying a voltage, anionic analytes are stacked at the sample/BGS boundary. Although the microchannel is coated with the neutral polymer, a faster EOF is generated by the low-I of the SM (Fig. 3b), which is predicted by the Debye-Hückel theory (*see* **Note 1**). Thus, both the focused analytes and the vacant SM move toward the cathode by the enhanced EOF. As the high-I BGS is introduced into the microchannel, the EOF is suppressed. When most of the SM is removed out from the cathodic end, the electric field in the BGS zone becomes higher. As a result, the electrophoretic velocity of the analyte exceeds the suppressed EOF rate, and the moving direction of the analytes is inversed (Fig. 3c). After the complete removal of the SM, the analytes are separated by CZE (Fig. 3d). Since both the sample enrichment and the SM removal occur simultaneously under the constant voltage in LVSEP, complicated voltage programs for the gated injection technique can be omitted.

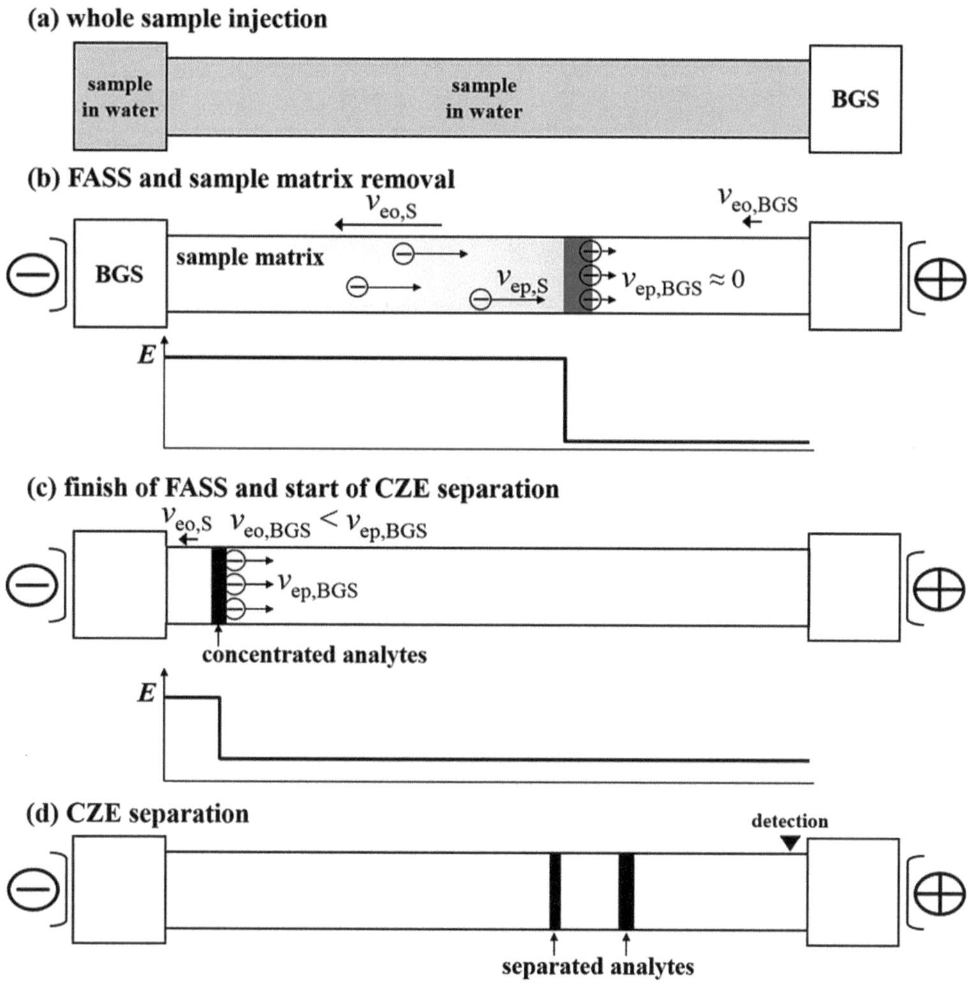

Fig. 3 Principle of LVSEP. $v_{eo,S}$ and $v_{eo,BGS}$ are the electroosmotic velocities in the sample and BGS zones, respectively

This modified LVSEP technique has been applied to analyze various biogenic compounds including oligosaccharides, cationic amines, enantiomers, proteins, and DNA [5, 6].

1.4 Sweeping

In the sweeping technique, the preconcentration is achieved by the migration velocity changes of analytes due to the retention by micelles in micellar electrokinetic chromatography (MEKC) [7]. Under a conventional sweeping condition, sodium dodecyl sulfate (SDS) is used as the pseudostationary phase to enrich neutral, hydrophobic, and/or cationic analytes. In CE, a long plug of a sample solution devoid of SDS is injected in a capillary filled with a BGS containing SDS as shown in Fig. 4a. Under a normal EOF condition, neutral or cationic analytes move faster in the sample zone mainly by EOF, while the SDS micelle in the BGS zone migrates slower due to highly negative charges of the micelle.

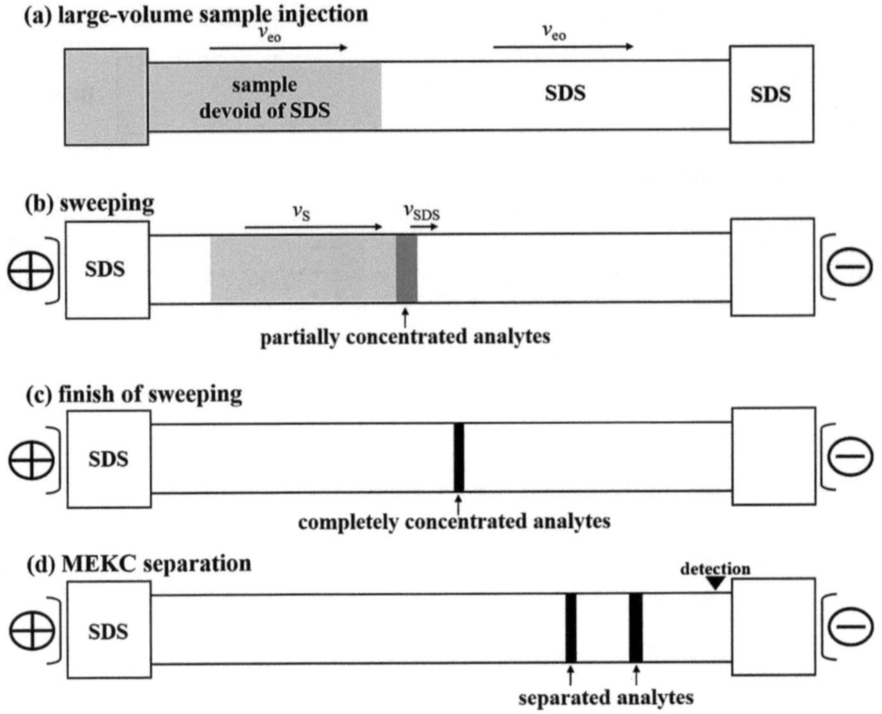

Fig. 4 Principle of sweeping. v_S and v_{SDS} are the apparent velocities in the sample and SDS-BGS zones, respectively

When the analytes are efficiently incorporated into the SDS micelle at the sample/BGS boundary, the migration velocities of the analytes are decreased as shown in Fig. 4b. As a result, the analytes are concentrated to a narrow zone (Fig. 4c). In MCE, the SEFs ranging from 100 to 500 can be achieved in spite of limited injection volumes of the samples. Generally, the applications of suppressed EOF conditions are better to enhance the sweeping efficiencies.

2 Materials

2.1 MCE Apparatus

All solutions are prepared with ultrapure water and analytical grade reagents.

1. MCE analysis of fluorescent samples is performed by using a home-made laser-induced fluorescence (LIF) detection scheme as shown in Fig. 5. A green laser beam (488 or 532 nm) for excitation is introduced to an inverted optical microscope and irradiated to a microchannel through an objective lens (×20, NA = 0.40). Alternatively, a ~100 W mercury lamp for excitation is introduced to the microscope and irradiated to the microchannel through a dichroic filter for passing a specific wavelength range and the objective lens. Fluorescence from

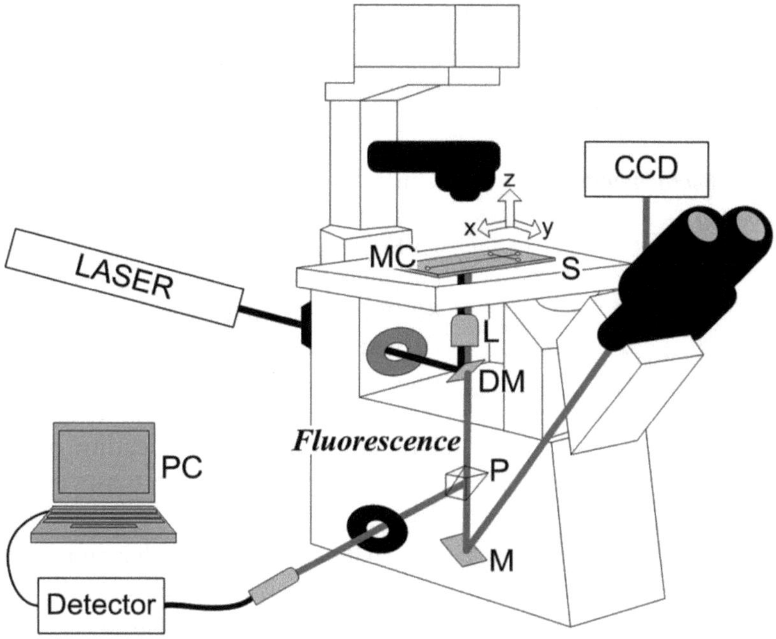

Fig. 5 Schematic of the microscope-based LIF detection system. *DM* dichroic mirror, *L* objective lens, *M* mirror, *P* prism, *S* 3D-stage, *MC* microchip

analytes collected by the same objective lens is passed through a dichroic filter and led to a photomultiplier, multichannel photodetector or CCD camera.

2. Microchips are fixed on the microscope stage by using a jig for adjusting two or four platinum wires, which ensures the repeatable positioning of the electrodes.
3. In the conventional MCE and the on-line sample preconcentration-MCE analyses, complicated voltage programs are performed via platinum electrodes connected to a multichannel-type high-voltage power supply. Conventional two-channel-type high-voltage power supply for CE can be employed to LVSEP in MCE.

2.2 Microchip Preparation

1. A poly(dimethyl siloxane) (PDMS) microfluidic device is fabricated by a conventional soft lithography technique [8] on the basis of PDMS base and curing reagents (Sylgard 184 from Dow Corning Toray) and a negative photoresist (SU8-50 from MicroChem). Typically, a cross-channel microchip consists of three 10-mm-long channels and a 50-mm-long separation channel (50–100 μm width × 50–100 μm depth). For LVSEP experiments, a microchip which has a single straight channel with a total separation channel length of 60 mm can be utilized.

2. After producing reservoir wells (2–3 mm diameter) with a piercer, the surface of the microchannel plate and a PDMS lid are activated by O_2 or air plasma. These activated substrates are directly bonded to each other.

2.3 Microchannel Coating

1. Under experimental conditions suppressing EOF and/or irreversible adsorption of analytes onto the channel surface, *e.g.*, LVSEP and biomolecule analysis, the fabricated PDMS microchannels are coated with poly(vinyl alcohol) (PVA, Mw 80,000, 88% hydrolyzed) by a heating passivation method [9]. At first, 2% PVA is introduced into the microchannel and then left for 15 min. The solution is removed out, and the microchip is heated at 110 °C for 15 min. The injection of the PVA solution and the heating of the microchip are repeated three times, where the temperature at the third heating is 140 °C.
2. Alternatively, a vacuum-drying method can be employed to the coating of polymers [10]. A 10 mM sodium dodecyl sulfate (SDS) solution is manually introduced into the whole microchannel by using a micro-syringe to hydrophilize the PDMS surface. After removing the SDS solution, the microchannel is loaded with 0.1% PVA solution. The microchip is put into a vacuum chamber under reduced pressure (<~1 Torr) with an oil rotary vacuum pump for 10 min (*see* **Note 2**).

2.4 FASS

1. A high-conductivity buffer solution is prepared as a BGS. Typically, 10–100 mM phosphate, Tris, acetate, or borate buffer (pH 2–10) is selected (*see* **Note 3**).
2. In the preparation of a standard sample solution, the standard reagents are dissolved in a diluted BGS. Typical dilution ratio is ranging from 10 to 1000 (*see* **Note 4**).
3. As for real samples, sample cleanup by SPE, dialysis, gel filtration, ion exchange, and so on is made prior to FASS for removing excess salts in the sample matrix.

2.5 LVSEP

1. A high-σ buffer solution is prepared as a BGS. In addition to conventional buffer electrolytes used in the FASS experiments, good buffer electrolytes such as HEPES, MES, and CHES are employed to suppress the Joule heating in LVSEP (*see* **Note 5**).
2. In the preparation of a standard sample solution, the standard reagents are dissolved in a diluted BGS or deionized water. The concentration of the electrolytes should be less than ~0.1 mM (*see* **Note 6**).
3. As well as FASS, the desalting pretreatment for saline real samples is carried out prior to LVSEP.

2.6 Sweeping

1. A buffer solution containing high-concentration surfactant is prepared as a BGS. The concentration of the surfactant should be higher than the critical micelle concentration (CMC). Typically, 10–100 mM SDS is added to the BGS (*see* **Note 7**).
2. In the preparation of a sample solution, the analytes are dissolved in the buffer solution devoid of the surfactant (*see* **Note 8**).

3 Methods

Prior to the MCE analysis, all of the solutions should be filtered by using 0.22 or 0.45 μm filters to avoid the clogging of the microchannels. Especially for high-concentration buffers, the used solutions are degassed by sonicating for 10–20 min to prevent bubble formations during the MCE operations. For the FASS and sweeping experiments, the gated injection technique is employed to introduce a long sample zone into the separation channel on the cross-channel chip. In the LVSEP measurements, on the other hand, the sample solution is manually injected into the straight channel.

3.1 General Operation in MCE

1. The microchannel is washed with methanol and water and finally conditioned with a BGS solution. Under a faster EOF condition, the channel that is washed with 0.1 M NaOH is used after methanol rinsing.
2. The BGS is introduced into the entire cross-channel by manually using a syringe.
3. The solutions remained in the reservoir wells are completely removed with a 10 μL micropipette.
4. Same volumes of the sample solution and the BGS, typically 1–10 μL, are poured into the reservoir S and other reservoirs, respectively (*see* **Note 9**).
5. The prepared microchip is placed on the X-Y stage of the microscope.
6. The X-Y and focus positions are adjusted to locate the channel region for the detection, typically at the point of 5–10 mm from the BW reservoir.
7. Platinum electrodes are inserted into each reservoir to provide electrical contact (*see* **Note 10**).
8. Inserted electrodes are connected to cables from a high-power voltage supply.
9. Electric fields are applied according to the voltage programs as summarized in Table 1.
10. At the same time with applying voltage, the record of the fluorescence intensity at the detection point is started.

Table 1
Typical voltage program in the gated injection technique on the cross-channel microchip under a faster EOF condition

		Applied voltage (V)			
	Time (s)	S	SW	B	BW
Loading	30	700	400	0	1500
Injection	1–20	1500	Float	Float	0
Separation	120	700	400	1500	0

11. Same washing procedure is repeated before starting a new experiment (*see* **Note 11**).
12. To estimate SEF, the peak height in the preconcentration experiment is compared with that obtained with a conventional MCE analysis using a pinched injection (PI) technique [11]. SEF is calculated from the following equation:

$$\mathrm{SEF} = \frac{C_{\mathrm{S,PI}}}{C_{\mathrm{S,PC}}} \times \frac{h_{\mathrm{PC}}}{h_{\mathrm{PI}}} \tag{1}$$

where C_S, h, PC, and PI are the sample concentration, the peak height, the on-line preconcentration experiment, and normal PI-MCE analysis, respectively (*see* **Note 12**).

3.2 FASS

1. After conditioning the microchannel, a high-conductivity BGS is introduced into the entire cross-channel.
2. The reservoir S is filled with a low-conductivity sample solution, and the BGS is poured into the B, BW, and SW reservoirs.
3. Four platinum electrodes are inserted into each reservoir, and the electric fields are applied to perform the gated injection. The injection time is typically 10–60 s (*see* **Note 13**).

3.3 LVSEP

1. After conditioning the microchannel, a low-I sample solution is introduced into the entire straight channel by manually using a syringe.
2. Both anodic and cathodic reservoirs are filled with same volumes of a high-I BGS (*see* **Note 9**).
3. For anionic analytes, a reversed electric field (detection end is anode), typically 100–500 V/cm, is applied through two platinum electrodes immersed in the two reservoirs (*see* **Note 14**).
4. The enriched and separated analytes are detected around the anodic channel end.

3.4 Sweeping

1. After conditioning the microchannel, the cross-channel is fully filled with a BGS containing high-concentration SDS.
2. A sample solution devoid of SDS is poured into the reservoir S, and the B, BW, and SW reservoirs are filled with the SDS-BGS (*see* **Note 9**).
3. A long plug of the sample solution is injected by the gated injection technique through four platinum electrodes inserted into each reservoir. The injection time is typically 10–90 s (*see* **Note 11**).

4 Notes

1. The electroosmotic mobility (μ_{eo}) depends on the zeta potential (ζ) of the microchannel surface. According to the Debye-Hückel theory, ζ can be approximated as a linear function of $I^{-1/2}$. Thus, μ_{eo} is proportional to $I^{-1/2}$. When the PVA-coated channel is filled with deionized water, actually, μ_{eo} is enhanced to ~5×10^{-4} cm^2/V·s, while with 25 mM BGS that is well suppressed to less than ~1×10^{-4} cm^2/V·s [4].
2. The vacuum-drying method is applicable to the coating of various polymers onto the channel surface of the PDMS microchips [10]. Immobilization of neutral polymers, including PVA, poly(vinylpyrrolidone) (PVP), cellulose derivatives, and so on, can suppress the EOF. To obtain reversed EOF, cationic polymers such as polyamines, polybrene, poly(ethyleneimine), and chitosan are coated by the vacuum-drying method.
3. The concentration of BGS affects the Joule heating. Higher Joule heating caused by a high salt concentration often reduces peak heights of enriched analytes.
4. Since a large degree of the changes in the migration velocity of the analytes at the sample/BGS boundary causes efficient enrichments, a large difference in the electric conductivity of the two zones gives higher SEFs. In FASS, therefore, the SEF is strongly affected by the conductivity ratio on the basis of the following equation:

$$l_{inj} = \gamma \cdot l_{stack} \tag{2}$$

where γ, l_{inj}, and l_{stack} are the conductivity ratio between the two zones, the length of the injected sample, and stacked zones, respectively.

5. To suppress the Joule heating, it is important to select buffer electrolytes. As a typical example, the Joule heating is more significant at the same buffer concentration as follows: phosphate >> borate > HEPES. In phosphate buffer, the electric current is ~fivefold higher than that in HEPES buffer at pH 7.

6. As well as FASS, the SEF value depends on the γ in LVSEP. In the LVSEP experiments, furthermore, the inversion position of the stacked analytes is also determined by the γ value [4]. When 25 mM HEPES buffer and deionized water are used as the BGS and the sample matrix, respectively, γ is approximately 200. Under the condition, the inversion of the concentrated analytes is observed at the position of 96% of the total channel length from the anodic end, while that is decreased to ~90% in the case of 0.2 mM HEPES sample matrix. Since the inversion position of the analytes is regarded as the starting point of the CZE separation, a longer separation length can be utilized at a higher γ condition, resulting in a better resolution.
7. In the sweeping, the SEF is determined by the retention factor (k) to the micelle in the BGS as follows [12]:

$$l_{\text{sweep}} = l_{\text{inj}}\frac{1}{1+k} \tag{3}$$

where l_{sweep} is the length of the swept zones. This means that efficiently incorporated analytes are well concentrated by the sweeping mechanism. Because ionic analytes are well incorporated by micelles having opposite charges, anionic SDS surfactant is effective especially for concentrating cationic analytes, whereas the use of cationic surfactants such as cetyltrimethylammonium bromide (CTAB) can enhance the SEFs of anionic analytes. Since the retention factor is linearly increased upon increasing the SDS concentration, furthermore, SEF can be improved by employing a BGS containing a higher concentration of pseudostationary phase.
8. In the sweeping technique, the preparation of a sample solution by using a low-conductivity matrix can improve the SEFs due to a dual preconcentration effect of FASS and sweeping.
9. Same volume loadings are needed to equate the heights of the liquid levels at the solution reservoirs. Different heights of the liquid levels cause a weak pressure flow. The pressure flow alters the migration time, the peak height, and the SEF relative to those obtained under the identical liquid-level condition.
10. To avoid the different heights of the liquid levels, the soaked portions of the electrodes are kept in the same lengths by using a jig for adjusting platinum wires.
11. The electrolysis of the sample during the application of the voltages causes the changes in the pH of the sample solution. Since the pH affects the injected analyte amounts, especially for weakly acidic and basic analytes, continuous measurements without sample replacements bring about irreproducible results. Hence, the sample and BGS should be replaced in every run.

12. In the PI-MCE analysis, a sample solution should be prepared by dissolving analytes in the BGS to avoid the preconcentration effect. In the sweeping experiment, however, a simple comparison of the peak height causes an underestimation since the fluorescence quantum yield of the fluorophore under the sweeping condition is often different from that under the conventional PI-MCE due to the absence or presence of the SDS micelle at the detection point. Thus, corrected sensitivity enhancement factor (SEF_{cor}) including the quantum yield difference should be used:

$$SEF_{cor} = \frac{C_{S,PI}}{C_{S,sweep}} \times \frac{h_{PI}}{h_{sweep}} \times \frac{Q_{SM}}{Q_{BGS}} \quad (4)$$

where Q_{SM} and Q_{BGS} are the fluorescence quantum yields in the sample matrix devoid of SDS and the BGS containing SDS, respectively.

13. Longer injection time usually gives a higher peak height of the stacked analytes. However, too much injected amounts of the analytes result in the peak broadenings and the reduction of the resolution. Hence, the injection time should be optimized with these adjustments.
14. Neutral polymer (*e.g.*, PVA and PVP) coatings onto the microchannel surface, where a cathodic EOF is well suppressed in BGS but drastically enhanced in a low-I sample matrix, are suitable for the LVSEP analysis of anions. Such EOF characteristics allow the preconcentration and the separation of anionic analytes. In the LVSEP analysis of cations, on the other hand, a weakly positive-charged channel surface is required to reverse the direction of EOF. For this purpose, the coating of polymer mixture of PVA and poly(allylamine) is recommended [13].

Acknowledgments

We would like to thank Dr. Kenji Sueyoshi (Osaka Prefecture University, Japan) and Dr. Takayuki Kawai (RIKEN, Japan) for collaborating the sweeping and LVSEP experiments, respectively. This work was supported in part by the Grant-in-Aid for Scientific Research (C) (No. 24550090 and 15K05527) from the Japan Society for the Promotion of Science (JSPS). This research was also supported by SENTAN, JST.

References

1. Jacobson SC, Koutny LB, Hergenroder R, Moore AW, Ramsey JM (1994) Microchip capillary electrophoresis with an integrated postcolumn reactor. Anal Chem 66:3472–3476
2. Burgi DS, Chien RL (1991) Optimization in sample stacking for high-performance capillary electrophoresis. Anal Chem 63:2042–2047
3. He Y, Lee HK (1999) Large-volume sample stacking in acidic buffer for analysis of small organic and inorganic anions by capillary electrophoresis. Anal Chem 71:995–1001
4. Kawai T, Sueyoshi K, Kitagawa F, Otsuka K (2010) Microchip electrophoresis of oligosaccharides using large-volume sample stacking with electroosmotic flow pump in single channel. Anal Chem 82:6504–6511
5. Kitagawa F, Kawai T, Otsuka K (2013) On-line sample preconcentration by large volume sample stacking with an electroosmotic flow pump (LVSEP) in microscale electrophoresis. Anal Sci 29:1129–1139
6. Kitagawa F, Kinami S, Takegawa Y, Nukatsuka I, Sueyoshi K, Kawai T, Otsuka K (2017) On-line coupling of sample preconcentration by LVSEP with gel electrophoretic separation on T-channel chips. Electrophoresis 38:380–386
7. Quirino JP, Terabe S (1998) Exceeding 5000-fold concentration of dilute analytes in micellar electrokinetic chromatography. Science 282:465–468
8. Duffy DC, McDonald JC, Schueller OJA, Whitesides GM (1998) Rapid prototyping of microfluidic systems in poly(dimethylsiloxane). Anal Chem 70:4974–4984
9. Wu D, Luo Y, Zhou X, Dai Z, Lin B (2005) Multilayer poly(vinyl alcohol)-adsorbed coating on poly(dimethylsiloxane) microfluidic chips for biopolymer separation. Electrophoresis 26:211–218
10. Kitagawa F, Nakagawara S, Nukatsuka I, Hori Y, Sueyoshi K, Otsuka K (2015) Simple and rapid immobilization of coating polymers on poly(dimethyl siloxane)-glass hybrid microchips by a vacuum-drying method. Anal Sci 31:1171–1175
11. Jacobson SC, Hergenroder R, Moore AW, Ramsey JM (1994) Precolumn reactions with electrophoretic analysis integrated on a microchip. Anal Chem 66:4127–4132
12. Quirino JP, Terabe S (1999) Sweeping of analyte zones in electrokinetic chromatography. Anal Chem 71:1638–1644
13. Kawai T, Ito J, Sueyoshi K, Kitagawa F, Otsuka K (2012) Electrophoretic analysis of cations using large-volume sample stacking with an electroosmotic flow pump using capillaries coated with neutral and cationic polymers. J Chromatogr A 1267:65–73

Chapter 5

Microchip Electrophoresis Containing Electrodes for Integrated Electrochemical Detection

Lucas Paines Bressan, Dosil Pereira de Jesus, Dulan Bandara Gunasekara, Susan Marie Lunte, and José Alberto Fracassi da Silva

Abstract

Microchip electrophoresis is a versatile separation technique. Electrochemical detection is suitable to apply to microdevices due to its easy integration to the fabrication process and good sensitivity and selectivity. Here we describe the procedures to prepare Pt band electrodes deposited on glass to couple to polydimethylsiloxane (PDMS) microchips aiming the separation and detection of nitrite using an isolated potentiostat.

Key words Amperometric detection, Polydimethylsiloxane, Isolated potentiostat, Gated injection, Reactive nitrogen species

1 Introduction

Micro total analysis systems (μTAS), also known as lab on a chip (or simply "microchips"), have revolutionized the way we perform chemical analysis. The term μTAS was introduced by Andreas Manz and colleagues in the beginning of the 1990s, and since then thousands of applications and developments have been reported [1–5]. Capillary electrophoresis (CE) is suitable to apply in microchip format (MCE) because high voltages applied for the separation can also be used to control the flow on the microchannels (electroosmotic flow, EOF).

Among other detection techniques, such as fluorescence and mass spectrometry, electrochemical detection has been used as detection strategy in microchip and presents some advantages such as high sensitivity and selectivity (can be tuned adjusting the electrode potential or through the modification of the electrode), easy integration to fabrication process, and easy miniaturization. Particularly, amperometric detection (AD) at constant potential can

Debashis Dutta (ed.), *Microfluidic Electrophoresis: Methods and Protocols*, Methods in Molecular Biology, vol. 1906, https://doi.org/10.1007/978-1-4939-8964-5_5, © Springer Science+Business Media, LLC, part of Springer Nature 2019

be easily implemented owing to the simple configuration and instrument control. Also, the response in AD is related to the electron transfer at the electrode-solution interface and does not depend on the volume of the detection cell. This is important considering the miniaturization of the detection system.

Unfortunately, when MCE is employed, the high voltages used for the separation cause serious interferences on the potential of the electrodes in AD. In extreme cases the electronic circuitry of the potentiostat can be damaged. To avoid this, some authors have reported the use of decouplers, electrode positioning outside the microchannels, dual channel configuration, and isolated potentiostats.

Here we describe how to configure a MCE for AD using Pt band electrodes positioned at the edge of the separation microchannel in a hybrid polydimethylsiloxane (PDMS) and glass microchip. Example of application to the separation of nitrite is also included.

2 Materials

2.1 Microchip Preparation

1. Silicon/SU-8 or metallic template (mold) (*see* Fig. 1 and **Note 1**).
2. Polydimethylsiloxane, PDMS, (Sylgard 184 Silicone Elastomer Kit, Dow Corning).
3. Vacuum desiccator.
4. Oven.

2.2 Electrode Preparation

1. Glass substrate with good planarity (*see* Fig. 1 and **Note 2**).
2. Positive photoresist.
3. Spin coater.
4. Hot plate.
5. Metal deposition system (sputtering or electron beam deposition).
6. Acetone.

2.3 Background Electrolyte (BGE) for Microchip Electrophoresis

1. Prepare the running buffer in ultrapure water by dissolving the needed amount of boric acid to give a 10-mM solution.
2. Adjust the pH 10–11 with sodium hydroxide solution.
3. Add tetradecyltrimethylammonium bromide (TTAB) to give a 2-mM concentration. TTAB act as an electroosmotic flow inverter.

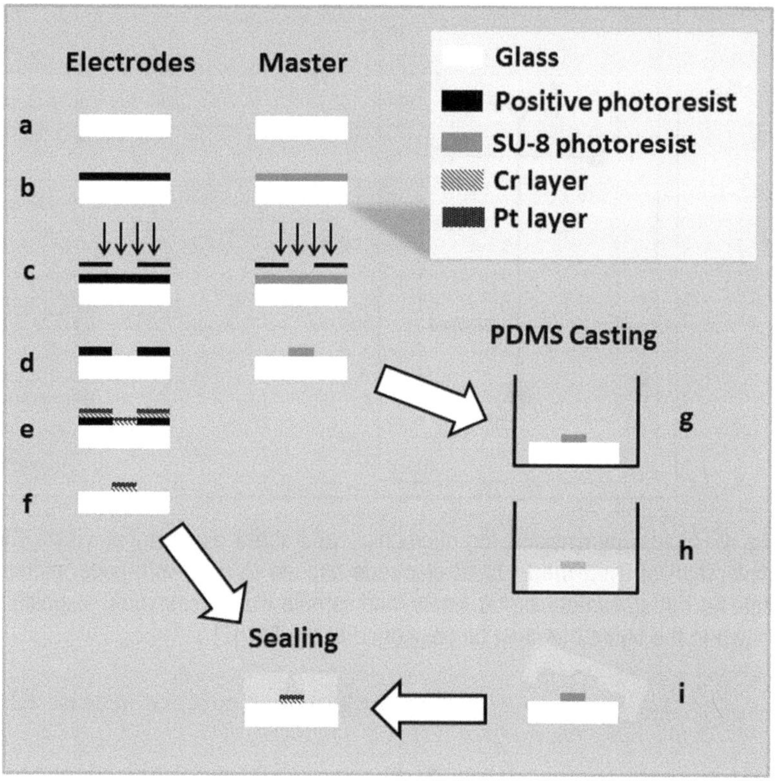

Fig. 1 Microchip preparation. Hybrid PDMS/glass microchips were used throughout this work, but other materials can also be applied (*see* **Note 5**). (**a**) Glass substrate; (**b**) Photoresist is applied by spinning; (**c**) Mask containing the design of the electrodes is positioned, and photoresist is exposed to UV light; (**d**) Photoresist is developed; (**e**) Metal deposition step, Cr is deposited first to improve adhesion; (**f**) Residual photoresist is removed leaving the electrode on the surface; (**g**) Template is placed in a support; (**h**) Polymer casting; (**i**) After curing the polymer is removed from the template. The last step concerns the sealing of the polymer microchip against the substrate containing the electrodes

2.4 Microchip Electrophoresis

1. Dual channel high-voltage power supply (HVPS).
2. Isolated potentiostat (Pinnacle Technologies, Lawrence, KS).
3. Ag/AgCl reference electrode (RE-5B, Bioanalytical Systems, Inc., West Lafayette, IN, USA).

3 Methods

3.1 Microchip Preparation [6]

1. Using a plastic cup, mix vigorously 10 parts of Sylgard 184 pre-polymer with 1 part (in mass) of cure agent (*see* **Note 3**).
2. Let the mixture in a desiccator under vacuum until all trapped air is removed.
3. Pour the mixture over the template for casting.

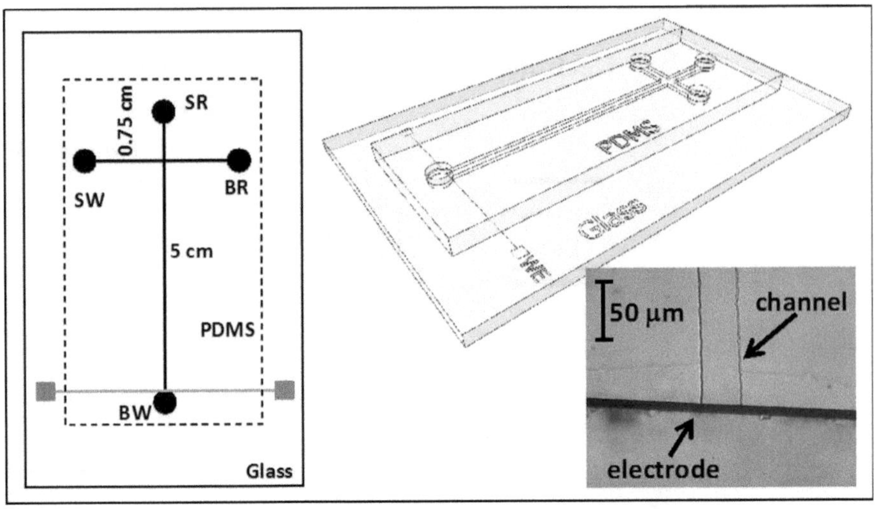

Fig. 2 Typical dimensions for microchips and three-dimensional view. The inset shows a microscope photograph of a 15-μm Pt band electrode aligned to a 40-mm-wide microchannel. BR, SR, BW, and SW state for buffer, sample, buffer waste, and sample waste reservoirs, respectively. Reference electrode is not shown in the figure but must be positioned at the BW [7]

4. Let the polymer cure for at least 12 h in a preheated oven at 70 °C.
5. Remove carefully the PDMS from the template, and drill access holes to the reservoirs using a 3-mm diameter biopsy puncher.
6. Align the edge of the separation channel with the band working electrode under a microscope. The working electrode should not be positioned more than 5 μm inside the separation channel (Fig. 2).
7. The PDMS must seal reversibly to a clean glass surface containing the working electrode. So, the preparation of the microchip must be conducted in an environment free of dust.

3.2 Electrode Preparation (Lift-Off Technique)

1. Clean the surface rigorously.
2. Deposit positive resist using a spinner.
3. Expose the positive photoresist to UV light using a mask containing the design of the electrodes.
4. Develop the photoresist. After this step, the surface of the substrate will be available for metal deposition (**step 5**).
5. The electrode deposition depends on the instrumentation and materials available. As an example, we have prepared 15-μm Pt band electrodes deposited on a 4-in. borosilicate glass (Precision Glass and Optics, Santa Ana, CA, USA) by sputtering. Usually, a Cr or Ti layer is deposited under the Pt to improve the adhesion of Pt on the glass surface [7–10]. Some vendors supply glasses already coated by Cr. An alternative to this

procedure is to etch area of electrode on the glass plate to make a groove before metal deposition [8]. The groove makes the procedure harder but provides better electrode stability compared to an electrode deposited directly on glass surface.

6. Remove the remaining photoresist.
7. Electrical contacts can be made using copper wires connected to the electrode through a conductive silver paint (Ted Pella, CA, USA).

3.3 Microchip Electrophoresis

1. Fill all reservoirs and then channels using the BGE and applying negative pressure in opposite ends of channels (*see* Fig. 3 and **Note 4**).
2. Connect the output cables of the HVPS to 1-cm Pt wires, and place them on the reservoirs as indicated in Fig. 2. Be sure that all cables are fixed to eliminate the risks of short circuit and electrical shock.
3. Check if there is electric current for all channels. If not, remove the microchip and check for bubbles.
4. Place the reference electrode at the buffer waste (BW) reservoir (*see* Fig. 2).
5. Connect the working and reference electrodes to the isolated potentiostat.
6. Replace the BGE at the sample reservoir by sample solution, and run the program for gated injection while acquiring the signal from the potentiostat.

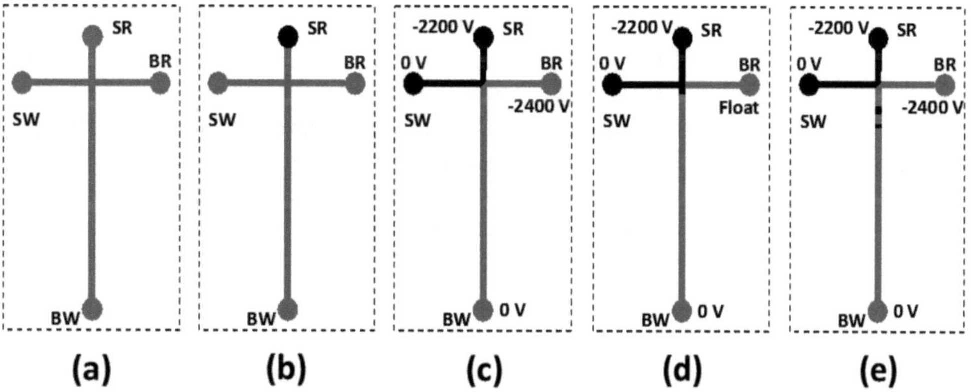

Fig. 3 Microchip operation. (**a**) All reservoirs and channels are filled with BGE; (**b**) Solution at the SR is replaced by the sample solution; (**c**) Voltage is applied to the reservoirs using a double channel HVPS (*see* **Note 6**); (**d**) The voltage at the BR is floated and part of the separation channel is filled with sample solution; (**e**) The gate is reestablished and the components of the sample are separated. Detection electrodes are not shown in the figure

4 Notes

1. The template must contain the design of the microchannels in high relief and can be prepared in any material that does not stick to the PDMS. We have successfully used silicon wafers as base material and SU-8 negative photoresist as a structural material prepared using photolithography protocols.
2. Optical grade or glass wafers such as Borofloat 33 or soda lime usually produce good results.
3. The total mass of PDMS depends on the area of the template and support and the desired thickness of the microchip. We typically use 10 g of prepolymer to 1 g of curing agent, which gives a thickness around 2 mm for a 4-inch silicon wafer used to prepare the template.
4. Sometimes, filling PDMS microchannels with aqueous solutions is difficult. Any trapped bubble must be removed before the application of high voltage. If the filling of PDMS microchannels is troublesome, try to fill them first using ethyl alcohol (ethanol). Take care to keep the levels of solution in all reservoirs balanced in order to avoid flow forced by hydraulic pressure differences.
5. Microchip made totally in borosilicate glass has also been used in our group with success. Unfortunately, the process of sealing glass parts is laborious and troublesome and needs to be performed in a clean room. Other polymers such as cyclic olefin copolymer (COC), poly(methyl methacrylate) (PMMA), polycarbonate (PC), and off-stoichiometric thiol-ene polymers (OSTE) can also be used as microchip substrates, but care must be taken to avoid cracks on the electrodes and loss of the electrical contact.
6. In practice, voltages can differ from the one indicated in Fig. 3. The voltage applied to the buffer reservoir must be higher than that applied to the sample reservoir to establish the gate. We have obtained good results using the voltages at buffer reservoir about 200 V higher. A deep discussion about the influence of several parameters on gate injection can be found in reference [11].

Acknowledgments

This work was supported by Conselho Nacional de Desenvolvimento Científico e Tecnológico (CNPq) grants 444514/2014-7 and 465389/2014-7, Fundação de Amparo à Pesquisa do Estado de São Paulo (FAPESP) grants 2014/50867-3 and 2013/22127-2, Coordenação de Aperfeiçoamento de Pessoal de Nível Superior (CAPES) grant CAPES-COFECUB 802-14, and National Institutes of Health (NIH). Authors thank Instituto Nacional de Ciência e Tecnologia em Bioanalítica (INCTBio).

References

1. Patabadige DEW, Jia S, Sibbitts J, Sadeghi J, Sellens K, Culbertson CT (2016) Micro total analysis systems: fundamental advances and applications. Anal Chem 88:320–338
2. Reyes DR, Iossifidis D, Auroux P-A, Manz A (2002) Micro total analysis systems. 1. Introduction, theory, and technology. Anal Chem 74:2623–2636
3. Culbertson CT, Mickleburgh T, Stewart-James SA, Sellens KA, Pressnall M (2014) Micro total analysis systems: fundamental advances and biological applications. Anal Chem 86:95–118
4. Kovarik ML, Ornoff DM, Melvin AT, Dobes NC, Wang Y, Dickinson AJ, Gach PC, Shah PK, Allbritton NL (2013) Micro total analysis systems: fundamental advances and applications in the laboratory, clinic, and field. Anal Chem 85:451–472
5. Kovarik ML, Gach PC, Ornoff DM, Wang YL, Balowski J, Farrag L, Allbritton NL (2012) Micro total analysis systems for cell biology and biochemical assays. Anal Chem 84:516–540
6. Gunasekara DB, Hulvey MK, Lunte SM, da Silva JAF (2012) Microchip electrophoresis with amperometric detection for the study of the generation of nitric oxide by NONOate salts. Anal Bioanal Chem 403:2377–2384
7. Gunasekara DB, Hulvey MK, Lunte SM (2011) In-channel amperometric detection for microchip electrophoresis using a wireless isolated potentiostat. Electrophoresis 32:832–837
8. Gunasekara DB, Siegel JM, Caruso G, Hulvey MK, Lunte SM (2014) Microchip electrophoresis with amperometric detection method for profiling cellular nitrosative stress markers. Analyst 139:3265–3273
9. Hulvey MK, Frankenfeld CN, Lunte SM (2010) Separation and detection of peroxynitrite using microchip electrophoresis with Amperometric detection. Anal Chem 82:1608–1611
10. Moreira NH, Almeida ALJ, Piazzeta MHO, de Jesus DP, Deblire A, Gobbi AL, da Silva JAF (2009) Fabrication of a multichannel PDMS/glass analytical microsystem with integrated electrodes for amperometric detection. Lab Chip 9:115–121
11. Zhang G, Du W, Liu B-F, Hisamoto H, Terabe S (2007) Characterization of electrokinetic gating valve in microfluidic channels. Anal Chim Acta 584:129–135

Chapter 6

Micellar Electrokinetic Chromatography

Braden C. Giordano, Ronald Siefert, and Greg E. Collins

Abstract

Micellar electrokinetic chromatography (MEKC) is a mode of capillary electrophoresis that allows for the separation of neutral molecules in an electric field. Typically, neutral molecules move with electroosmotic flow (EOF) or bulk flow during electrophoretic separations resulting in no temporal resolution between mixtures of neutral analytes. Inclusion of surfactant micelles in the separation buffer allows for the separation of neutral analytes from one another through association with the micelle. Here we outline the implementation of MEKC for the separation of neutral molecules using a mixture of nitroaromatic explosives and their degradation products serving as a test analyte mixture.

Key words Electrophoresis, Preconcentration, Explosives

1 Introduction

Capillary electrophoresis is a separation technique wherein analytes are separated based upon differences in their charge-to-size ratio in an electric field. While neutral molecules all carry the same charge-to-size ratio of zero, separation of neutral analytes is possible through the addition of surfactant micelles [1] to the separation background electrolyte. Briefly, in the absence of micelles, neutral analytes migrate with the velocity of electroosmotic flow (EOF)—the bulk flow of liquid through a capillary when an electric field is applied across the capillary. Upon addition of surfactant micelles, an equilibrium is established where a given analyte is either migrating at the velocity of EOF or at the velocity of the micelle. A given analyte's migration or elution time is a function of the analyte's affinity for the micelle.

This manuscript will outline how to prepare sample matrix and background electrolytes for MEKC-based separation, performing MEKC on a mixture of neutral analytes and highlight the use of on-line sample preconcentration to improve sensitivity and separation performance in MEKC.

Debashis Dutta (ed.), *Microfluidic Electrophoresis: Methods and Protocols*, Methods in Molecular Biology, vol. 1906, https://doi.org/10.1007/978-1-4939-8964-5_6, © Springer Science+Business Media, LLC, part of Springer Nature 2019

2 Materials

Chemicals used in the method were all ACS reagent grade. Water (18.2 MΩ-cm) was acquired from an EMD Millipore Milli-Q Advantage A10 system. Solvents were HPLC grade. Trinitrotoluene (TNT), 1,3,5-trinitrobenzene (1,3,5-TNB), 2,4-dinitrotoluene (2,4-DNT), and 2-amino-4,6-dinitrotoluene (2-Am-4,6-DNT) standards were obtained from AccuStandard (New Haven, CT).

2.1 Solution Preparation (See Note 1)

1. Capillary Conditioning Solution: 1.00 M NaOH. The 1.00 M NaOH stock solution was prepared by weighing 2.00 g of NaOH (CAS 1310-73-2, FW 40.00 g/mol), then quantitatively transferring the NaOH to a 50.00 mL volumetric flask, and adding 18.2 MΩ-cm water to the mark. A pipet was used to transfer 1600 μL of 1.00 M NaOH to a vial (*see* **Note 2**).
2. Capillary Rinse Solution: 18.2 MΩ-cm water. A pipet was used to transfer 1600 μL of 18.2 MΩ-cm water to a vial.
3. Separation Background Electrolyte (BGE): 200 mM sodium cholate with 20 mM Tris(hydroxymethyl)amino-methane (Tris) buffer and 10% methanol. A 500 mM sodium cholate stock solution was prepared by weighing 10.76 g of sodium cholate (CAS 361-09-1, FW 430.55 g/mol), then quantitatively transferring the sodium cholate to a 50.0 mL volumetric flask, and adding 18.2 MΩ-cm water to the mark. A 500 mM Tris stock solution was prepared by weighing 3.03 g of Tris (CAS 77-86-1, FW 121.14 g/mol), then quantitatively transferring the Tris to a 50.0 mL volumetric flask, and adding 18.2 MΩ-cm water to the mark. BGE was prepared from these stock solutions by adding 2400 μL of 500 mM sodium cholate, 240 μL of 500 mM Tris, 600 μL of methanol, and 2760 μL of 18.2 MΩ-cm water which resulted in 6000 μL of BGE. The solution could be scaled appropriately for larger volumes if needed (*see* **Note 3**).
4. Sample: Analyte samples were prepared by serial dilution of standards from AccuStandard. The concentration for all standards was 1000 μg/mL in 50/50 methanol/acetonitrile. In all examples contained herein, final analyte concentration is 5 μg/mL and was prepared via serial dilution in the indicated sample matrix (*see* **Note 4**).

3 Methods

3.1 Capillary Preparation

1. A newly prepared capillary should be installed per manufacturer's recommendations and conditioned appropriately. In the case of the MEKC separations described herein, a new capillary

is prepared and installed in a capillary cartridge with a total length of 60 cm and an effective length of 50 cm (*see* **Note 5**). The capillary is conditioned by alternating flushes of 1 M NaOH and 18.2 MΩ-cm water. Each flush is 2 min at a pressure of 20 psi. This process typically takes 20 min.

3.2 Separation Protocol

This section outlines the steps necessary to perform a single MEKC-based separation. All separations described in this manuscript follow these steps with the only variable being the sample, sample matrix, or sample time. It is important to note that most commercial capillary electrophoresis systems allow for injection on either side of a capillary.

1. Rinse capillary with capillary conditioning solution (1 M NaOH) for 1 min at 50 psi (*see* **Note 6**).
2. Rinse capillary with capillary rinse solution (18.2 MΩ-cm water) for 1 min at 50 psi.
3. Rinse capillary with separation background electrolyte (20 mM Tris, 200 mM sodium cholate, 10% v/v methanol, pH 10.0) for 4 min at 20 psi.
4. Inject sample at 1 psi for a desired amount of time (*see* **Note 7**).
5. Inject separation background electrolyte for 10 s at 1 psi (*see* **Note 8**).
6. Replace both inlet and outlet vials with vials containing separation background electrolyte and apply desired separation voltage (*see* **Notes 9 and 10**).
7. Perform separation for a sufficient time such that all analytes of interest are observed.

3.3 Establish the Separation Window

The separation window is the region of the electropherogram where one would expect to see neutral analytes. One side of the separation window is the migration time of electroosmotic flow, and the second is the time associated with the migration of the micelle. The following section outlines the method by which the separation window can be determined.

1. Prepare a sample where the sample matrix is the separation background electrolyte and the analyte of interest is Sudan III (*see* **Note 11**).
2. Perform a separation as indicated in Subheading 3.2.
3. Due to the hydrophobic nature of Sudan III, it serves as a micelle marker. The peak associated with Sudan III is used to determine the mobility or separation time of the micelles in the sample plug. Note the migration time of EOF (t_{EOF}) and the migration time of Sudan III used to mark the mobility of the micelle (t_{mc})—neutral analytes will fall between these two

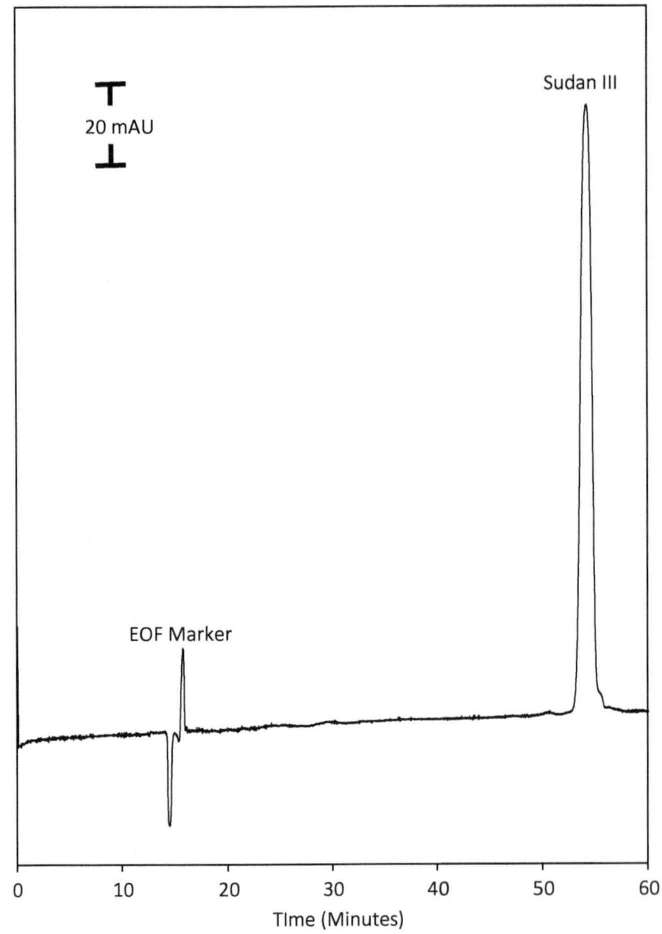

Fig. 1 Determination of separation window using Sudan III in the sample plug. Background electrolyte is 20 mM Tris, 200 mM sodium cholate, 10% v/v MeOH, pH 10.0. Sample was 100 μM Sudan III in separation background electrolyte and was injected for 16 s at 1 psi for a sample plug length of 1.6 cm. Separation voltage of 15 kV with an operating current of 43 μA. Capillary is 50 μ i.d., 60 cm length (50 cm effective length)

points in the electropherogram. Analytes with low affinity for the micelle will migrate closer to t_{EOF}, and those with high affinity for the micelle will migrate closer to t_{mc} (*see* **Notes 12 and 13**). The results of a typical separation window determination experiment are shown in Fig. 1.

3.4 Separation Performance: Sample Matrix Is Separation Background Electrolyte

Upon the establishment of the separation window, it is now appropriate to begin analysis of the analytes of interest. A test mixture of four nitroaromatic explosives and degradation products is being used to illustrate typical MEKC-based separations. The analytes used are 1,3,5-trinitrobenzene (1,3,5-TNB), 2,4,6-trinitrotoluene

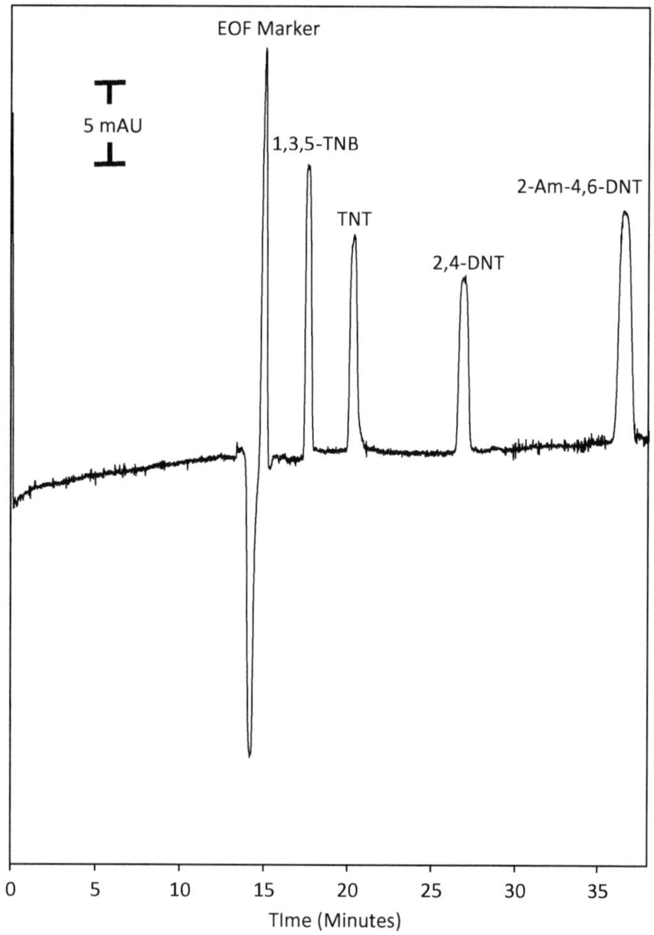

Fig. 2 Separation of four nitroaromatic compounds. Background electrolyte is 20 mM Tris, 200 mM sodium cholate, 10% v/v methanol, pH 10.0. Sample matrix is background electrolyte. Sample was injected for 16 s at 1 psi for a sample plug length of 1.6 cm. Separation voltage of 15 kV with an operating current of 43 μA. Capillary is 50 μ i.d., 60 cm length (50 cm effective length)

(TNT), 2,4-dinitrotoluene (2,4-DNT), and 2-amino-4,6-dinitrotoluene (2-Am-4,6-DNT).

1. Prepare a sample where the sample matrix is separation background electrolyte.
2. Perform separation as indicated in Subheading 3.2. A representative separation of these four analytes is shown in Fig. 2.

3.5 Separation Performance: Sample Matrix Is NaCl

One benefit associated with capillary electrophoresis is the ability to perform on-line or "in-capillary" analyte preconcentration. This functionality enables sweeping [2–4] or high-salt stacking [5] to be extended to MEKC. In either case, neutral analyte preconcentration requires a discontinuity between the sample plug and the

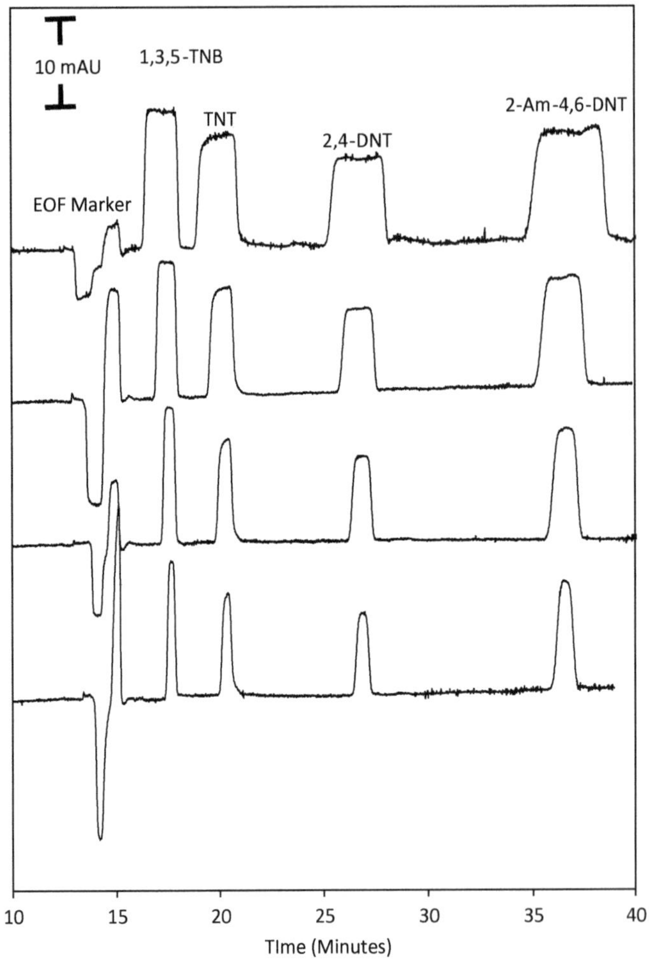

Fig. 3 Separation of four nitroaromatic compounds. Background electrolyte is 20 mM Tris, 200 mM sodium cholate, 10% v/v methanol, pH 10.0. Sample matrix is background electrolyte. From the bottom to the top, sample plug lengths are 1.6 cm, 2.4 cm, 4.0 cm, and 6.5 cm. Separation voltage of 15 kV with an operating current of 43 μA. Capillary is 50 μ i.d., 60 cm length (50 cm effective length)

separation background electrolyte—or put more plainly, the surfactant in the separation background electrolyte cannot be in the sample matrix.

On-line preconcentration allows for longer sample injections with minimal impact on peak width. The consequence is peak height increases with increasing sample volume, thus improving sensitivity without the detriment of disrupting selectivity. Figure 3 illustrates the effects of increasing injection plug length with a sample matrix that is separation background electrolyte. Clearly, with increasing injection plug length, the analyte peaks get wider with no increase in peak height.

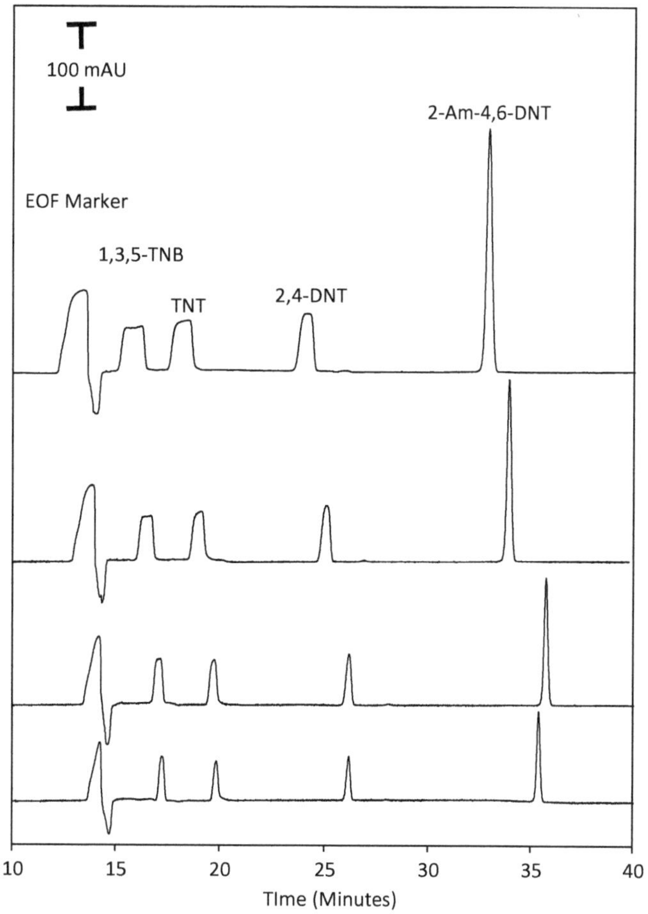

Fig. 4 Separation of four nitroaromatic compounds. Background electrolyte is 20 mM Tris, 200 mM sodium cholate, 10% v/v methanol, pH 10.0. Sample matrix is 180 mM NaCl. From the bottom to the top, sample plug lengths are 1.6 cm, 2.4 cm, 4.0 cm, and 6.5 cm. Separation voltage of 15 kV with an operating current of 43 μA. Capillary is 50 μ i.d., 60 cm length (50 cm effective length)

In order to implement on-line preconcentration:

1. Prepare a sample where the sample matrix does not contain the surfactant present in the separation background electrolyte (*see* **Note 14**). In the example shown in Fig. 4, the sample matrix is 180 mM NaCl.
2. Perform separation as indicated in Subheading 3.2. Representative separations for 1.6, 2.4, 4.0, and 6.5 cm injection plugs are shown in Fig. 4. A summary of peak heights and peak widths at half height is presented in Table 1 (*see* **Note 15**).

Table 1
Summary of peak height and peak width at half heaight for non-preconcentration and on-line preconcentration sample injections

		Sample matrix is BGE		Sample matrix is 180 mM Nacl	
Analyte	Sample plug length (cm)	Peak height (mAU)	Peak width at half height (min)	Peak height (mAU)	Peak width at half height (min)
1,3,5-TNB	1.6	1.857	0.375	3.657	0.274
	2.4	1.858	0.574	3.731	0.423
	4.0	1.868	0.945	3.686	0.695
	6.5	1.903	1.580	3.629	1.120
TNT	1.6	1.385	0.454	3.251	0.242
	2.4	1.397	0.669	3.738	0.362
	4.0	1.48	1.114	4.038	0.595
	6.5	1.543	1.850	4.12	0.974
2,4-DNT	1.6	1.121	0.600	3.624	0.203
	2.4	1.161	0.909	4.269	0.276
	4.0	1.131	1.509	4.688	0.421
	6.5	1.226	2.560	4.741	0.700
2-Am-4,6-DNT	1.6	1.463	0.800	7.473	0.181
	2.4	1.515	1.211	10.608	0.202
	4.0	1.516	2.040	15.162	0.232
	6.5	1.565	3.470	20.251	0.302

4 Notes

1. All sample volumes presented herein are appropriate for the commercially available capillary electrophoresis system used in our laboratory. Sample volume needs will obviously change as a function of an end user's instrumentation.
2. MEKC-based separations can be performed under either basic or acid pH. Typically, when using a basic separation buffer, the capillary is conditioned with base. Conversely, if one was using an acidic separation buffer, the capillary would be conditioned with an acidic solution such as 1 M HNO_3 or 1 M HCl prior to rinsing with water.
3. Sodium cholate is one of many common surfactants used in MEKC-based separations. Choice of surfactant and surfactant concentration can depend on a number of factors, including analytes of interest, sample matric composition, and desired conductivity of separation background electrolyte. Separation optimization may require extensive evaluation of background electrolyte composition.

4. The use of the NaCl sample matrix promotes analyte preconcentration at the boundary between the sample plug and the separation background electrolyte. This mode of preconcentration can be referred to as sweeping [2–4] and/or high-salt stacking [5].
5. Capillaries used in electrophoresis are coated with polyimide to provide mechanical stability to the fused silica. A small window (approximately 1 mM in length) is burned into the capillary serving as a detection window for UV absorbance or laser-induced fluorescence detection. Since most commercial capillary electrophoresis instrumentation allows for injection on either side of the capillary, it is useful to define the effective length of the capillary in order to orient the user to which side of the capillary injections are taking place. In the work described herein, the long side of the capillary is being used for separation, with that side being referred to as the inlet side of the capillary. The short side of the capillary will be referred to as the outlet side of the capillary. In all instances where hydrodynamic flow is used to move liquid through the capillary, the vial of liquid being moved is at the inlet side, and a vial filled with 18.2 MΩ-cm water is placed on the outlet side.
6. A capillary flush at this pressure and time replaced the capillary volume approximately five times. A general rule of thumb is that all capillary rinsing/flushing should replace the total capillary volume between 5 and 20 times.
7. Injection time can vary based upon the presence of on-line sample preconcentration. In the absence of preconcentration, a condition met when the sample matrix is the separation background electrolyte, injection plug length should not exceed 5% of the effective length of the capillary. In the work described herein, the effective length of the capillary is 50 cm; thus a sample plug should not exceed 2.5 cm long.
8. A small plug of BGE should be injected behind the sample plug to ensure that during the early moments of a separation, analyte does not inadvertently migrate out of the capillary.
9. Both vials are replaced with fresh vials of separation background electrolyte to minimize the effects of capillary action associated with vials at the inlet and outlet side of the capillary filled to different levels. For example, a partially filled outlet vial may result in slight hydrodynamic flow of material from the inlet toward the outlet.
10. The magnitude of the separation voltage will determine the total separation time. The limit to applied voltage is often Joule heating (heating of the liquid in the capillary as current passes through) which can contribute to migration time irreproducibility and peak broadening due to increased diffusion. The

applied voltage at which excess Joule heating occurs can be determined with a simple experiment where the operational current is measured at different applied voltages. Deviation from linearity is indicative of Joule heating occurring, and operational currents should be set below the voltage at which this deviation occurs.

11. Sudan III is a hydrophobic dye demonstrated to have a very high affinity for both cholate and SDS micelles. Sudan III can be prepared as a 10 mM solution in methanol and added to surfactant-containing solutions.
12. The EOF marker is the result of trace amounts of methanol in the injected sample plug.
13. Charged analytes may have an affinity for a micelle, but will not necessarily migrate between t_{eof} and t_{mc}.
14. There are a number of factors that contribute to the choice of sample matrix composition in regard to on-line sample preconcentration and are beyond the scope of this manuscript. Several articles highlight the importance of sample matrix composition and sample plug length as it relates to on-line preconcentration when considering sensitivity optimization needs [2–15].
15. On-line sample preconcentration efficiency is directly related to an analyte's affinity for a micelle. The more it retained by the micelle, the greater the extent of on-line preconcentration. The ability to preconcentrate will eventually plateau for all analytes as sample plug length increases.

References

1. Terabe S, Otsuka K, Ichikawa K, Tsuchiya A, Ando T (1984) Electrokinetic separations with micellar solutions and open-tubular capillaries. Anal Chem 56:111–113
2. Quirino JP, Kim JB, Terabe S (2002) Sweeping: concentration mechanism and applications to high-sensitivity analysis in capillary electrophoresis. J Chromatogr A 965:357–373
3. Quirino JP, Terabe S (1998) Exceeding 5000-fold concentration of dilute analytes in micellar electrokinetic chromatography. Science 282:465–468
4. Quirino JP, Terabe S (1999) Sweeping of analyte zones in electrokinetic chromatography. Anal Chem 71:1638–1644
5. Palmer J, Munro NJ, Landers JP (1999) A universal concept for stacking neutral analytes in micellar capillary electrophoresis. Anal Chem 71:1679–1687
6. Palmer J, Burgi DS, Landers JP (2002) Electrokinetic stacking injection of neutral analytes under continuous conductivity conditions. Anal Chem 74:632–638
7. Liu ZY, Sam P, Sirimanne SR, Mcclure PC, Grainger J, Patterson DG (1994) Field-amplified sample stacking in micellar electrokinetic chromatography for on-column sample concentration of neutral molecules. J Chromatogr A 673:125–132
8. Giordano BC, Newman CID, Federowicz PM, Collins GE, Burgi DS (2007) Micelle stacking in micellar electrokinetic chromatography. Anal Chem 79:6287–6294
9. Quirino JP, Terabe S (1998) On line concentration of neutral analytes for micellar electrokinetic chromatography. 5. Field enhanced sample injection with reverse migrating micelles. Anal Chem 70:1893–1901

10. Quirino JP, Terabe S (1998) On-line concentration of neutral analytes for micellar electrokinetic chromatography—IV. Field-enhanced sample injection. J Chromatogr A 798:251–257
11. Quirino JP, Otsuka K, Terabe S (1998) On-line concentration of neutral analytes for micellar electrokinetic chromatography—VI. Stacking using reverse migrating micelles and a water plug. J Chromatogr B 714:29–38
12. Quirino JP, Terabe S (1998) On-line concentration of neutral analytes for micellar electrokinetic chromatography. 3. Stacking with reverse migrating micelles. Anal Chem 70:149–157
13. Kim JB, Otsuka K, Terabe S (2001) On-line sample concentration in micellar electrokinetic chromatography with cationic micelles in a coated capillary. J Chromatogr A 912: 343–352
14. Monton MRN, Otsuka K, Terabe S (2003) On-line sample preconcentration in micellar electrokinetic chromatography by sweeping with anionic-zwitterionic mixed micelles. J Chromatogr A 985:435–445
15. Monton MRN, Quirino JP, Otsuka K, Terabe S (2001) Separation and on-line preconcentration by sweeping of charged analytes in electrokinetic chromatography with nonionic micelles. J Chromatogr A 939:99–108

Chapter 7

Microchip Isotachophoresis: Analysis of Pharmaceuticals

Marián Masár and Jasna Hradski

Abstract

Microchip isotachophoresis (μITP) is a miniaturized version of conventional isotachophoresis (ITP) characterized by low sample and buffer consumption and reduced waste production. μITP with universal conductivity detection is suitable for quantitative analysis of relatively simplified samples that contain analyte(s) at relatively high concentration, e.g., pharmaceutical preparations. Here we describe in detail a principle of μITP in terms of reaching highly precise results. A practical use of μITP is shown on the analyses of various pharmaceutical preparations for content of major constituents including active pharmaceutical ingredients as well as pharmaceutical counterions. The pharmaceuticals are treated only minimally prior to the ITP run on a microchip with coupled channels and sample injection channel with 0.9 μL volume. Developed method is suitable for rapid (analysis time up to 10 min), precise (less than 1% RSD of analyte zone length), and accurate (recovery of 98–101%) determination of major pharmaceutical ingredients using a method of internal standard for data evaluation.

Key words Microchip isotachophoresis, Microchip with coupled channels, Contact conductivity detection, Pharmaceuticals, N-acetylcysteine, Buserelin acetate

1 Introduction

Microchip isotachophoresis (μITP), as one of the microchip electrophoresis (MCE) techniques, has a significant position among miniaturized separation techniques. This is mainly due to the self-sharpening effect which results in sharp zone boundaries in steady-state and its concentration effect which can be accompanied with removal of matrix constituents [1]. Isotachophoresis (ITP) was first introduced in miniaturized format for the separation of herbicides on a single-channel microchip [2]. The popularity of ITP not only in miniaturized format is expressed by a number of review articles published periodically [3–5] or specifically on some topic related to the ITP analysis [1, 6]. Even though ITP is most widely used in combination with some other (electro)separation techniques, its concentration and sample cleanup abilities make it suitable for further development not only for multidimensional applications but also as a stand-alone technique.

Debashis Dutta (ed.), *Microfluidic Electrophoresis: Methods and Protocols*, Methods in Molecular Biology, vol. 1906, https://doi.org/10.1007/978-1-4939-8964-5_7,

μITP is based on the same principles as conventional ITP, i.e., separation is carried out in a discontinuous buffer system, when an electric field gradient from the leading (L) to the terminating (T) buffer is formed. L buffer contains an ion with higher effective mobility than the effective mobility of the most mobile ion in the sample, and T buffer contains an ion with lower effective mobility than the effective mobility of the least mobile ion in the sample. Sample is introduced between L and T buffers, and during the separation, ions exhibiting mobilities intermediate to those of L and T ions form discrete zones. After reaching a steady-state condition, all ITP zones migrate with the same velocity. Based on the Kohlrausch regulating function, the analyte is present only in its own zone at a concentration c_A, so-called limiting concentration, which depends on the concentration of L ion, as well as the mobilities of analyte, L ion, and counterion (Eq. (1)) [7, 8].

$$c_A = c_L \frac{\mu_L + \mu_R}{\mu_L} \frac{\mu_A}{\mu_A + \mu_R} \frac{z_L}{z_A} \tag{1}$$

where c_L is the concentration of L ion; μ_L, μ_R, and μ_A are the effective mobilities of L ion, counterion, and analyte, respectively; and z_L and z_A are the ionic charges of L ion and analyte, respectively.

Once the limiting concentration is reached, the analyte forms a plateau (regular zone), and this type of ITP is called a plateau mode ITP. This mode is characterized by a stepwise response from a universal conductivity detector where the qualitative parameter is the relative step height (RSH; Eq. (2)) and the quantitative parameter is the zone length of the analyte. If the analyte is present at low concentration, the limiting concentration is not reached. In this case, the analyte is concentrated between two regular zones and forms a peak on isotachopherogram. This type of ITP is called peak mode ITP [8]. In this work, only the plateau mode ITP was employed.

Compared to conventional ITP, μITP has several advantages, such as shorter analysis time, lower consumption of sample and buffer solutions, and thus lower waste production. These properties are the result of miniaturization, i.e., shortening of the separation path and reducing the internal diameter of the channels. On the other hand, main disadvantages of miniaturized electrophoretic systems, in general, include technologically complex manufacturing process as well as requirement of sensitive detection systems. Despite these shortcomings, μITP has found its applicability in clinical, pharmaceutical, environmental, and food analysis [4, 9–15].

In order to improve physicochemical properties of pharmaceuticals, e.g., thermal stability or solubility, these are often prepared in a salt form. In such case, precise quantitation of all pharmaceutical constituents is required to confirm production of pharmaceuticals with proper stoichiometry. This is a reason why nowadays methods are being developed not only for determination of active pharmaceutical ingredient but also counterions and impurities in pharmaceuticals [16]. Usually small inorganic or organic ions are used as pharmaceutical counterions, e.g., chloride, sulfate, or acetate, for basic active pharmaceutical ingredients [17].

In the recent years, a growing interest in the analysis of pharmaceuticals by μITP or by MCE, in general, is evident [9, 12]. Applicability of μITP as a miniaturized electrophoretic technique in the analysis of pharmaceutical macroconstituents is shown in this study on (1) active pharmaceutical ingredient, N-acetylcysteine in mucolytic preparations [18], and (2) pharmaceutical counterion, acetate in cancer treatment drug buserelin acetate [19]. N-acetylcysteine is an antioxidant which has found a wide use in pharmaceutical industry as mucolytic, antiviral, antitumor, and anti-inflammatory agent [20, 21]. Buserelin is widely used in treatment of hormone-dependent cancers [22].

1.1 Qualitative and Quantitative Analysis

Quality in plateau mode ITP is given as

$$\mathrm{RSH_A} = \frac{h_\mathrm{A} - h_\mathrm{L}}{h_\mathrm{T} - h_\mathrm{L}} \tag{2}$$

where $\mathrm{RSH_A}$ is the relative step height of the analyte and h_A, h_L, and h_T are the step heights of analyte, L ion, and T ion, respectively (*see* Fig. 1).

Two types of quantitative methods can be employed in the evaluation of ITP measurements based on the zone length measurement. First, the simpler one, is called external calibration method (ECAL; Eq. (3)).

$$t_{\mathrm{A,ECAL}} = k\, c_{\mathrm{A,smpl}} \tag{3}$$

where $t_{\mathrm{A,ECAL}}$ is the time needed for the zone of the analyte to pass through the detector, i.e., zone length of the analyte, k is the slope of the calibration line, and $c_{\mathrm{A,smpl}}$ is the concentration of the analyte in the injected sample.

To reduce run-to-run fluctuations of various experimental conditions, and thus to reach highly precise determination, it is preferable to use internal standard method (ISTD; Eq. (4)).

$$t_{\mathrm{A,ISTD}} = \omega \frac{c_{\mathrm{A,smpl}}}{c_{\mathrm{ISTD,smpl}}} \tag{4}$$

where $t_{\mathrm{A,ISTD}}$ is the relative time of passage of the analyte zone through the detector, i.e., zone length of the analyte corrected on the zone length of the ISTD, ω is the slope of the calibration line using the ISTD, and $c_{\mathrm{ISTD,smpl}}$ is the concentration of the ISTD in the injected sample.

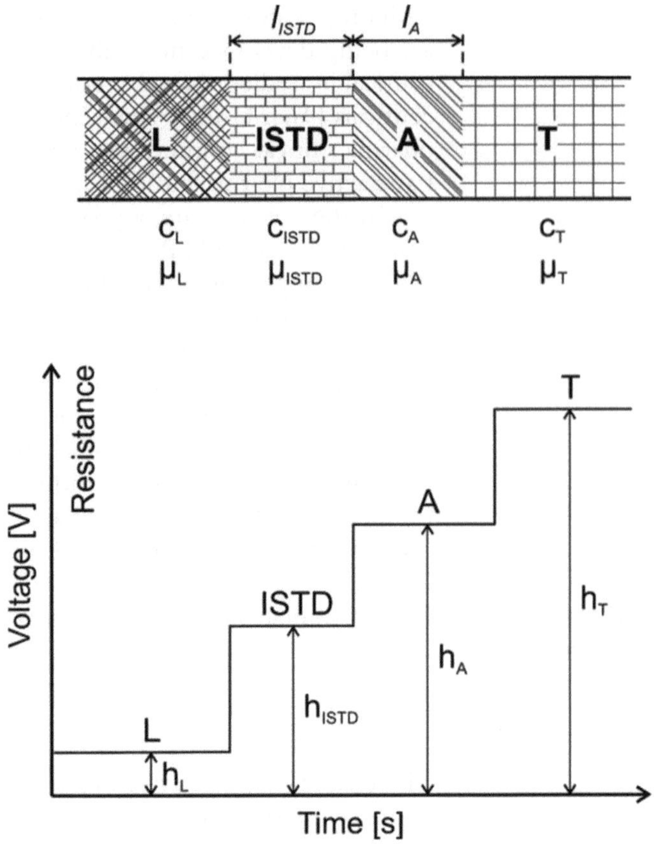

Fig. 1 ITP qualitative and quantitative parameters. *A* analyte; *ISTD* internal standard; *L*, *T* leading and terminating ions, respectively; *l* zone length; *c* concentration of the constituent in its ITP zone; μ effective mobility; *h* step height

The ISTD should fulfil several requirements in terms of (1) stability during the analysis, (2) quantitative recovery, and (3) electromigration properties, given that it has to migrate between L and T ions and it has to be resolved from the analyte and any other constituents present in the sample.

2 Materials

2.1 Instruments and Components

1. MCE analyzer with the electrolyte (*see* **Note 1**) and electronic (*see* **Note 2**) units connected to the PC (*see* Fig. 2).
2. IonChip™ 3.0 poly(methyl methacrylate) microchip with coupled separation channels (Merck, Darmstadt, Germany) and integrated conductivity sensors (*see* Fig. 2 and [23]).
3. MicroITP, ver. 2.4 data acquisition and analysis software (Merck) for Microsoft Windows. The software is used to

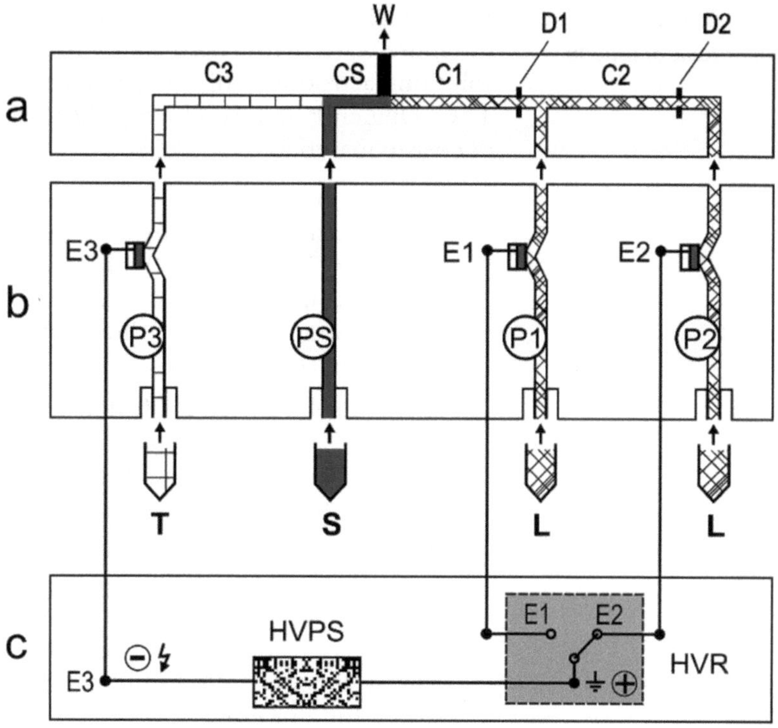

Fig. 2 MCE equipment with conductivity detectors used for the μITP separations. (**a**) Microchip with coupled channels: *C1*, first separation channel filled with L buffer (4.5 μL volume; 59 × 0.2–0.5 × 0.14–0.2 mm [length × width × depth]); *C2*, second separation channel filled with L buffer (4.3 μL volume; 56 × 0.2–0.5 × 0.14–0.2 mm); *CS*, sample injection channel with 0.9 μL volume; *C3*, third channel filled with T buffer; *D1*, *D2*, conductivity sensors placed in the C1 and C2, respectively; *L*, *T*, leading and terminating buffers, respectively; *S*, sample; *W*, waste channel. (**b**) Electrolyte unit: *P1*, *P2*, *P3*, *PS*, peristaltic micropumps for filling C1, C2, C3, and CS channels with corresponding solutions, respectively; *E1*, *E2*, driving electrodes for the C1 and C2 separation channels, respectively; *E3*, driving electrode connected to a high-voltage pole of HVPS. (**c**) Electronic unit: *HVPS*, high-voltage power supply (0–50 μA, 0–7 kV); *HVR*, high-voltage relay. Driving current passed between E3 and E2

control the electronic unit and to perform data acquisition and analysis (*see* **Note 3**).

4. Magnetic stirrer with Teflon stir bar.
5. The pH meter with combined glass electrode (*see* **Note 4**).
6. Test tube shaker for sample homogenization.
7. Ultrasonic bath for degassing the buffer solutions.
8. Small laboratory centrifuge (5000 × *g*).
9. Disposable syringe filters of 0.8 μm pore size.
10. Single-channel mechanical pipettes with a variable volume (10–200 μL and 100–1000 μL) and adequate tips.

2.2 Reagents and Stock Solutions

Prepare all the solutions using analytical grade reagents and fresh ultrapure water with resistivity of 18 MΩ cm at 25 °C. Prepare all solutions at room temperature and store them in refrigerator at 4 °C. Diligently follow all waste disposal regulations when disposing waste materials.

1. A 1% (w/v) stock solution of electroosmotic flow suppressor [24] (methylhydroxyethylcellulose 30,000; MHEC): Add 1 g of MHEC into a 150–200 mL glass beaker filled with 100 mL of water, and stir using magnetic stirrer until dissolved (*see* Subheading 2.1). Store the solution in a plastic bottle for a maximum of 3 months.
2. A stock solution of 1000 mg/L acetate used for calibration: Weigh 57.6 mg of sodium acetate trihydrate into a 25 mL volumetric flask, and add water to reach the level of the etched line. Mix well or use ultrasonic bath. The stock solution can be stored for a month.
3. A stock solution of 1000 mg/L succinate used for calibration as ISTD: Weigh 25.2 mg of succinic acid into a 25 mL volumetric flask, and add water to reach the level of the etched line. Mix well or use ultrasonic bath. The stock solution can be stored for a month.
4. A stock solution of 1000 mg/L N-acetylcysteine used for calibration: Weigh 25.2 mg of N-acetyl-L-cysteine into a 25 mL volumetric flask, and add water to reach the level of the etched line. Mix well or use ultrasonic bath. The stock solution can be stored for a month.
5. A stock solution of 1000 mg/L 2-aminobenzoate used for calibration as ISTD: Weigh 25.2 mg of 2-aminobenzoic acid into a 25 mL volumetric flask, and add water to reach the level of the etched line. Mix well or use ultrasonic bath. The stock solution can be stored for a month.
6. A stock solution of 2% (v/v) detergent for microchip maintenance: Pipet 1 mL of Extran® MA 03 (Merck) into a 50 mL volumetric flask and fill up with water. The stock solution can be stored for 3 months.

2.3 Preparation of ITP Buffers (See Note 5)

Choose suitable L and T ion based on the mobility values of the studied analyte and ISTD (*see* Subheading 1).

1. L buffer: Weigh 104.8 mg of L-histidine monohydrochloride monohydrate into a 50 mL volumetric flask filled with 20 mL of water. Add 77.6 mg of L-histidine, 5 mL of 1% (w/v) MHEC (*see* **step 1** in Subheading 2.2 and **Note 6**), and fill up with water to a volume of 50 mL. Homogenize and degas the solution and measure its pH (6.1 ± 0.05) (*see* **Note 7**).
2. T1 buffer for more mobile analytes: Weigh 87.1 mg of hexanoic acid (*see* **Note 8**) into a 50 mL volumetric flask filled with

20 mL of water. Add 155.2 mg of L-histidine, 5 mL of 1% (w/v) MHEC (*see* **step 1** in Subheading 2.2 and **Note 6**), and fill up with water to a volume of 50 mL. Homogenize and degas the solution and measure its pH (6.0 ± 0.05) (*see* **Note** 7).

3. T2 buffer for less mobile analytes: Weigh 97.6 mg of 2-(N-morpholino)ethanesulfonic acid into a 50 mL volumetric flask filled with 20 mL of water. Add 93.1 mg of L-histidine, 5 mL of 1% (w/v) MHEC (*see* **step 1** in Subheading 2.2 and **Note 6**), and fill up with water to a volume of 50 mL. Homogenize and degas the solution and measure its pH (6.2 ± 0.05) (*see* **Note** 7).

2.4 Preparation of Calibration Solutions

Choose suitable ISTD (*see* Subheading 1.1) based on the mobility values of studied analytes.

1. Prepare calibration solutions of acetate (20–100 mg/L; *see* **step 2** in Subheading 2.2) with ISTD (100 mg/L succinate; *see* **step 3** in Subheading 2.2) and T1 buffer (*see* **step 2** in Subheading 2.3):
   (a) 20 mg/L acetate and 100 mg/L succinate calibration solution: Pipet 20 μL of stock solution of acetate, 100 μL of stock solution of succinate, 100 μL of T1 buffer, and 780 μL of water into a 1.5 mL vial. Mix well and 1 day process only.
   (b) 40 mg/L acetate and 100 mg/L succinate calibration solution: Pipet 40 μL of stock solution of acetate, 100 μL of stock solution of succinate, 100 μL of T1 buffer, and 760 μL of water into a 1.5 mL vial. Mix well and 1 day process only.
   (c) 60 mg/L acetate and 100 mg/L succinate calibration solution: Pipet 60 μL of stock solution of acetate, 100 μL of stock solution of succinate, 100 μL of T1 buffer, and 740 μL of water into a 1.5 mL vial. Mix well and 1 day process only.
   (d) 80 mg/L acetate and 100 mg/L succinate calibration solution: Pipet 80 μL of stock solution of acetate, 100 μL of stock solution of succinate, 100 μL of T1 buffer, and 720 μL of water into a 1.5 mL vial. Mix well and 1 day process only.
   (e) 100 mg/L acetate and 100 mg/L succinate calibration solution: Pipet 100 μL of stock solution of acetate, 100 μL of stock solution of succinate, 100 μL of T1 buffer, and 700 μL of water into a 1.5 mL vial. Mix well and 1 day process only.
2. Prepare calibration solutions of N-acetylcysteine (20–100 mg/L; *see* **step 4** in Subheading 2.2) with ISTD (100 mg/L 2-aminobenzoate; *see* **step 5** in Subheading 2.2) and T2 buffer (*see* **step 3** in Subheading 2.3). Repeat the process described in **step 1** of Subheading 2.4 using given solutions.

3 Methods

3.1 Preparation of Pharmaceutical Samples

1. Prepare stock solutions of pharmaceutical samples: Fully dissolve selected pharmaceutical preparations in water to approx. 1–20 mg/mL concentrations. Mix well or use ultrasonic bath. The stock solution can be stored at 4 °C for 1 week.
2. Prepare sample solutions for analysis: Dilute the stock solution of the pharmaceutical sample in 10% (v/v) T solution (*see* **steps 2** or **3** in Subheading 2.3). When ISTD method is used, add also appropriate volume of stock solution of ISTD corresponding to approx. 100 mg/L concentration in the sample solution. Mix well or use ultrasonic bath. The total dilution factor should be from 10 to 100, depending on the analyte concentration in the sample.
3. Filter the sample solutions using syringe filters (*see* Subheading 2.1) into the 1.5 mL vials. Mix well and 1 day process only.

3.2 ITP Running Protocol

The procedure to perform μITP separations is shown in Table 1.

1. Fill four 1.5 mL vials with water and insert the inlet capillaries connected to the micropumps into the vials.
2. Wash the microchip channels using the micropumps controlled by MicroITP software (*see* Procedure 1 in Table 1). The cleaning procedure is applied prior to the first run of the day.
3. Filter the buffer solutions (*see* Subheading 2.3) using syringe filters (*see* Subheading 2.1) into the vials (*see* **Note 9**).
4. Replace the vials filled with water at inlet capillaries with vials containing buffer (L and T) and sample (T buffer used in a blank run, calibration solution, and real sample) solutions.
5. Enter the "Filling method" in MicroITP software (*see* Table 2).
6. Enter the "Running method" in MicroITP software (*see* Table 3).
7. Set the required number of repetitive runs in MicroITP software and start the run (*see* **Note 10**).
8. Replace the sample vial with water vial and wash the sample channel (*see* Procedure 4 in Table 1). This procedure is applied between samples.
9. Replace the water vial with the vial containing next sample and repeat the procedure from **steps 7–9**.
10. Clean and wash the microchip channels (*see* Procedure 5 and 6 in Table 1) after the last run of the day.

Table 1
Microchip procedure

Procedure	Channel C1	C2	C3	CS	Time [s]	Note
1. Wash	Water	Water	Water	Water	300	
2. Fill	L buffer	L buffer	T buffer	Sample	120	*see* Table 2
3. Separation	L buffer[a]	L buffer[a]	T buffer[a]	Sample[a]	600	*see* Table 3
4. Wash	L buffer[a]	L buffer[a]	T buffer[a]	Water	30	
5. Clean	Detergent	Detergent	Detergent	Detergent	300	
6. Wash	Water	Water	Water	Water	600	

For abbreviations, *see* Fig. 2
[a]Peristaltic micropumps turned off

Table 2
Filling method

Step	Pump turned on	Time [s]	Rpm
1	P1, P2, P3, PS	30	20
2	P1, P2, P3	10	10
3	P2	20	5
4	P1	20	5
5	P3	20	5
6	PS	20	5

Each step is followed by 5 s relaxation time (all pumps turned off) for a complete standstill of the hydrodynamic flow
Rpm rotations per minute. For other abbreviations, *see* Fig. 2

Table 3
Running method

Step	Time [s]	Current [μA]	Current direction	Data acquisition
1	100	20	E3 → E2	–
2	500	20	E3 → E2	D1, D2[a]

For abbreviations, *see* Fig. 2
[a]Both detectors can be used for data evaluation (*see* **Note 11**)

3.3 Data Analysis

Ideal isotachopherograms used for evaluation of RSH values and zone lengths are shown in Fig. 3.

1. Evaluate the isotachopherogram of a blank run: The migration window defined by chosen L and T ions should be clear (*see* **Note 12**). If there are zones present on the

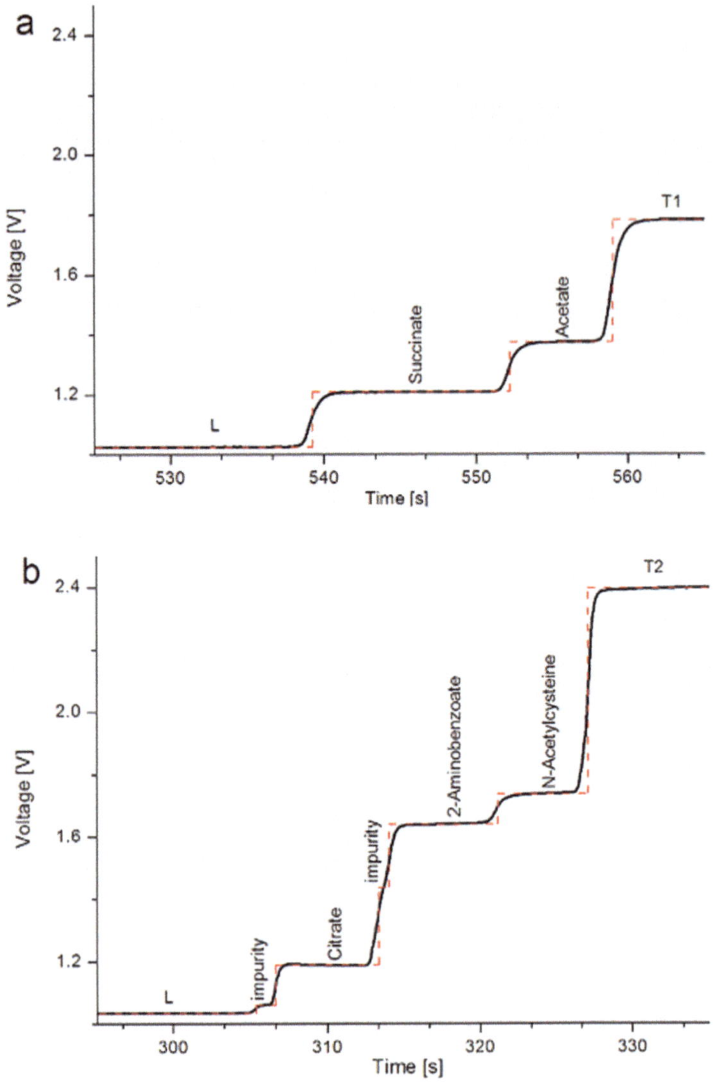

Fig. 3 μITP analysis of major constituents in pharmaceuticals. (**a**) Isotachopherogram from the analysis of 20-fold diluted stock solution of Buserelin acetate (17.7 g/L) and 100 mg/L of succinate (ISTD) in 10% (v/v) T1 buffer. Response from D2 was used for data evaluation. (**b**) Isotachopherogram from the analysis of 100-fold diluted stock solution of Solmucol 200 (10 g/L) and 100 mg/L of 2-aminobenzoate (ISTD) in 10% (v/v) T2 buffer. Response from D1 was used for data evaluation. Dashed line represents an ideal graph used for data evaluation. For other experimental conditions, *see* Subheading 3.2 and Table 3; for sample pretreatment, *see* Subheading 3.1. *L, T* leading and terminating ions, respectively

isotachopherogram, identify them by calculating the RSH values according to Eq. (2) or by overlapping the isotachopherograms of blank and calibration sample. Measure the zone lengths corresponding to the RSH values of analyte and ISTD (*see* **Note 13**) from D1 or D2 (*see* **Note 11**), if present in the blank sample.

2. Evaluate the isotachopherograms of a calibration samples: Identify the zones of analyte and ISTD by calculating the RSH values according to Eq. (2) or by overlapping the isotachopherograms of blank and calibration sample. Measure the zone length of analyte and ISTD (*see* **Note 13**) from D1 or D2 (*see* **Note 11**). Subtract the zone length corresponding to the analyte and/or ISTD in the blank sample if necessary (*see* **Note 14**).
3. Construct the calibration curves based on the least square linear regression analysis: Use analyte concentration on X-axis and its zone length on Y-axis, i.e., ECAL method, according to Eq. (3). Alternatively, use ISTD method, according to Eq. (4), when zone length ratio is plotted versus concentration ratio. Calculate slope, intercept, and correlation coefficient for calibration curves.
4. Evaluate the isotachopherograms of real pharmaceutical samples: Isotachopherograms from ITP separations of main components in different pharmaceuticals are shown in Fig. 3. Identify the zones of analyte as well as ISTD by calculating the RSH values according to Eq. (2) or by overlapping the isotachopherograms with the calibration sample, when the position is not clear. Measure the zone lengths of the analyte and ISTD (*see* **Note 13**) from D1 or D2 (*see* **Note 11**). Subtract the zone length corresponding to the analyte and/or ISTD in the blank sample if necessary.
5. Calculate the concentration: Use equations of calibration curves (*see* **step 3**) for calculation of the analyte concentration by ECAL (Eq. (3)) and ISTD (Eq. (4)) methods (*see* **Note 15**). The total sample dilution factor should be taken into account in the calculation (*see* **step 2** in Subheading 3.1).

4 Notes

1. An electrolyte unit consists of microchip holder, four peristaltic micropumps, and three membrane driving electrodes (E1–E3). For mutual connections of these parts and their connection to the inlets of the microchip channels, Teflon capillaries are used. The rollers of the micropumps are used to close the corresponding inlets of the microchip channels, when not in use. An excess of the solutions pumped through the microchip channels is removed into a waste container connected to a permanently open outlet from the microchip (*see* Fig. 2). The membrane driving electrodes are used to eliminate disturbances due to bubble formation during the separation [23].
2. The high-voltage power supply (max 50 μA and 7 kV) of the electronic unit is used for delivering the driving voltage of

required polarity to the electrodes (E1–E3) placed between the corresponding inlets of the microchip channels (C1–C3) and the peristaltic micropumps (P1–P3). Here, the driving electrode E3 is permanently connected to the high-voltage pole (negative, in this case) of the power supply, while the driving electrodes E1 and E2 are connected to its positive pole via a high-voltage relay (*see* Fig. 2). The electronic unit also includes the measuring electronics of the contact conductivity detectors, controls the micropumps in the preparation step of the run, and interfaces the MCE analyzer to a PC.

3. A data acquisition rate of 25 points per second should be used.
4. A two-point calibration of the pH meter should be used (pH 4 and 7).
5. Prepare buffer solutions using analytical grade reagents and fresh ultrapure water and store them at 4 °C for a week. Filter the solutions through disposable syringe filters of 0.8 μm pore size prior to the use.
6. Graduated plastic syringe should be used to measure the volume of viscous solution of MHEC (*see* Subheading 2.2).
7. Use only small volume of buffer (approx. 5 mL) for pH measurement. Do not pour back the used buffer into the original stock.
8. Use hexanoic acid of highest purity ($\geq$ 99%).
9. Discard the first part of the filtrate (approx. 1 mL). One vial (1.5 mL) filled with the buffer solution can be used for about 15 runs.
10. The ITP runs are performed with the same sample (*see* Procedures 2 and 3 of Table 1).
11. Use preferably the isotachopherogram acquired from the response of D1 for data evaluation. The isotachopherogram from the response of D2 is used alternatively when the constituents are not separated completely in D1.
12. When reagents of the highest purity are used for preparation of buffer solutions, migration position of the analyte and ISTD should be clear.
13. The zone length of the constituents is measured by automated data evaluation using the first derivative of the signal from conductivity detector D1 or D2 (*see* Fig. 2).
14. Apply this procedure to each of the calibration sample.
15. The use of a suitable ISTD significantly improves the precision (from six to eight times lower RSD of the corrected zone lengths) of determination of the analyte.

Acknowledgments

This work was supported by the Slovak Research and Development Agency (APVV-17-0318) and the Slovak Grant Agency for Science (VEGA 1/0340/15). The financial support of Merck (Darmstadt, Germany) is also acknowledged.

References

1. Smejkal P, Bottenus D, Breadmore MC et al (2013) Microfluidic isotachophoresis: a review. Electrophoresis 34:1493–1509
2. Walker PA III, Morris MD, Burns MA et al (1998) Isotachophoretic separations on a microchip. Normal Raman spectroscopy detection. Anal Chem 70:3766–3769
3. Malá Z, Gebauer P, Boček P (2017) Analytical capillary isotachophoresis after 50 years of development: recent progress 2014–2016. Electrophoresis 38:9–19
4. Malá Z, Gebauer P, Boček P (2015) Recent progress in analytical capillary isotachophoresis. Electrophoresis 36:2–14
5. Malá Z, Gebauer P, Boček P (2013) Recent progress in analytical capillary isotachophoresis. Electrophoresis 34:19–28
6. Breadmore MC, Wuethrich A, Li F et al (2017) Recent advances in enhancing the sensitivity of electrophoresis and electrochromatography in capillaries and microchips (2014–2016). Electrophoresis 38:33–59
7. Boček P (1981) Analytical isotachophoresis. In: Boschke FL (ed) Analytical problems. Topics in current chemistry, vol 95. Springer, Berlin, pp 131–177
8. Sádecká J, Masár M (2014) Electrophoresis | capillary isotachophoresis. In: Reedijk J (ed) Reference module in chemistry, molecular sciences and chemical engineering. Elsevier, Waltham
9. Nuchtavorn N, Suntornsuk W, Lunte SM et al (2015) Recent applications of microchip electrophoresis to biomedical analysis. J Pharm Biomed Anal 113:72–96
10. Štěpánová S, Kašička V (2017) Analysis of proteins and peptides by electromigration methods in microchips. J Sep Sci 40:228–250
11. Shang FJ, Guihen E, Glennon JD (2012) Recent advances in miniaturisation—the role of microchip electrophoresis in clinical analysis. Electrophoresis 33:105–116
12. Castro ER, Manz A (2015) Present state of microchip electrophoresis: state of the art and routine applications. J Chromatogr A 1382:66–85
13. Cui F, Rhee M, Singh A et al (2015) Microfluidic sample preparation for medical diagnostics. Annu Rev Biomed Eng 17:267–286
14. Tetala KKR, Vijayalakshmi MA (2016) A review on recent developments for biomolecule separation at analytical scale using microfluidic devices. Anal Chim Acta 906:7–21
15. Chen L, Prest JE, Fielden PR et al (2006) Miniaturised isotachophoresis analysis. Lab Chip 6:474–487
16. Štěpánová S, Kašička V (2014) Determination of impurities and counterions of pharmaceuticals by capillary electromigration methods. J Sep Sci 37:2039–2055
17. Paulekuhn GS, Dressman JB, Saal C (2007) Trends in active pharmaceutical ingredient salt selection based on analysis of the Orange book database. J Med Chem 50:6665–6672
18. Rudašová M, Masár M (2016) Precise determination of N-acetylcysteine in pharmaceuticals by microchip electrophoresis. J Sep Sci 39:433–439
19. Hradski J, Drusková Chorváthová M, Bodor R et al (2016) Quantitative aspects of microchip isotachophoresis for high precision determination of main components in pharmaceuticals. Anal Bioanal Chem 408:8669–8679
20. Atkuri KR, Mantovani JJ, Herzenberg LA et al (2007) N-Acetylcysteine-a safe antidote for cysteine/glutathione deficiency. Curr Opin Pharmacol 7:355–359
21. Poole PJ, Black PN (2001) Oral mucolytic drugs for exacerbations of chronic obstructive pulmonary disease: systematic review. Br Med J 322:1271–1274
22. Limonta P, Marelli MM, Mai S et al (2012) GnRH receptors in cancer: from cell biology to novel targeted therapeutic strategies. Endocr Rev 33:784–811
23. Kaniansky D, Masár M, Bodor R et al (2003) Electrophoretic separations on chips with hydrodynamically closed separation systems. Electrophoresis 24:2208–2277
24. Kaniansky D, Masár M, Bielčíková J (1997) Electroosmotic flow suppressing additives for capillary zone electrophoresis in a hydrodynamically closed separation system. J Chromatogr A 792:483–494

Chapter 8

Microfluidic Free-Flow Isoelectric Focusing with Real-Time pI Determination

Stefan Nagl

Abstract

Free-flow electrophoresis (FFE) may be used for continuous and preparative separation of a wide variety of biomolecules. Isoelectric focusing (IEF) provides for the separation of compounds according to their isoelectric point (pI). Here we describe a microfluidic chip-based protocol for the fabrication, application, and optical monitoring of free-flow isoelectric focusing (FFIEF) of proteins and peptides on the microscale with optical surveillance of the microscopic pH gradient provided by an integrated pH sensing layer. This protocol may be used with modifications also for the FFIEF of other biomolecules and may serve as template for the fabrication of microfluidic chips with integrated fluorescent or luminescent pH sensor layers for FFE and other applications.

Key words Free-flow electrophoresis, Isoelectric focusing, Fluorescent pH sensor, Isoelectric point determination, Multispectral imaging

1 Introduction

Preparative and continuous separations of proteins, nucleotides and their complexes, organelles, or whole cells are being enabled by free-flow electrophoresis (FFE). Although the separation efficiencies are not as high as in other forms of electrophoresis, due to its continuous operation, FFE is particularly useful for sample preparation and pre-separation prior to other procedures. Conventional FFE is typically carried out in rather large devices, but the method has been successfully miniaturized onto microfluidic platforms in various implementations [1–4].

One of the possible separation modes on these platforms is free-flow isoelectric focusing (FFIEF), providing for the migration of zwitterionic compounds to their isoelectric point (pI) [5–9]. One challenge is the monitoring of the pH gradient and pI determination in these miniaturized platforms. In conventional free-flow IEF, it is usually sufficient to measure the pH at each

Debashis Dutta (ed.), *Microfluidic Electrophoresis: Methods and Protocols*, Methods in Molecular Biology, vol. 1906, https://doi.org/10.1007/978-1-4939-8964-5_8, © Springer Science+Business Media, LLC, part of Springer Nature 2019

outlet using a pH microelectrode, but for microfluidic chips with fewer outlets and small channels, other solutions are needed.

As in gel and capillary electrophoresis, isoelectric point markers (certain fluorescent dyes or labeled proteins) may be employed, but these are not well-suited for FFE because they contaminate sample fractions and they follow the typically smaller electrophoretic resolutions in FFE. We have developed an approach that uses noninvasive thin film fluorescent pH sensor layers in FFIEF microfluidic chips. They can display the pH gradient on the microscale and determine the pI of focused fractions noninvasively, almost in real time (reviewed in [10]).

In this protocol, largely based on the work in [11], all steps that are needed to carry out an FFIEF of protein or other mixtures on a microfluidic chip with an integrated fluorescent pH sensor layer are described, starting from the fabrication of a suitable microfluidic FFE chip with an integrated sensing coating, fluidic connections, experimental conditions for sample preparation, fluidic and optical setup, and the quantitative readout of the results which are discussed along with variations of these methods that we found possible and useful for certain samples and analytes.

2 Materials

Prepare all aqueous solutions using ASTM type 1 water or better, and use analytical grade chemicals (unless indicated otherwise).

2.1 Buffers and Solutions

1. Ampholyte solution: 0.1% (w/w) ampholyte pH 4–7 and 50 ppm (w/w) Tween 20 (*see* **Note 1**).
2. 200 mM bicarbonate buffer pH 9.0.
3. 150 mM PBS buffer pH 7.4.
4. Dye stock solution: 1 mg Atto 425-NHS in DMSO (*see* **Note 2**).
5. Britton-Robinson buffers (BRBs, 40 mM boric acid, 40 mM phosphoric acid, and 40 mM acetic acid, titrated with 200 mM NaOH), one each from pH 3 to pH 10 in steps of 1 pH unit.
6. Anolyte solution: 20 mM H_2SO_4.
7. Acidic sheath flow: 20 mM H_2SO_4.
8. Catholyte solution: BRB pH 10.
9. Cathodic sheath flow: BRB pH 10.

2.2 Equipment

1. Inverted fluorescence microscope with stable holding mechanism for glass substrates that fits the microfluidic chip and a monochrome CCD or CMOS camera with good sensitivity for quantitative readout.

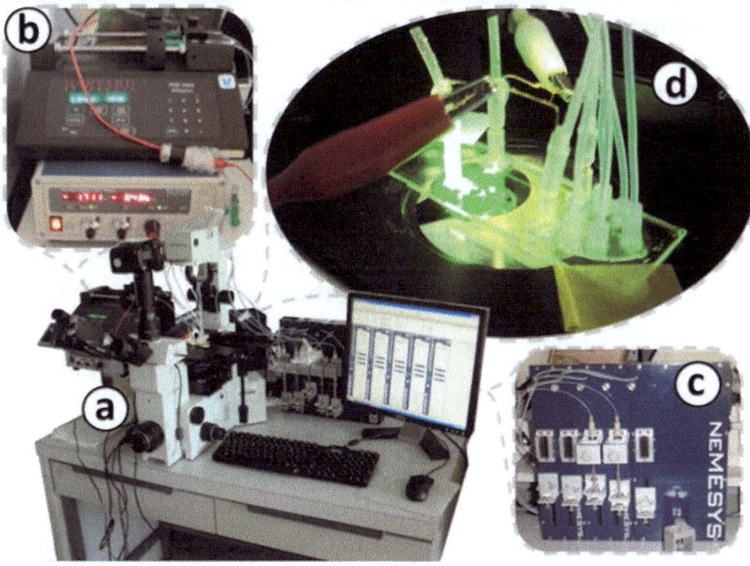

Fig. 1 Picture of an experimental setup for μFFIEF with integrated pH observation. (**a**) Inverted fluorescence microscope with fluidic and electrical connections, (**b**) high-voltage source and syringe pump according to Subheading 2.2.6. (**c**) High precision syringe pump according to Subheading 2.2.5. (**d**) Close-up of the microfluidic chip, fluidic and electric connections. From ref. 16

2. A low magnification microscope objective, suitable for fluorescence (*see* Fig. 1; *see* **Note 3**).
3. LED excitation at 420 nm and 660 nm (*see* **Note 4**), with appropriate filter sets for blue emission (label channel) and NIR emission (pH sensor channel), respectively. Capability to monitor two optical channels in the emission pathway (*see* **Note 5**).
4. High-voltage DC power supply, with at least 500 V, and up to 1 mA, or better.
5. Precision pulsation-free syringe pump system with six syringe channels for flow rates in the nL/min to μL/min regime, independently adjustable (*see* **Note 6**).
6. Additional syringe pump(s) for at least two channels in the nL/min to μL/min regime (*see* **Note 7**).
7. PTFE tubings of 1 mm inner diameter and 2 mm outer diameter.
8. Silicone tubings of 2 mm inner diameter and 4 mm outer diameter.

3 Methods

3.1 Microfluidic Chip

1. Prepare a photomask as displayed in Fig. 2 in a CAD program. The outer dimensions should correspond to the size of the employed microscope glass slides (approx. 75 × 25 mm). The separation bed is 10 mm in width and 20 mm in length. All the inlet and outlet channels including the serpentine reactor structure are 300 μm in width except the visibly thicker channels flanking the separation bed that are 1 mm in width each.
2. Print the photomask at 3600 dpi resolution with a permanent ink onto an approx. 100 μm thick transparency slide (*see* **Note 8**).
3. Drill fluidic access holes of 1–2 mm diameter into microscope glass slides corresponding to the inlet and outlet positions of the photomask (*see* **Note 9**). Dust off with an air or nitrogen stream. Clean these slides according to **step 4**, and modify according to **steps 5** and **6**, and use for **step 11**. Do not use for **steps 7–10**.
4. Clean borosilicate glass microscope slides (*see* **Note 10**), with fresh piranha solution (*caution*: very aggressive and corrosive solution, handle with care, hot when freshly mixed) in a stainless steel or fluoropolymer tray, rinse with methanol and water, and dry with nitrogen or air.
5. Immerse clean and dry glass slides in a 5 mM solution of (3-methacryloyloxypropyl) trichlorosilane (TPM) in *n*-heptane and trichloromethane (3:1, w/w) for 2 min.
6. Rinse with *n*-heptane and water (*see* **Note 11**).
7. Prepare a mixture of 85.00% (w/w) acryloylmorpholine, 14.78% (w/w) polyethylene glycol diacrylate PEG-DA700 (*see* **Note 12**), 0.20% (w/w) 2,2-dimethoxy-2-phenylacetophenone (DMPA) photoinitiator, and 0.02% (w/w) of the pH probe PBI (*see* **Note 13**).
8. Spread 60 μL of the prepolymer mixture of **step** 7 as a thin layer between an acrylated glass slide (treated according to **steps 5** and **6**) and a non-acrylated top glass slide (cleaned as in **step 3** but not further modified). Avoid air bubbles.

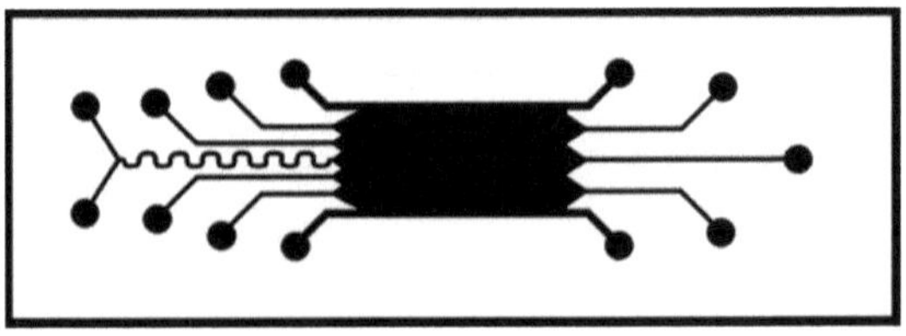

Fig. 2 Photomask for FFIEF chip manufacture. Print at the size of a glass microscope slide

9. Photopolymerize the assembly via exposure to 15 mW/cm^2 365 nm UV light for 15 s (*see* **Note 14**).
10. Remove the top glass slide.
11. Apply 60 μm thick adhesive transfer foil as ca. 5 × 5 mm pieces on the edges of the structure (*see* **Note 15**) as a spacer between the sensing layer and the top plate to be applied in the next step.
12. Spread 100 μL polyethylene glycol diacrylate PEG-DA250 (note the different molecular weight to **step** 7) containing 0.10% (w/w) 2,2-dimethoxy-2-phenylacetophenone (DMPA) photoinitiator in between the glass slide with a PBI-PEG-PAM layer and a cover glass slide with access holes prepared according to step 3 and cleaned according to **steps 4** and **5**.
13. Attach the photomask prepared in **steps 1** and **2** on top of the assembly (*see* **Note 16**).
14. Photopolymerize the assembly via exposure to 15 mW/cm^2 365 nm UV light for 1.3 s (*see* **Note 14**).
15. Attach silicone tubing connected to a mild vacuum via the fluidic access holes on top of the microfluidic chip to remove unpolymerized PEG-DA from the separation bed and the channels.
16. Illuminate again with the same settings for 1.3 s (*see* **Note 14**).
17. Store microfluidic chips in the dark, in a clean, low humidity environment.

3.2 Electrophoretic Setup

1. Prepare 1 cm long pieces of the silicone tubings, and glue them onto the inlets and outlets of the microfluidic chip using Elastosil E43 silicone adhesive. Avoid penetration of the glue into the chip inlets/outlets.
2. Put PTFE tubing inside silicone tubings at the chip inlets and outlets (*see* **Note 17**).
3. Pinch a copper wire into the tubings at the outer inlets and outlets (depicted in orange in Fig. 3f), and fix with silicone adhesive. Pull a short piece of silicone tubing over the pinching hole, and also fix with silicone adhesive (*see* **Note 17**).
4. Connect copper wires to HV power supply; ground the assembly (*see* **Note 18**).

3.3 pH Calibration

1. Dilute BRBs from pH 3 to pH 10 to match the conductivity of the ampholyte solution (*see* **Note 19**).
2. Focus with the microscope onto the evaluation area near the outlet (depicted in Fig. 1).

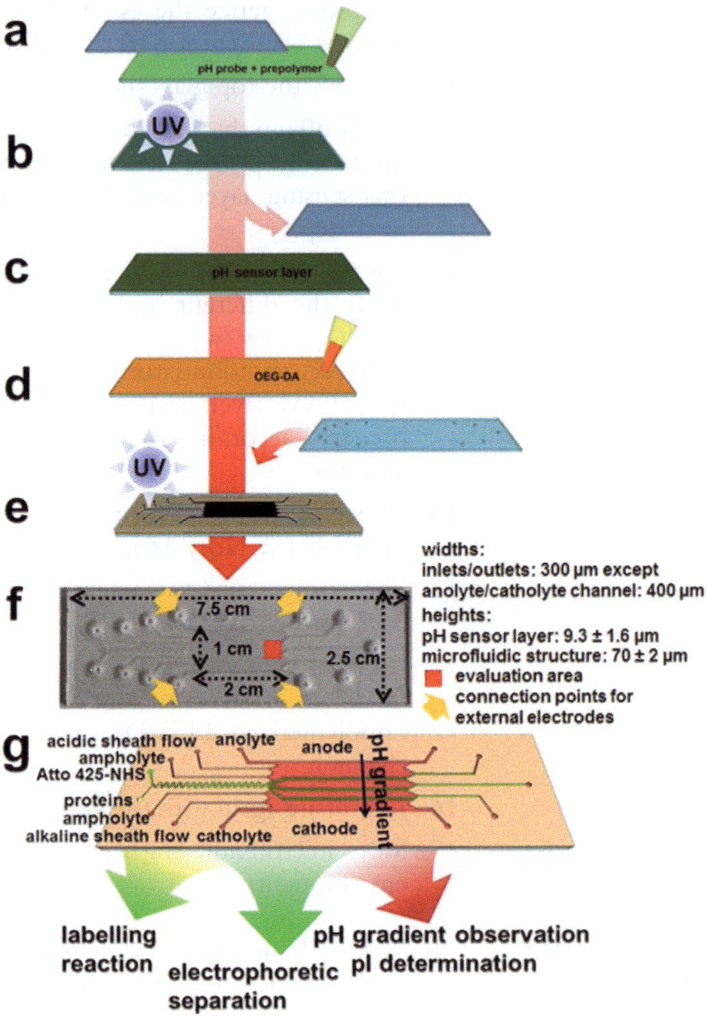

Fig. 3 Overview of the manufacturing procedure and layout of microfluidic chips for free-flow isoelectric focusing, on-chip labeling and integrated pH sensing. (**a**) Application of pH sensing matrix prepolymer onto acrylated glass, sealing with cover glass, (**b**) photopolymerization of the pH sensor layer, (**c**) removal of cover glass, (**d**) application of PEG-DA onto the sensor layer, (**e**) addition of cover glass with access holes, masked photopolymerization of the microfluidic structure, (**f**) completed microfluidic chip with added relevant length scales. Red is a typical evaluation area near the end of the separation bed and orange the connection points for external electrodes; (**g**) is a schematic of the chip and channel layout. From ref. 10 with permission from the Royal Society of Chemistry

3. Prepare eight 5 mL syringes for the syringe pump system in Subheading 2.2.6.
4. Fill the syringes with BRBs from pH 3 to pH 10 in steps of 1 pH unit.

5. Connect the syringe with BRB pH 3 with the tubing of a chip inlet.
6. Fill the microfluidic chip with BRB pH 3 using a flow rate of 50 μL/min.
7. Record the fluorescence of the area with the camera. Adjust the exposure so that there is no saturation at any position (*see* **Note 20**).
8. Repeat **steps 5–7** for every BRB from pH 4 to 10.
9. Construct a calibration curve according to the readings at each pH relating the fluorescence of the pH sensor channel to a distinct pH value.

3.4 Isoelectric Focusing

1. Prepare labeling buffer: 9:1 (v/v) 200 mM bicarbonate buffer pH 9.0 and 150 mM PBS buffer pH 7.4 (should yield a pH of 8.3).
2. Dissolve Atto 425 stock solution into labeling buffer to a final concentration of 10 μM (dye solution).
3. Dissolve solid biological sample or dilute a sample solution into ampholyte solution (*see* **Note 21**).
4. Prepare six 1 mL glass syringes for the syringe pump system in Subheading 2.2.5.
5. Fill the syringes and attach on the pump from left to right with the following solutions:

   BRB pH 10 (cathodic sheath flow)
   Ampholyte solution
   Sample solution
   Dye solution
   Ampholyte solution
   20 mM H_2SO_4 (acidic sheath flow)
   (*see* **Note 22**)
6. Connect syringe outlets with corresponding tubings according to the layout in Fig. 1 (*see* **Note 23**).
7. Prepare two 5 mL syringes for the syringe pump system in Subheading 2.2.6.
8. Fill the syringes with BRB pH 10 (catholyte) and 20 mM H_2SO_4 (anolyte), respectively (*see* **Note 22**).
9. Connect the additional syringes with corresponding tubings according to the layout in Fig. 1 (*see* **Note 23**).
10. Adjust the syringe outflow to the following settings and start the flow from the reservoirs (*see* **Note 24**):

    Catholyte: 7 μL/min
    Cathodic sheath flow: 1 μL/min
    Ampholyte solution: 1 μL/min

Sample solution: 0.3 μL/min
Dye solution: 0.1 μL/min
Ampholyte solution: 1 μL/min
Acidic sheath flow: 1 μL/min
Anolyte: 7 μL/min

11. Wait until the fluorescence of the label can be observed in the label channel at the observation area near the outlet of the separation bed (several minutes) (*see* **Note 25**).
12. Switch on the electric field at a low voltage, and simultaneously observe the fluorescence of the product stream in the label channel.
13. Raise the voltage slowly, and observe the separation of the product stream. Optimize the voltage according to the visually displayed separation in the label channel. Focused analytes should follow a straight path in the separation bed near the outlet and not deflect perpendicular to the flow direction (*see* **Note 26**).

3.5 Isoelectric-Point Determination

1. Record the fluorescence in the label channel and in the pH sensor channel around the separated products.
2. Integrate the fluorescence recordings of both spectral channels within a distance of around 200 to 500 μm in the axis that is parallel to the flow direction.
3. Use the calibration curve to translate values in the pH sensor channel to local pH values.
4. Fit the resulting pH curve to account for small local deviations (*see* Fig. 4; *see* **Note 27**).

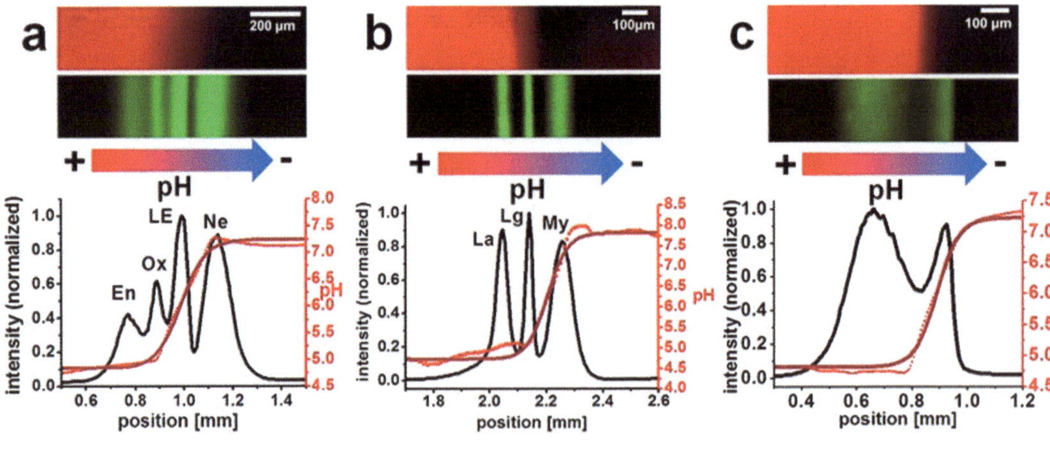

Fig. 4 Example FFIEF with simultaneous pH observation (false-colored fluorescence images, green, label channel; red, pH sensor channel), (**a**) neuropeptide separation, endothelin (En), oxytocin (Ox), leucine-enkephalin (LE), and neurotensin (Ne); (**b**) protein separation, lactalbumin (La), lactoglobulin (Lg), and myoglobin (My); (**c**) tryptic digest of physalaemin separation. From ref. 10

5. Overlay the data of the pH sensor channel with the fluorescence of the label channel (*see* Fig. 4).
6. Determine the peaks of the isoelectrically focused products, and read the local pH at the same position from the *y*-axis of the plot in the pH sensor channel (*see* **Note 28**).

3.6 Sample Collection and Analysis

1. Observe the flow direction of focused, separated products.
2. If the desired fraction is migrating into more than one outlet, manipulate the flow velocities in the sheath flow and ampholyte channels at the optimized voltage to steer the fractions fully into one outlet.
3. Collect the fraction at the outlet of the tubing of the respective outlet via an attached Eppendorf vial or similar.
4. If required the collected fractions can be further analyzed, e.g., via mass spectrometry, ampholyte may need to be removed from the fraction before analysis.

4 Notes

1. Other ampholytes may be used; however in our experience some ampholytes, from some manufacturers, work better than others. Also the pH sensitivity range of the employed pH sensor layer has to match with the pH range of the separation.
2. If the commercial pH sensor layer as suggested in **Note 13** is used, the choice of labeling dye has to be adjusted accordingly as to avoid spectral overlap. Not all reactive labeling dyes may work well with the protocol and chip platform described here. An alternative is the use of intrinsic biomolecule fluorescence as described in ref. 11, but this requires an elaborate optical setup.
3. The observed area on the microscope should cover at least 1 mm^2; the results in Fig. 4 were obtained with a 2.5×, NA 0.12 objective.
4. If other pH probes or biomolecule labels are used as noted in **Notes 2** and **13**, excitation sources and filters need to be adjusted accordingly.
5. Either manual or automatic or a multispectral beam splitter such as the photometrics dual/quad view.
6. The performance of the syringe pumps is critical for μFFIEF to work properly. Pumps that we found to perform sufficiently include CETONI neMESYS and Fluigent MFCS systems; systems with related specifications from other manufacturers exist. The Harvard PHD 2000 or related pumps that are very common in microfluidics labs are insufficient (*see* also **Note 7**).

7. For the outer electrolyte channels, Harvard PHD 2000 or related systems are sufficient.
8. This is a relatively simple and economic way of producing photomasks that we have found to be sufficient for μFFE chips with horizontal dimensions of the channels on the order of several hundred micrometers. Other types of photomasks may be used as well with this procedure.
9. Good results were achieved with a powder blaster of 50 μm grain size, but this step can be executed via different technologies.
10. In principle any silicate-based glass may be used, but for best results, glass slides with low autofluorescence throughout the visible and NIR region are recommended.
11. Measure contact angle if desired to confirm cleanliness of the original glass substrate and successful modification with TPM.
12. Here I use the commercial name polyethylene glycol diacrylate (PEG-DA) to aid the reader in finding suppliers. In several publications where we employed these prepolymers for FFIEF chip manufacture [11–14], I referred to them in the chemically more correct term oligoethylene glycol diacrylate (OEG-DA) since the chain length of the ethylene glycol units is very short, but they are the same substances.
13. The pH probe PBI (a perylene bisimide) is, to the best of my knowledge, not yet commercially available but may be prepared according to ref. 15, alternatively the sensor layer composition of ref. 13 using commercially available BCECF-dextran nanoparticles as the pH probe may be employed with the same procedure. The choice of pH probe also depends on the pI range to be observed. pH probes for other pI ranges in FFIEF are described in ref. 14.
14. The optimum polymerization time depends heavily on the exact characteristics of the employed light source and illumination arrangement, we used a SÜSS FE mask aligner, but other 365 nm illumination sources also yielded good results.
15. We used 467MP tape from 3 M.
16. Take care that the access holes in the glass match with the photomask. A dedicated mask aligner may be used or observation on a microscope. If done manually, the photomask may be fixed on the assembly via a water drop, a clip, or a small amount of glue on the edges.
17. An alternative is to directly glue female Luer connectors onto the chip entry and exit points. Other diameters, materials, or adhesives will work in a similar fashion. The important issue is to ensure a tight, leakage-free connection between the in-/

outlet and the tubing. If required a short, second piece of silicone tubing can be wrapped around the PTFE tubing at the end, or the connection may be sealed with silicone glue.

18. See the supplementary info of ref. 13 for a more detailed discussion and pictures.
19. This is because the optical pH sensor layers are affected to a small degree by ionic strength. If no conductivity meter is available, dilute the BRBs to 10 mM, the error introduced that way will be small.
20. The layer should yield an overall homogeneous fluorescence, with possibly some small defects; if not, check layer, illumination, and fluidics. Set the exposure time so that the fluorescence is around 70–80% of maximum intensity at the pH with the highest fluorescence (pH 3).
21. This depends on the sample; final concentration should be in the same range as the dye concentration (10 μM).
22. Ensure air bubble-free filling of all syringes. Fill slowly, if necessary drain and fill again.
23. Use Luer or similar connectors if the diameters of syringe outlets and tubings do not match. Tubings should connect tightly to the tubings on the microfluidic chip inlets, if they do not, use different ones.
24. The flow rate should be adjusted and optimized for each sample and sets of experiments. The numbers here are from the protein separation of ref. 11 and to be understood as a rough guideline.
25. Sometimes the stream may be located outside this area on the upper or lower side before the electrical field is switched on. If it cannot be found, record or lock the prior position and move the observation area on the microscope perpendicular to the separation bed, and then return to original position.
26. Optimum voltage depends heavily on the nature of the sample molecules. Usually for peptides and other small molecules, the optimum voltage is in the lower volt range, whereas it can be several hundred volts for proteins and other large compounds.
27. Usually a sigmoidal fit works well, but other functions can be considered. Observe visually if the fit is a good representation of the actual pH gradient; if not, try other parameters.
28. Use a chromatographic peak fitting algorithm if necessary. Usually it is sufficient to take the position of the maximum in the label channel.

References

1. Kohlheyer D, Eijkel JCT, van den Berg A, Schasfoort RBM (2008) Miniaturizing free-flow electrophoresis—a critical review. Electrophoresis 29:977–993
2. Agostino FJ, Krylov SN (2015) Advances in steady-state continuous-flow purification by small-scale free-flow electrophoresis. Trends Anal Chem 72:68–79
3. Novo P, Janasek D (2017) Current advances and challenges in microfluidic free-flow electrophoresis, a critical review. Anal Chim Acta 991:9–29
4. Johnson AC, Bowser MT (2018) Micro free flow electrophoresis. Lab Chip 18:27–40
5. Xu Y, Zhang CX, Janasek D, Manz A (2003) Sub-second isoelectric focusing in free flow using a microfluidic device. Lab Chip 3:224–227
6. Kohlheyer D, Besselink GAJ, Schlautmann S, Schasfoort RBM (2006) Free-flow zone electrophoresis and isoelectric focusing using a microfabricated glass device with ion permeable membranes. Lab Chip 6:374–380
7. Song YA, Chan M, Celio C, Tannenbaum SR, Wishnok JS, Han J (2010) Free-flow zone electrophoresis of peptides and proteins in PDMS microchip for narrow pI range sample prefractionation coupled with mass spectrometry. Anal Chem 82:2317–2325
8. Walowski B, Hüttner W, Wackerbarth H (2011) Generation of a miniaturized free-flow electrophoresis chip based on a multilamination technique—isoelectric focusing of proteins and a single-stranded DNA fragment. Anal Bioanal Chem 401:2465–2471
9. Cheng LJ, Chang HC (2014) Switchable pH actuators and 3D integrated salt bridges as new strategies for reconfigurable microfluidic free-flow electrophoretic separation. Lab Chip 14:979–987
10. Nagl S (2017) Micro free-flow isoelectric focusing with integrated optical pH sensors. Eng Life Sci 18(2):114–123.
11. Herzog C, Poehler E, Peretzki AJ, Borisov SM, Aigner D, Mayr T, Nagl S (2016) Continuous on-chip fluorescence labelling, free-flow isoelectric focusing and marker-free isoelectric point determination of proteins and peptides. Lab Chip 16:1565–1572
12. Poehler E, Herzog C, Lotter C, Pfeiffer SA, Aigner D, Mayr T, Nagl S (2015) Label-free microfluidic free-flow isoelectric focusing, pH gradient sensing and near real-time isoelectric point determination of biomolecules and blood plasma fractions. Analyst 140:7496–7502
13. Jezierski S, Belder D, Nagl S (2013) Microfluidic free-flow electrophoresis chips with an integrated fluorescent sensor layer for real time pH imaging in isoelectric focusing. Chem Commun 49:904–906
14. Herzog C, Beckert E, Nagl S (2014) Rapid isoelectric point determination in a miniaturized preparative separation using jet-dispensed optical pH sensors and micro free-flow electrophoresis. Anal Chem 86:9533–9539
15. Aigner D, Borisov SM, Petritsch P, Klimant I (2013) Novel near infra-red fluorescent pH sensors based on 1-amino perylene bisimides covalently grafted onto poly(acryloyl)morpholine. Chem Commun 49:2139–2141
16. Jezierski S (2013) Mikrofluidische Freifluss-Elektrophorese mit integrierten optischen Sensoren. Ph.D. Thesis, Universität Leipzig, Germany

Chapter 9

Nanochannel Gradient Separations

Michael A. Startsev and David W. Inglis

Abstract

Gradient-based electrophoretic separations enable simultaneous separation and concentration of molecules. Compared with conventional injection-based separations, they enable enrichment of low-concentration analytes from larger sample volumes that are not limited by an injection volume. We have demonstrated that a nanochannel, connecting two chemically different reservoirs, can maintain a stationary chemical gradient while trapping biomolecules and effectively averaging out many of the complex physicochemical hydrodynamics that would broaden the bands in a meso- or microscale capillary. Here we describe chemical and physical methods that enable this work.

Key words Nanofluidic, Fabrication, Conductivity gradient, pH gradient

1 Introduction

Micro- and nanofluidic techniques have had a significant impact on the fields of analytical chemistry and biomolecular sciences in general. The advantages of reducing the scale of critical elements within instrumentation include faster reaction, separation and detection times, reduced sample consumption, and increased resolution and precision. A persistent challenge faced by all detection methods is sensitivity. The reduced sample volumes associated with micro- and nanotechniques clash with this challenge because the absolute number of molecules in nano- and femto-liter samples can approach zero.

Electrophoretic gradient-based separation methods were pioneered in the 1980s using chromatographic matrices to purify and enrich proteins and biomolecules [1]. All gradient methods use counteracting forces to create stable equilibrium positions or traps. The locations of these traps depend on the physical and chemical properties of the molecules being trapped. An example of this type of trap is electric field gradient focusing [2]. Such methods use a channel (column) where the electric field changes from one end to the other but the average fluid velocity does not.

Debashis Dutta (ed.), *Microfluidic Electrophoresis: Methods and Protocols*, Methods in Molecular Biology, vol. 1906, https://doi.org/10.1007/978-1-4939-8964-5_9,

At one end of the column, the electrophoretic force will dominate and push the molecule into the column. At the other end, where the electric field is weaker, the fluidic drag force will dominate, also pushing the molecule into the column. These forces trap and enrich the molecule at a particular location.

There are a wide variety of gradient methods. In this work we will describe methods used to create conductivity and pH gradients in nanochannels.

2 Materials

2.1 Chip Fabrication

We have used two types of silica-based nanofluidic devices.

2.1.1 Silicon Device Fabrication

1. 3", p-type 1–10 Ωcm, <100> wafer with 50 nm of thermal oxide.
2. AZ 1518 photoresist.
3. Metal photomask.
4. March PX-250 plasma system.
5. SU-8 resist.
6. Office tape.
7. Sandblaster containing 75 μm alumina particles (Carlsbad CA, USA).
8. Detergent.
9. Acetone.
10. Remover PG (MicroChem Corp., Newton MA, USA).
11. Potassium hydroxide.
12. Isopropyl alcohol.
13. Hydrofluoric acid.
14. Borofloat 33 wafers.

2.1.2 Fused Silica Device Fabrication

1. Deep reactive-ion etching (DRIE) system.
2. SU-8 etch mask.
3. All materials from Subheading 2.1.1.

2.2 Physical Experimental Setup

Running these devices requires:

1. A bench power supply.
2. Platinum or Ag-AgCl electrodes.
3. An inverted fluorescence microscope.
4. Two different buffer solutions which create the end points of the gradient.
5. A fluorescent analyte.

2.3 Buffers

1. Tris fluorescein (TF) buffer: 333 μM Tris, 167 μM fluorescein acid, and 50 mM NaCl.
2. High-conductivity gradient buffer: standard phosphate-buffered saline containing 141 mM NaCl.
3. Low-conductivity gradient buffer: standard phosphate-buffered saline, diluted with deionized water to 7–24 mM NaCl.
4. pH gradient buffers: citric acid buffers of pH values between 2.6 and 7.0 were first prepared with the conductivities between 1 and 4.87 mS/cm by mixing specific volumes of 0.1 M citric acid solution with 0.2 M disodium orthophosphate (Na_2HPO_4). These were then brought to equal conductivity of 4.60 ± 0.02 mS/cm by adding NaCl or deionized water as needed. Protein from storage was added to both the high and low pH buffers.

2.4 Biomolecules

1. R-phycoerythrin.
2. DyLight 488-labelled streptavidin.
3. PNA Probe: A single-stranded 18-mer PNA probe (probe 1) (Invitrogen Life Technologies) with Alexa555 label at the 5' end.
4. A single-base mismatch strand of PNA probe with the same fluorescent label was also generated (probe 2).
5. DNA oligo: a single-stranded DNA oligonucleotide identical to the 16s ribosomal RNA sequence (oligo) was purchased from Integrated DNA Technologies (Australia). The oligo was pre-labelled at the 3' end with Alexa 488. The sequences of PNA probe 1, PNA probe 2, and ribosomal RNA mimic oligo can be found in [3].

3 Methods

The silicon devices were made in such a way that the nanochannel was etched into a thick layer of high-quality thermal oxide (Fig. 1). These were lidded with borosilicate wafers [3–6]. Other devices were made entirely out of fused silica [7]. This process produced more robust devices with rectangular microchannel cross sections (as opposed to trapezoidal for wet-etched silicon). However, the fused silica process is more expensive.

3.1 Silicon Device Fabrication

1. Acquire a 3", p-type 1–10 Ωcm, <100> wafer with 50 nm of thermal oxide.
2. Pattern the microchannels onto this wafer using standard AZ1518 photolithography and a metal photomask.

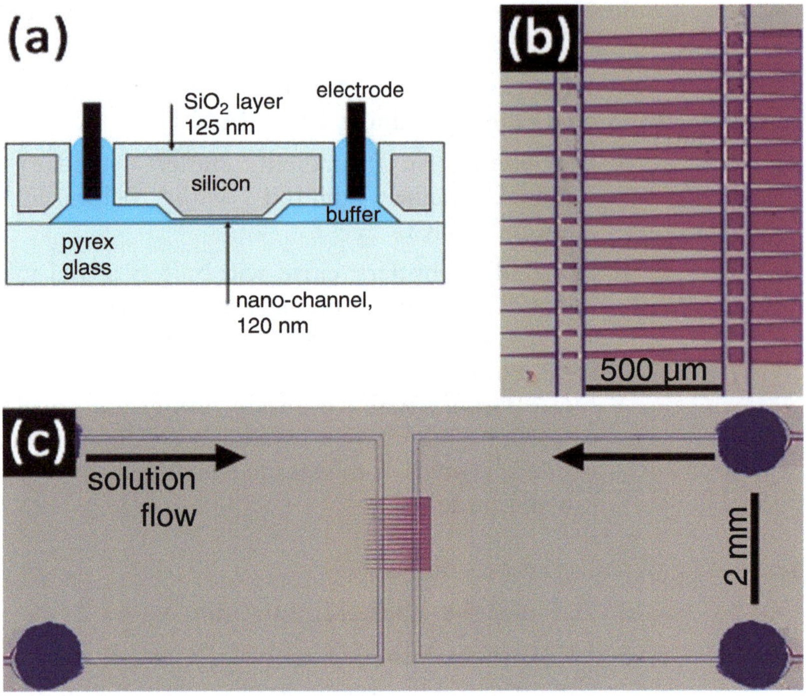

Fig. 1 Representative silicon device used in this work. (**a**) Cross section showing protective oxide layer. (**b**) Top view of the tapered nanochannels (horizontal), connecting two vertical microchannels. (**c**) Wide view of the entire chip

3. Etch the pattern by CF_4/O_2 plasma etching in a March PX-250 plasma system (2 sccm O_2, 16 sccm CF_4, 300 mTorr, 150 W) with 8" square electrodes. The wafer is placed on the positive electrode with a grounded electrode placed 25 mm above (*see* **Note 1**).
4. After etching through the 50 nm oxide layer, coat the entire front side of the wafer with approximately 50 µm of SU-8 resist, and then office tape (*see* **Note 2**).
5. Use a dental sandblaster containing 75 µm alumina particles from Danville Materials (Carlsbad CA, USA) to create holes for backside fluid and electrode ports.
6. Sequentially wash in detergent, acetone, and Remover PG (MicroChem Corp., Newton MA, USA) to remove sand particles, SU-8, and photoresist.
7. Etch clean wafers for 23 min in a stirred, 70 °C solution of 14% w/v potassium hydroxide and 17% v/v isopropyl alcohol.
8. After etching microchannels to between 5 and 10 µm, remove the remaining oxide by hydrofluoric acid. 130 nm of new thermal oxide was then grown (*see* **Note 3**).

9. Pattern nanochannels connecting a pair of microchannels, and etch to depths ranging from 85 to 120 nm by photolithography followed by CF_4/O_2 plasma etching.
10. Bond Borofloat 33 wafers to these silicon wafers using the reverse RCA (Radio Corporation of America) procedure; piranha clean followed by RCA2 then RCA1 on both wafers [8] (*see* **Notes 4** and **5**).
11. Seal fluid ports with tape, and then dice the wafers as needed.
12. Used devices were stored in DI water.

3.2 Fused Silica Device Fabrication

Fabrication of fused silica devices was mostly the same, except that microchannels were created using deep reactive-ion etching (DRIE) with an SU-8 etch mask. This etch mask was removed by baking in air at 900 °C for 5 h. Nanochannels were patterned using AZ photolithography and the same O_2, CF_4 plasma etching described earlier. Sandblasting of through holes was identical to the silicon devices. After cleaning, the wafer was bonded to a blank fused silica wafer using the same reverse RCA procedure and annealed for 12 h at 1050 °C in air.

4 Experimental Procedure

4.1 Nanochannel Conductivity Gradient Focusing

1. Place the device onto an inverted microscope, and initially view with an objective ring light or other broad-spectrum lamp placed below the chip.
2. Focus on the channels using scattered light from the micro- and nanochannel features (*see* **Note 6**).
3. Wet new devices with DI water from one side of each microchannel (*see* **Note** 7).
4. Introduce high- and low-conductivity buffers, and allow a few minutes for them to flow through the microchannels (*see* **Note 8**). As shown in Fig. 2, we place the high-conductivity buffer and analytes into the bottom microchannel (*see* **Note 9**).
5. With electrodes in place, apply 2 to 5 V between the opposing microchannels with the positive end at the high-salt electrode (*see* **Note 10**).
6. Fluorescent bands of concentrated biomolecules form near the low-salt end of the nanochannel. Bands take seconds to minutes to become observable depending on bulk concentration and the applied voltage.

4.2 Nanochannel pH Gradient Focusing

1. For pH gradient focusing, introduce a low pH buffer (typical pH 2–3) into the anodic microchannel and a more neutral buffer (typical pH 5–7) into the cathodic microchannel. Both

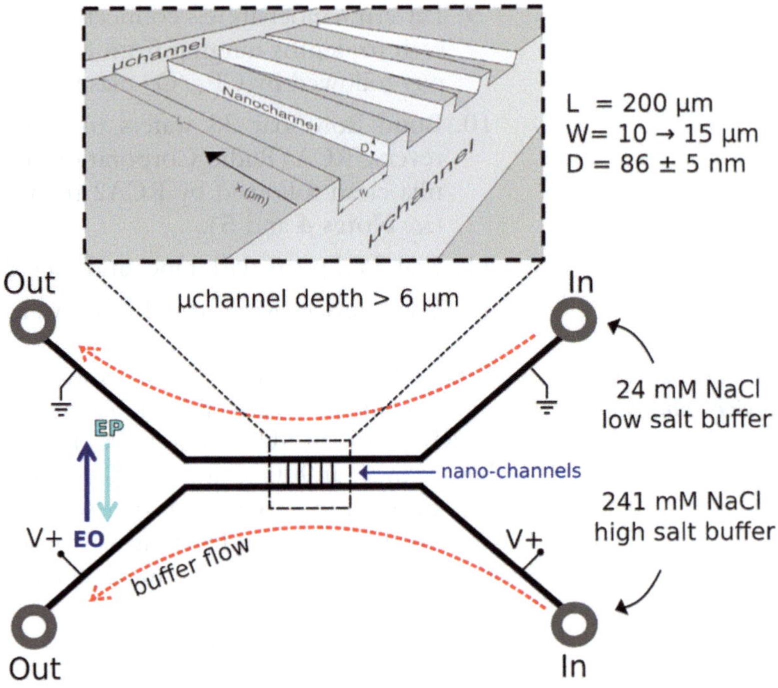

Fig. 2 Schematic of the nanochannel device used for conductivity gradient focusing. The inset describes exemplary dimensions. The low- and high-salt buffers are introduced to the ground and V+ microchannel inlets, respectively. Within the nanochannels, electrophoretic flow is directed toward the positive electrode, and electroosmotic flow is directed toward the negative electrode

buffers should have the same conductivity, and both microchannels should contain analyte.

2. Now develop focused analyte by applying voltage. Focused bands are less intense in this technique compared to those realized with conductivity gradients and tend to become unstable with applied voltages above 3 V (*see* also **Notes 11–13**).

5 Notes

1. We observed etch rates of approximately 9 and 100 nm/min on thermal oxide and photoresist, respectively.
2. This tape and unexposed resist is to protect the wafers during sandblasting of holes for backside fluid and electrode ports.
3. It was important to grow a layer of oxide after sandblasting, and one that was thicker than the nanochannel was deep to electrically insulate the fluidic channel from the semiconducting silicon.

4. Between each step the wafers were rinsed thoroughly in running deionized water.
5. The sealed wafers were bonded by annealing at 350 °C for 12 h.
6. Devices were viewed with either a 10× water-immersion objective or a 40×/0.6 air objective with a 0–2 mm correction collar. This objective is particularly useful for viewing micro- and nanofluidic devices with varying lid thicknesses.
7. The entire device wets in a few minutes.
8. The effect of nanochannel taper is minor and is described in [5].
9. It is important to maintain flow in the microchannels as this prevents buildup of high salt at the anodic end of the nanochannel. This flow can be passively or actively controlled as described in Subheading 3.
10. We use four electrodes to reduce potential drop in the microchannels (Fig. 2).
11. To avoid bleaching, the nanochannels should not be continuously illuminated but rather spot-checked to monitor accumulation progress. In all work images were captured using a cooled monochrome CCD, mercury lamp, and appropriate filters. To simplify quantification of different molecules, it is important to use non-overlapping filter sets and molecules with well-separated emission spectra.
12. We define the concentration enhancement (CE) factor as the nanochannel band intensity (I_{Nano}), divided by the intensity of the microchannel containing the analyte (I_{Micro}), times the microchannel depth (D_{Micro}), divided by the nanochannel depth (D_{Nano}).

$$\mathrm{CE} = \frac{I_{\mathrm{Nano}} D_{\mathrm{Micro}}}{I_{\mathrm{Micro}} D_{\mathrm{Nano}}}$$

    Setting the applied voltage to 0 releases the focused band, allowing it to rapidly broaden. Applying a few negative volts releases the band and back flows low-conductivity buffer into the channel, quickly flushing out biomolecules and darkening the image.

13. Channels are flushed with buffer then deionized water before storing devices underwater in individual vials. Devices were reused after flowing a detergent solution followed by 10% bleach and a thorough rinse with water. Fused silica devices were occasionally cleaned by baking in air at 900 °C overnight.

References

1. O'Farrell PH (1985) Separation techniques based on the opposition of two counteracting forces to produce a dynamic equilibrium. Science 227(4694):1586–1589
2. Koegler WS, Ivory CF (1996) Focusing proteins in an electric field gradient. J Chromatogr A 726 (1–2):229–236
3. Startsev MA, Ostrowski M, Goldys EM, Inglis DW (2017) A mobility shift assay for DNA detection using nanochannel gradient electrophoresis. Electrophoresis 38:335–341
4. Inglis DW, Goldys EM, Calander NP (2011) Simultaneous concentration and separation of proteins in a nanochannel. Angew Chem Int Ed 50(33):7546–7550
5. Hsu WL, Inglis DW, Jeong H, Dunstan DE, Davidson MR, Goldys EM, Harvie DJE (2014) Stationary chemical gradients for concentration gradient-based separation and focusing in nanofluidic channels. Langmuir 30(18):5337–5348
6. Hsu WL, Harvie DJE, Davidson MR, Jeong H, Goldys EM, Inglis DW (2014) Concentration gradient focusing and separation in a silica nanofluidic channel with a non-uniform electroosmotic flow. Lab Chip 14(2014):3539–3549
7. Startsev MA, Inglis DW, Baker MS, Goldys EM (2013) Nanochannel pH gradient electrofocusing of proteins. Anal Chem 85(15):7133–7138
8. Kern W (1990) The evolution of silicon wafer cleaning technology. J Electrochem Soc 137 (6):1887–1892

Chapter 10

Paper-Based Electrophoresis Microchip as a Powerful Tool for Bioanalytical Applications

Cyro L. S. Chagas, Thiago M. G. Cardoso, and Wendell K. T. Coltro

Abstract

This chapter describes the development of paper-based microchip electrophoresis (pME) devices for the separation of clinically relevant compounds. pME were fabricated by laser cut and thermal lamination process using polyester pouches. In addition, hand-drawn pencil electrodes were integrated to the device to perform capacitively coupled contactless conductivity detection (C^4D). Finished device costs less than US$ 0.10 and did not require either sophisticated instrumentation or clean room facilities. Furthermore, pME is lightweight, easy to handle, flexible, and robust. pME-C^4D device revealed an excellent capacity to separate BSA and creatinine in less than 150 s with baseline resolution. The device proposed in this chapter has proven to be a good alternative as a platform for the diagnosis of diseases from renal disorders such as diabetes mellitus and heart disease.

Key words Paper electrophoresis, Bovine serum albumin, Creatinine, Pencil electrodes, Contactless conductivity detection, Biomolecules, Clinical diagnosis, Kidney failure

1 Introduction

Paper is a substrate that presents several advantages as low cost, lightweight, easy transport and manipulation, global affordability, and biocompatibility. In the 1940s, electrophoresis was first reported on paper substrates to the separation of biological compounds as proteins, amino acids, and peptides. In the pioneering studies, the instrument used as high-voltage power supply and detection system did not present power rating and sensitivity. For this reason, separations were proceeded using high concentration of the analytes. In addition, the first separations exhibited low resolution and longer analysis time [1–4]. Because of these problems, the analysis of these biomolecules was substituted by techniques that require modern instrumentation as high-performance liquid chromatography (HPLC) or need large volume of reagents and analyte as gel electrophoresis. These features have made the traditional paper electrophoresis unpopular until recent years and

Debashis Dutta (ed.), *Microfluidic Electrophoresis: Methods and Protocols*, Methods in Molecular Biology, vol. 1906, https://doi.org/10.1007/978-1-4939-8964-5_10,

pushed the use of this substrate toward conventional analytical techniques such as filtration and clinical applications such as pregnancy tests [5]. In 2007, Whitesides's research group developed the microfluidic paper-based analytical devices (μPADs) as a bioanalytical platform to measure glucose and protein concentration levels [6]. μPADs bring the idea of minimum reagent waste as well as operational simplicity [7], revealing the necessity of improving the portability and reducing the cost of analysis [6, 8–10].

The advent of μPADs has contributed to the return of paper as a platform for electrophoresis separations with the aim of rapid and low-cost analysis [11]. Currently, the paper can be found with different thickness, porous dimensions, and chemical modification, and these characteristics helped to the return of the paper as raw material to fabrication-fast of analytical devices; other important points were that high-voltage source and detector system are better, more robust, and reduced size. In 2014, ca. 7 years after the proposal of μPADs, the first publications bringing back the paper as a separation platform for biomolecules were reported. The separation of lysine, serine, and aspartic acid using an on-column wireless electrogenerated chemiluminescence detector and the separation of fluorescent molecules and serum proteins using low voltage in a device coupled with fluorescence detection were successfully reported by Ge et al. and Luo et al., respectively [12, 13]. These two reports showed several improvements on paper electrophoresis, but parameters such as sample injection control, separation performance, and robustness still need to be better studied to improve the use of this platform for routine analysis. In 2016, Xu and coworkers developed a paper-based electrophoresis device designed in a cross-shaped format. In their study, the electrokinetic injection was used to introduce sample through the floating mode. The separation of two different organic dyes was performed within 10 min and monitored using cell phone camera, but it was also possible to accomplish the separation by naked eye [14]. After that, Wu and coworkers developed a methodology to concentrate and electrokinetically separate bovine hemoglobin and cytochrome C using a paper analytical device [11].

Electrochemical detection is quite attractive to be integrated with electrophoresis microchips due to its low instrumental cost, high sensitivity, and possibility to analyze multiple compounds without derivatization requirement. Electrodes can be fabricated on paper using different protocols including screen printing, sputtering, microwires, or hand drawing with graphite pencil [15–19]. Amperometry is one of the most sensitive electrochemical detection methods, and it was already explored to monitor chromatographic separations on paper devices [16, 17]. In 1998, da Silva and coworkers and Zemann and coworkers published, independently, the development of a capacitively coupled contactless conductivity detector (C^4D) for electrophoresis [20, 21]. C^4D

offers several advantages over "contact modes." The positioning of electrodes without physical contact with samples prevents problems associated with bubble generation, electrode fouling, and electrical interference when high voltage for electrophoresis is applied [18–20, 22–24].

In this chapter, we focused in the description of a recent development of fully disposable paper-based microchip electrophoresis (pME) devices to promote the separation of bovine serum albumin and creatinine using C^4D as detection system [25]. The interest in these molecules is due to the clinical importance in the diagnosis of renal diseases. The pME device was integrated with electrodes developed using pencil drawn in office paper [26].

Several developing countries have problems to build a good health structure for your population, because they limit financial support to expensive analysis techniques. However, the miniaturization of analytical techniques during the last 30 years has offered modern analytical instrumentation with low consumption of reagents and samples, low analysis time, and portability which enables the use for point-of-care testing. Diabetes mellitus and cardiovascular diseases are the major concerns of the World Health Organization because a study in 2012 showed these diseases are among the four main causes of death in the world [27]. The concentrations of protein in blood and urine are parameters that require special attention because they show some disturb as an abnormal protein concentration in the serum or an excess of protein that is named proteinuria [28]. The human bodies can present disturb without showing visible and noticeable symptoms as high blood pressure, cardiovascular disease, and diabetes mellitus. These cited disturbs cause kidney lesion that generate the proteinuria. Colorimetry and immunochemistry assays are most common techniques used for detecting proteinuria. Strips test is the most used colorimetric tool, but this method presents low accuracy and provides only a semiquantitative result. In addition, the sample's condition and the incorrect care of operator during the analysis make it possible to obtain false-positive results. However, the immunochemistry test presents good analytical parameters as high sensitivity, specificity, and reproducibility, but it is expensive and requires ambient controlled condition, instrumentation, and facilities [29–32].

2 Materials

2.1 Paper-Based Electrophoresis Microchips

1. Computer with graphical software installed.
2. Chromatographic paper (Whatman® Grade 01 CHR, Maidstone, Kent, UK).
3. Polyester thermal laminating pouches.
4. Paper punch.

5. Micropipette tips.
6. Stylet.
7. Epoxy glue.
8. Thermal laminator machine.
9. CO_2 laser engraver.

2.2 Hand-Drawn Pencil Electrodes

1. Computer with graphical software installed.
2. Office LaserJet printer.
3. Office paper sheet.
4. Pencil grade 2B (STABILO, model 8008, Heroldsberg, Germany).
5. Adhesive tape.

2.3 C^4D Detection

1. Commercial or lab-made C^4D system.
2. Function generator (Tektronix, model AFG1022, Beaverton, Oregon, EUA).
3. Computer with software for operational control and data acquisition.

2.4 Electrophoresis Analysis of Biomolecules

1. Bipolar high-voltage sequencer (eDAQ, Denistone East, Australia).
2. Background electrolyte (BGE): mixture of 20 mM lactic acid and 2 mM histidine (pH 3.1).
3. Biomolecule stock solutions: 1 mM bovine serum albumin (BSA) and creatinine 10 mM.
4. Artificial serum samples (Doles, Goiânia, Goiás, BR).

3 Methods

3.1 Fabrication of pME Devices

1. The first step involves the layout drawing of the paper device (*see* **Note 1**). Initially, the design must be projected through a graphical software. Figure 1a shows the geometry chosen for the fabrication of paper channels. The cross-shaped layout is similar to those often used in the development of electrophoresis devices fabricated in different platforms. Injection and separation channels were defined with 1 mm wide and 26 and 60 mm long, respectively. The channel height was delimited by the chromatographic paper thickness, i.e., 100 μm.
2. Once the layout was established, paper channels were cut in a CO_2 laser machine (*see* **Note 2**). The cutting parameters must be adjusted according to the available laser power.

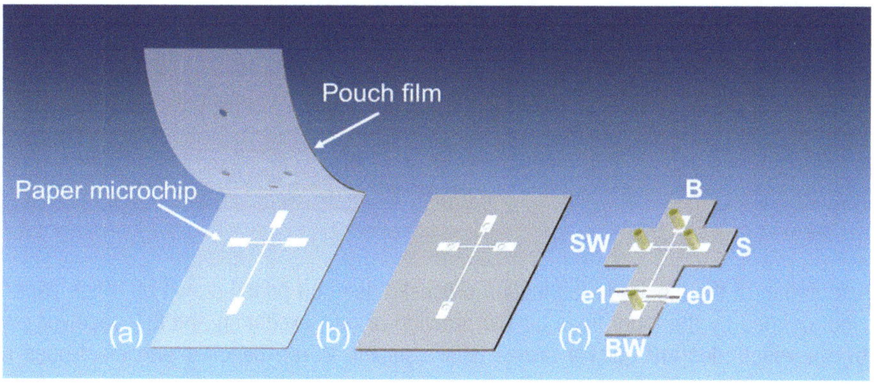

Fig. 1 Fabrication process of paper-based microchip electrophoresis (pME) with integrated electrodes for C^4D measurements. Images (**a–c**) show the positioning of paper microchip between the two layers of pouch film, the laminated device, and the resulting fully integrated pME-C^4D device containing pencil-drawn electrodes and solution reservoirs. In (**c**), the labels S, SW, B, BW, and e0 and e1 mean reservoirs for sample, sample waste, buffer, buffer waste, and excitation and receiver electrodes, respectively. Reprinted from ref. 25 with permission

3. After the cutting step, paper channels were laminated through a thermal laminator. First, the paper microchip was placed between two sheets of polyester for lamination (Fig. 1a). A marking of the position of the reservoirs was defined with the aid of a permanent marker. The paper was removed, and the perforation of the polyester was performed on the markings made previously through a paper punch in a circular geometry with 2 mm diameter. The paper was reintroduced between the two sheets of polyester, aligned with the holes, and, then, laminated at temperature of 160 °C (Fig. 1b) (*see* **Note 3**).
4. Micropipette tips of 200 μL were used for the create reservoirs for solutions. The tips were cut in half using a stylet, centered on perforations, made and glued on the polyester surface with bicomponent epoxy resin (Fig. 1c) (*see* **Note 4**).

3.2 Fabrication of Hand-Drawn Pencil Electrodes

1. For manufacturing the detection electrodes, it was necessary to define the layout through a graphical software. The geometry of the electrodes was defined in an antiparallel orientation, where each electrode was 20 mm long and 2 mm wide. The distance between the electrodes was 1 mm (*see* **Note 5**).
2. The drawn layout was printed on office paper surface using a LaserJet printer (Fig. 2a) to define the borders for controlling the area to be painted with a graphite pencil.
3. After printing, conductive surfaces were successfully created with the pencil drawing (Fig. 2b) until the entire area was visually filled (Fig. 2c).

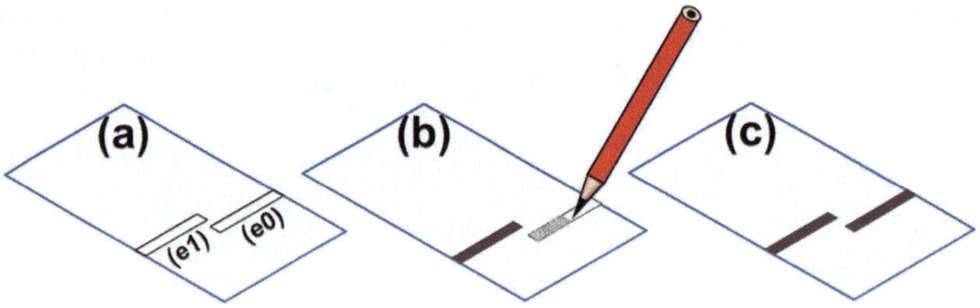

Fig. 2 Representation of the electrode fabrication process based on pencil drawn on office paper. In (**a**), we have the layout defined by the toner lines for the drawing of (e0) excitation and (e1) receiver electrodes on paper. In (**b**), the electrodes are quickly drawn with a pencil. In (**c**), pencil-drawn electrodes ready to be coupled to pME

4. The pair of electrodes were cut from the sheet of paper with scissors and glued with the adhesive tape under the pME device.

3.3 Electrophoresis and C⁴D Detection Procedure for BSA and Creatinine Analysis

1. A stock solution of 200 mM lactic acid was prepared daily by adding 438 μL of lactic acid (85%) in a 25 mL volumetric flask and diluting with ultrapure water. A stock solution of 100 mM histidine was prepared daily by adding 388 mg of L-histidine in a 25 mL volumetric flask and diluting with ultrapure water. BGE was prepared by mixing 1 mL of lactic acid stock solution and 200 μL of histidine stock solution in a 10 mL volumetric flask, and the volume was completed with ultrapure water.
2. The channels of the microchip were preloaded with BGE. Approximately 300 μL of the BGE was added to the buffer, sample, and sample waste reservoirs, leaving the buffer waste reservoir empty. This procedure was performed so that all the air present in the pores of the paper could be removed from the channel. After complete filling of the channels, 300 μL of the BGE was added to the buffer waste reservoir.
3. The device was attached to the platform with adhesive tape. The contacts between the electrodes, the function generator, and the detector were made using metallic alligator clips.
4. The high-voltage platinum electrodes for application and reading of electric potential were introduced into the device reservoirs. Potentials were applied to sample and buffer reservoirs, while potential readings were made in sample and buffer waste reservoirs.
5. The channel conditioning was performed by applying 1 kV for 10 min in both channels. After the conditioning of the channels, the BGE present in the reservoirs was renewed.

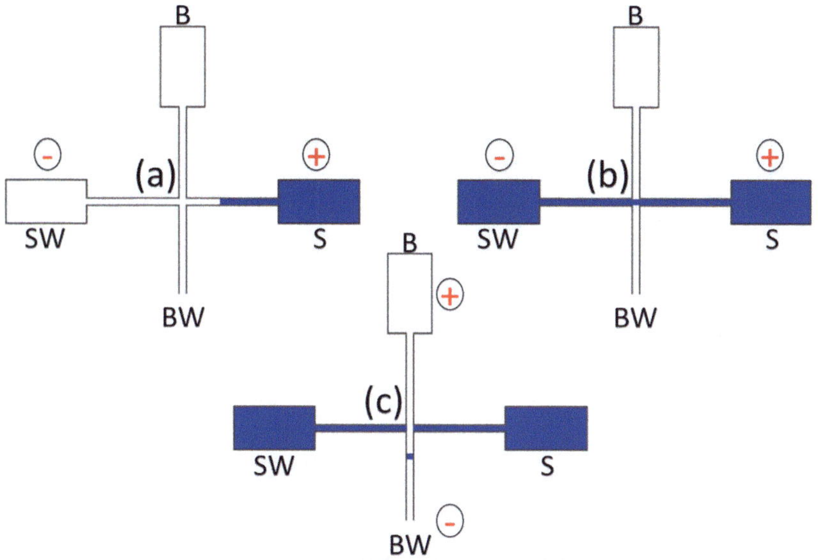

Fig. 3 Representation of the floating injection mode in the pME device. In (**a**), the electric potential is applied in the injection channel, and the sample begins to flow toward the sample waste reservoir. In (**b**), the injection channel was completely filled with the sample. In (**c**), the potential applied in the injection channel is switched off and immediately connected in the separation channel; the BGE loads the sample that was present at the intersection of the channels toward the detection point

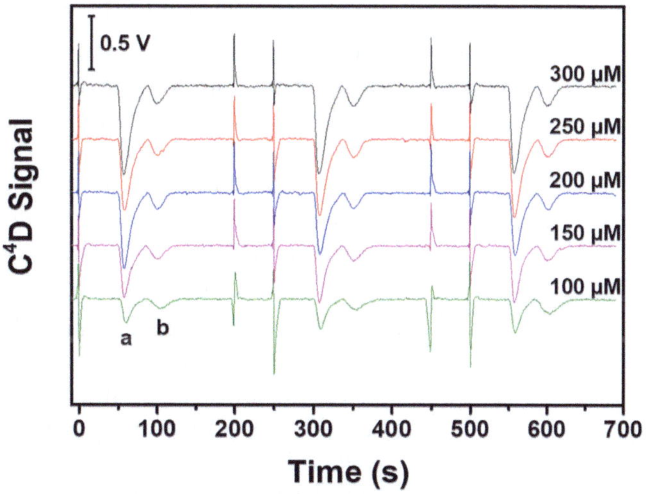

Fig. 4 Sequential electropherograms showing the separation of (**a**) BSA and (**b**) creatinine at five concentration levels. Reprinted from ref. 25, with permission

6. After preconditioning, the BGE solutions added on reservoirs were renewed prior to analysis. For the sample reservoir, BGE was replaced by sample solution (300 μL). Then, injections of the BSA and creatinine and serum samples were electrokinetically performed using the floating injection mode. The potential was applied during 50 s in the injection channel to ensure

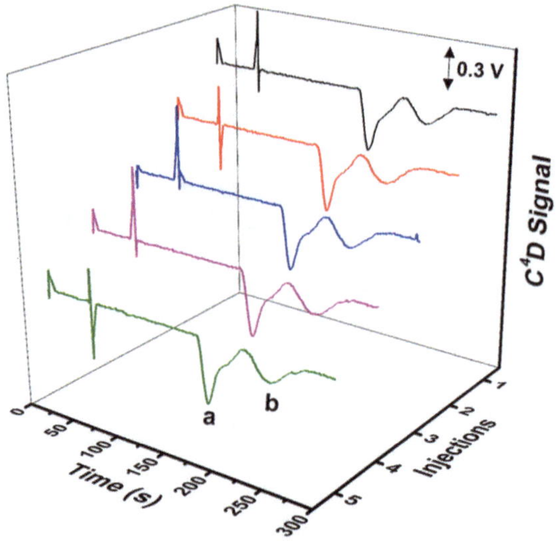

Fig. 5 Electropherograms showing five analyses of (**a**) BSA and (**b**) creatinine in artificial serum sample using paper-based electrophoresis microchips

that the sample reached the intersection of the channels (Fig. 3a). Then, the application of potential in the injection channel was switched to the separation channel (Fig. 3b). The amount of sample injected corresponds to the volume defined at the intersection of the channels (Fig. 3c).

7. Stock solutions of BSA and creatinine standards were prepared for the analysis. To prepare the stock solution of BSA (1 mM), 330 mg of the BSA standard was weighed and transferred into a 5 mL volumetric flask, completing the volume with ultrapure water (*see* **Note 6**). The stock solution of creatinine (10 mM) was made by weighing 28 mg of the creatinine standard and transferring into a 25 mL volumetric flask, completing the volume with ultrapure water.
8. To obtain the analytical curve, BSA and creatinine were mixed at five different concentrations (100 to 300 μM each). The standards were injected three consecutive times at a potential of 2.3 kV for 50 s and separated using a potential of 2.5 kV for 200 s (Fig. 4). After the injections of each concentrations of the standard, all reservoir solutions were renewed, and BGE was placed in the sample reservoir. Again, the potential was switched on in the injection channel to ensure that previous standard was removed from the channel, thus preventing one pattern from being contaminated with another concentration of standard.
9. After all injections of the standards, a sample of artificial human serum was injected (Fig. 5). Due to the high conductivity of the serum in relation to the standards, a lower separation potential (2.3 kV) was used (*see* **Note 7**).

4 Notes

1. The device layout can be made in various shapes and sizes, while respecting the size of the polyester sheet.
2. If a CO_2 laser cutter is not available, the paper can be cut through a cutoff printer.
3. To facilitate lamination and alignment of the device with the perforations, it is best to not separate the two sheets of polyester.
4. Avoid the use of glue in excess, once it can spread and fill partially the channels.
5. The electrodes can be made in several different geometries.
6. During the preparation of BSA solution, it is recommend to keep a slow stirring to prevent the foaming formation.
7. A strategy for quantifying biological compounds in complex samples is to use the standard addition method.

References

1. Kunkel HG, Tiselius A (1951) Electrophoresis of proteins on filter paper. J Gen Physiol 35:89–118
2. Jencks WP, Jetton MR, Durrum EL (1955) Paper electrophoresis as a quantitative method. Serum proteins. Biochem J 60:205–215
3. Mejbaum-Katzenellenbogen W, Dobryszycka WM (1959) New method for quantitative determination of serum proteins separated by paper electrophoresis. Clin Chim Acta 4:515–522
4. Righetti PG (2005) Electrophoresis: the march of pennies, the march of dimes. J Chromatogr A 1079:24–40
5. Ehrenkranz JRL (2002) Home and point-of-care pregnancy tests: a review of the technology. Epidemiology 13:S15–S18
6. Martinez AW, Phillips ST, Butte MJ et al (2007) Patterned paper as a platform for inexpensive, low-volume, portable bioassays. Angew Chem Int Ed Engl 46:1318–1320
7. Nanthasurasak P, Cabot JM, See HH et al (2017) Electrophoretic separations on paper: past, present, and future-a review. Anal Chim Acta 985:7–23
8. Tomazelli Coltro WK, Cheng CM, Carrilho E et al (2014) Recent advances in low-cost microfluidic platforms for diagnostic applications. Electrophoresis 35:2309–2324
9. Cate DM, Adkins JA, Mettakoonpitak J et al (2015) Recent developments in paper-based microfluidic devices. Anal Chem 87:19–41
10. Santhiago M, Nery EW, Santos GP et al (2014) Microfluidic paper-based devices for bioanalytical applications. Bioanalysis 6:89–106
11. Wu Z-Y, Ma B, Xie S-F et al (2017) Simultaneous electrokinetic concentration and separation of proteins on a paper-based analytical device. RSC Adv 7:4011–4016
12. Ge L, Wang S, Ge S et al (2014) Electrophoretic separation in a microfluidic paper-based analytical device with an on-column wireless electrogenerated chemiluminescence detector. Chem Commun 50:5699–5702
13. Luo L, Li X, Crooks RM (2014) Low-voltage origami-paper-based electrophoretic device for rapid protein separation. Anal Chem 86:12390–12397
14. Xu C, Zhong M, Cai L et al (2016) Sample injection and electrophoretic separation on a simple laminated paper based analytical device. Electrophoresis 37:476–481
15. Carvalhal RF, Kfouri MS, de Oliveira Piazetta MH et al (2010) Electrochemical detection in a paper-based separation device. Anal Chem 82:1162–1165
16. Shiroma LY, Santhiago M, Gobbi AL et al (2012) Separation and electrochemical detection of paracetamol and 4-aminophenol in a

paper-based microfluidic device. Anal Chim Acta 725:44–50

17. Dossi N, Toniolo R, Piccin E et al (2013) Pencil-drawn dual electrode detectors to discriminate between analytes comigrating on paper-based fluidic devices but undergoing electrochemical processes with different reversibility. Electroanalysis 25:2515–2522
18. Pumera M (2007) Contactless conductivity detection for microfluidics: designs and applications. Talanta 74:358–364
19. Coltro WKT, Lima RS, Segato TP et al (2012) Capacitively coupled contactless conductivity detection on microfluidic systems-ten years of development. Anal Methods 4:25–33
20. Fracassi da Silva JA, do Lago CL (1998) An oscillometric detector for capillary electrophoresis. Anal Chem 70:4339–4343
21. Zemann AJ, Schnell E, Volgger D et al (1998) Contactless conductivity detection for capillary electrophoresis. Anal Chem 70:563–567
22. Wang J, Pumera M, Collins G et al (2002) A chip-based capillary electrophoresis-contactless conductivity microsystem for fast measurements of low-explosive ionic components. Analyst 127:719–723
23. Pumera M, Wang J, Opekar F et al (2002) Contactless conductivity detector for microchip capillary electrophoresis. Anal Chem 74:1968–1971
24. Kubáň P, Hauser PC (2015) Contactless conductivity detection for analytical techniques-developments from 2012 to 2014. Electrophoresis 36:195–211
25. Chagas CLS, de Souza FR, Cardoso TMG et al (2016) A fully disposable paper-based electrophoresis microchip with integrated pencil-drawn electrodes for contactless conductivity detection. Anal Methods 8:6682–6686
26. Chagas CLS, Costa Duarte L, Lobo-Júnior EO et al (2015) Hand drawing of pencil electrodes on paper platforms for contactless conductivity detection of inorganic cations in human tear samples using electrophoresis chips. Electrophoresis 36:1837–1844
27. Organization WH and Organization WH The top 10 causes of death. 2012. http://www.who.int/mediacentre/factsheets/fs310/en/index1.html
28. Burtis C (2011) Tietz Fundamentos da Química Clínica. Elsevier Health Sciences, Amsterdam
29. Newman DJ, Thakkar H, Gallagher H (2000) Progressive renal disease: does the quality of the proteinuria matter or only the quantity? Clin Chim Acta 297:43–54
30. Wu MT, Lam KK, Lee WC et al (2012) Albuminuria, proteinuria, and urinary albumin to protein ratio in chronic kidney disease. J Clin Lab Anal 26:82–92
31. Lezaic V (2015) Albuminuria as a biomarker of the renal disease. In: Patel VB (ed) Biomarkers in kidney disease. Springer, Dordrecht, pp 1–18
32. Merrill AE, Khan J, Dickerson JA et al (2016) Method-to-method variability in urine albumin measurements. Clin Chim Acta 460:114–119

Chapter 11

Band Broadening Theories in Capillary Electrophoresis

Sandip Ghosal

Abstract

In capillary electrophoresis (CE), analytes are separated along the axis of a single microcapillary by virtue of their differential migration in an applied electric field. CE can also be performed in channels etched on solid substrates such as glass or PDMS and can be integrated into a microfluidic chip with a complex network of electric and fluidic circuits. The measure of quality of a CE instrument is resolution which is limited fundamentally by mixing due to various physical processes. The theoretical limit on the best separation that can be achieved is set by molecular diffusion, which is inevitable. The goal is to eliminate or minimize the other sources of dispersion by design. This chapter provides an overview of the various mechanisms of band broadening and the mathematical results that make it possible to estimate their relative contributions.

Key words Electrophoresis, Electroosmosis, Taylor dispersion, Zeta potential, Debye layer

1 Introduction

Electrophoretic separation relies on the fact that many macromolecules, including the biologically important ones, DNA, RNA, and proteins, carry an electric charge in aqueous solution. In an external electric field, the molecule migrates with a velocity (u_{ep}) that is proportional to the field strength (E), provided that the electric field strength is not too high. The electrophoretic mobility (*see* **Note 1**) of the molecule is its velocity per unit field strength, $\mu_{ep} = u_{ep}/E$. The electrophoretic mobility of a molecule depends not only on its size and charge but also on the ionic composition of the surrounding buffer. Electrophoresis relies on the fact that μ_{ep} provides a fingerprint of the molecular species and can be used to characterize it.

The calculation of μ_{ep} for a particle of a given shape and charge is a difficult fundamental problem in electrokinetics that has received much attention [1–3]. An interesting result of general validity is due to Morrison [4]. Morrison showed that in the limit where the Debye layer is very thin relative to particle dimensions

Debashis Dutta (ed.), *Microfluidic Electrophoresis: Methods and Protocols*, Methods in Molecular Biology, vol. 1906,
https://doi.org/10.1007/978-1-4939-8964-5_11,

and the zeta potential (ζ) is uniform, the particle moves without rotation at a velocity

$$u_{\text{ep}} = \frac{\epsilon \zeta}{\eta} E \tag{1}$$

irrespective of its shape. Here ϵ and η are, respectively, the electric permittivity and viscosity of the surrounding electrolyte, and E is the uniform applied electric field. One consequence of this result is that linear polyelectrolytes (e.g., DNA) that may be modeled as a uniformly charged rod (of radius a and charge density λ per unit length) have an electrophoretic mobility independent of length and therefore cannot be separated in free solution. Indeed, if the curvature of the rod is much larger than its radius, as is usually the case, its zeta potential is given by [5].

$$\zeta = \frac{\lambda}{2\pi\kappa a\ \epsilon} \frac{K_0(\kappa a)}{K_1(\kappa a)} \tag{2}$$

which is independent of the length of the polyelectrolyte. Here K_n ($n = 0{,}1$) is the order n modified Bessel function of the second kind, and κ^{-1} is the Debye length. In order to separate linear polyelectrolytes by size, one needs to fill the capillary with some sieving medium, although "free solution" techniques such as ELFSE (end-labeled free solution electrophoresis) are also used [6, 7].

In the general case where the Debye length (κ^{-1}) is not necessarily small in comparison with typical particle dimensions (a), that is, κa is not necessarily large compared to unity, the electrophoretic velocity depends on particle shape. There is a substantial body of literature devoted to calculating μ_{ep} for a spherical particle [2]. For a spherical particle, we have Eq. 1 when $\kappa a \gg 1$ (since this result is independent of particle shape), but

$$u_{\text{ep}} = \frac{2}{3} \frac{\epsilon \zeta}{\eta} E \tag{3}$$

in the opposite limit of thick double layers, $\kappa a \ll 1$. Asymptotic results for charged disks have been presented by Stone and Sherwood [8]. In this chapter, we do not discuss the problem of calculating μ_{ep} from first principles; instead, we will suppose that this quantity is fixed and given for each charged species. The band broadening problem that we discuss here pertains to the variation of concentration of the analyte on scales much larger than the size of individual particles.

The basic idea of CE and associated experimental techniques are discussed in textbooks devoted to the subject [9–11]. In brief, there are a number of different modes of CE:

1. *Capillary zone electrophoresis (CZE)*: is the simplest CE mode. The analyte is injected at one end and is carried toward the

opposite end by bulk electroosmotic flow. Each species moves slightly faster or slower than the bulk flow speed due to its unique electrophoretic mobility. Thus, the sample separates into zones and is detected at the opposite end of the capillary, usually with a UV absorbance photodetector. If the capillary is not prefilled with a sieving medium, then we have free solution CZE (FSCZE).

2. *Capillary isoelectric focusing (CIEF)*: is used to separate amphoteric molecules (e.g., amino acids and proteins). Here in addition to applying a voltage across the capillary, a pH gradient is maintained along it. Each analyte species migrates to its own isoelectric point where its net velocity is zero. CIEF capillaries are usually chemically coated to neutralize any electroosmotic flow that would also occur and to prevent adsorption of cationic components to capillary walls.
3. *Capillary isotachophoresis (CIETP)*: may only be used if the mixture of ions being separated all have the same sign. If there are N species with electrophoretic mobilities $\mu_1^{(ep)} < \mu_2^{(ep)} < \cdots < \mu_N^{(ep)}$, then the sample is injected between a leading buffer of mobility $\mu_\infty^{(ep)} > \mu_N^{(ep)}$ and a trailing buffer of mobility $\mu_0^{(ep)} < \mu_1^{(ep)}$. The constancy of the current $I = \sigma AE$ (σ is the buffer conductivity, A the capillary cross-sectional area, and E the local electric field) implies that the electric field is reduced in zones containing ions of higher electrophoretic mobility where the conductivity is higher. In the steady state, the leading and trailing electrolyte and each one of the analyte species migrate at one and the same rate (hence the name isotachophoresis) and pass a fixed detector in order of their mobilities [12–14].
4. *Capillary electrochromatography (CEC)*: is more of a chromatographic rather than an electrophoretic technique. It is mentioned here as the underlying bulk flow is usually electroosmotic. Here the capillary contains a second "stationary phase," often in the form of coated beads that have varying degrees of affinity for the component species in the analyte. Thus, in accordance with the well-known chromatographic principle, individual species migrate with an average speed depending on its affinity toward the stationary phase.
5. *Micellar electrokinetic capillary chromatography (MECC or MEKC)*: was invented [15, 16] to separate neutral molecules that would normally not separate by electrophoretic methods. In MEKC the buffer contains a high concentration of a detergent such as SDS (sodium dodecyl sulfate) above its critical micellar concentration (CMC). These micelles carry a large negative charge and have a high electrophoretic velocity. Depending on its solubility, each analyte molecule spends

part of the time trapped in a micelle and is therefore mobile and part of the time in solution when its migration velocity is zero. The principle of separation is identical to that of CEC except that the mobile phase is these micelles and the stationary phase is the surrounding buffer.

The band broadening theories we discuss here pertain to FSCZE, though the ideas could be generalized to other CE modes.

2 Materials

This chapter is a discussion of theoretical results relating to band broadening and is generally not dependent on any specific buffer composition or preparation method. Often, we need just a few buffer properties (e.g., permittivity, viscosity, etc.) as parameters in the theoretical model. In some instances, we need to assume a specific kind of background electrolyte (e.g., a symmetric binary electrolyte), and these restrictions are pointed out at the appropriate point in the text. Most often, these restrictions are introduced in order to make progress in the mathematical description of the problem and may not always correspond to the specific conditions of laboratory experiments. The results are nevertheless still useful if one desires an estimate of band broadening, and high accuracy is not necessarily a priority.

3 Methods

3.1 Elementary Considerations

If a voltage is applied across an electrolyte-filled glass capillary, there is a bulk flow of fluid along the capillary. This electroosmotic flow (EOF) velocity is uniform across the capillary cross section (plug flow) and given by the formula

$$u_{\mathrm{eo}} = -\frac{\epsilon\zeta}{\eta} E \tag{4}$$

where ϵ and η are, respectively, the electric permittivity and viscosity of the electrolyte, ζ is the zeta potential of the glass wall, and E is the applied electric field [14]. Equation 4 is known as the Helmholtz-Smoluchowski formula and is valid provided that the Debye length is much smaller than the capillary width. If an analyte species is now introduced, clearly it is advected at a velocity, $u_{\mathrm{eo}} + u_{\mathrm{ep}}$, where u_{ep} is its electrophoretic velocity relative to stationary fluid and, simultaneously, diffuses along the capillary by an amount determined by its molecular diffusivity (D). Thus, the concentration along the capillary, $c(x, t)$, evolves according to the one-dimensional advection-diffusion equation

$$\frac{\partial c}{\partial t} + \left(u_{eo} + u_{ep}\right)\frac{\partial c}{\partial x} = D\frac{\partial^2 c}{\partial x^2} \tag{5}$$

which admits exact analytical solutions. If the initial concentration distribution is a Gaussian with variance σ_0^2, then the variance after time "t" is

$$\sigma^2 = \sigma_0^2 + 2Dt \tag{6}$$

and the centroid of the peak is advected with velocity $u_{eo} + u_{ep}$. Thus, the time between injection and detection is $t = L/(u_{eo} + u_{ep})$ where L is the capillary length. If the initial peak width is considered negligible, then

$$\sigma^2 \approx 2Dt = \frac{2DL}{\left(u_{eo} + u_{ep}\right)} = \frac{2DL}{\left(\mu_{eo} + \mu_{ep}\right)E} = \frac{2DL^2}{\left(\mu_{eo} + \mu_{ep}\right)V} \tag{7}$$

where V is the applied voltage and μ_{ep}, μ_{eo} are, respectively, the electrophoretic and electroosmotic mobility. This equation is often written in dimensionless form as

$$N \equiv \frac{L^2}{\sigma^2} = \frac{\left(\mu_{eo} + \mu_{ep}\right)V}{2D} \tag{8}$$

where the dimensionless quantity N, called "the number of theoretical plates," is a measure of the separation quality. The higher the N, the more efficient the separation. Sometimes, the dimensional quantity

$$H = \frac{\sigma^2}{L} = \frac{L}{N} \tag{9}$$

known as the "plate height" is used in place of N. Clearly, H has dimensions of length. A notable aspect of Eq. 8 is the fact that N is proportional to the applied voltage but is independent of the capillary length. In practice, there is a limit to how short a capillary one may use or how high a voltage is permissible, due to the production of Joule heat in the system. This is because shorter capillaries would have lower electrical resistances and therefore produce a greater amount of heat at a fixed voltage. The limitations imposed due to Joule heat are discussed in the following section.

A quantity related to N is the resolution, R_s, of a CE system. It is defined as the distance between neighboring peaks at the detector divided by four times the peak standard deviation. The resolution is related to N [17, 18] by

$$R_s = \frac{\sqrt{N}}{4}\Delta\mu\bar{\mu} \tag{10}$$

Here $\Delta\mu$ is the difference in electrophoretic mobility between the two species represented by neighboring peaks, and $\bar{\mu}$ is their average electrophoretic mobility. Thus, two peaks can be resolved only if $\Delta\mu > 4\bar{\mu}/\sqrt{N}$. In practice, N is often several thousands.

3.2 Anomalous Dispersion

The value of N calculated in Eq. 8 presumes an advection velocity that is uniform over the capillary cross section. This, however, represents an ideal "diffusion-limited" separation, $N = N_{\text{ideal}}$. In reality, there are often a multitude of additional effects that result in N being less than N_{ideal}; we call this "anomalous" dispersion. An interesting analogy may be drawn with the science of designing optical instruments where one strives for "diffraction limited" optics by minimizing all sources of optical aberrations. In CE we seek a "diffusion-limited" system by trying to minimize all sources of anomalous dispersion. In this section, we discuss some common sources of anomalous dispersion and present methods for estimating their effects quantitatively.

3.2.1 Thermal Broadening

The flow of electric current through the capillary results in the production of Joule heat. The heat produced per unit time is V^2/R, R being the resistance across the capillary. Since N is proportional to voltage, a high operational voltage is desirable. In fact, the applied voltage is often in the range of kilovolts for which Joule heat is a serious concern. Temperature differences due to Joule heat could lead to convection in the fluid that would obliterate the signal altogether or lead to channel blockage due to vapor lock. The actual temperature rise is determined by a balance of Joule heat production and heat dissipation from the capillary walls. This is one of the reasons why CE requires microcapillaries; in fact, the Joule heat problem was a major impediment to the development of CE until capillaries less than 200 μm internal diameter became available. Joule heating can be minimized by increasing the capillary resistance R by (1) using longer capillaries, (2) reducing capillary diameter, and (3) using buffers of low electrical conductivity. With each one of these strategies, one runs into other anomalous dispersion sources that we discuss later. Thus, achieving high N is always a fine balancing act, and for this one must understand in depth the transport physics in the capillary.

Assuming that the heat generation in the capillary is sufficiently controlled so that none of the abovementioned catastrophic events happen, the analyte band would nevertheless show anomalous broadening. This is due to the fact that Joule heating leads to a radial temperature variation since the heat sources are uniformly distributed in the fluid volume but cooling is by heat conduction from the capillary walls. The fluid viscosity is related to temperature by Sutherland's formula

$$\eta = A \exp(B/T) \tag{11}$$

where T is the absolute temperature and A, B are constants for a given fluid. According to Eq. 1, the electrophoretic migration speed of molecules varies inversely with the fluid viscosity η. Thus, analyte molecules near the wall migrate with a slightly different speed than those at the center. This differential migration leads to anomalous band broadening known as thermal broadening. Since the electroosmotic slip velocity in Eq. 4 depends only on the value of the viscosity at the wall, radial temperature variations only change the bulk electroosmotic flow speed without introducing any shear in the flow [19]. Thus, in the case of thermal broadening, there are no contributions to dispersion due to perturbations of the electroosmotic flow. The variation of the average electroosmotic flow can however be used to measure the temperature rise in a capillary. Based on such measurements, Burgi et al. [20] produced the following useful semiempirical formula for estimating the core temperature of the capillary from the measured power

$$T \text{ (in K)} = 11.5 \times (\text{Power in W}) + 297.9 \qquad (12)$$

The magnitude of the thermal broadening effect can be calculated from the temperature rise, and formulas for doing this have been presented by Davis [21], who considers the radial variation of not only the viscosity but other transport parameters as well. These expressions are somewhat complicated and require numerical integration of profile functions. We do not reproduce them here but refer the reader to Davis' paper. Explicit expressions for the plate height have also been provided by Knox [19]. These formulas are somewhat simpler due to simplifying assumptions made in the analysis. The simplest expression for plate height has been provided by Grushka et al. [22]:

$$H = \frac{2D}{u} + \frac{R_1^6 E^4 C_b^2 B^2 \Lambda^2 u}{24D\left(8k_1 T_1^2 - E^2 \Lambda C_b R_1^2 B\right)^2} \qquad (13)$$

where D is the solute diffusivity, u is the solute migration velocity averaged over the cross section, R_1 is the inner diameter of the capillary, E is the field strength, C_b is the buffer concentration, B is the constant in Sutherland's formula (Eq. 11), Λ is the equivalent conductance of the buffer, k_1 is the buffer thermal conductivity, and T_1 is the absolute temperature at the inside wall of the capillary. Grushka et al. assumed that the temperature differential between core and wall was small and that the variation of viscosity with temperature was the sole contributing factor in thermal broadening. Similar calculations were provided by Andreev and Lisin [23]. Other researchers have addressed the thermal broadening problem incorporating various levels of detail in their mathematical models [22, 24]. In addition to the obvious radial dependence on temperature, axial variations in temperature could also occur due to various inhomogeneities in the capillary. Such variations could induce pressure gradients and lead to band broadening. It is not clear whether such axial variations are present and if so whether they do cause significant dispersion. Convective motion of fluid in the capillary is also possible. These are open areas for investigation.

3.2.2 Geometric Dispersion Due to Channel Curvature

Much of the modern interest in CE systems is in the context of integrating CE into more general "lab-on-a-chip" microfluidic devices. Unlike tabletop CE capillaries that could be a meter in length, practical considerations prevent microfluidic chips of this size. The only way that a long CE channel can be imprinted on a microfluidic chip of modest footprint is by introducing turns along the channel. Various geometric designs such as spirals and zigzag patterns have been considered. If a microfluidic channel of uniform cross section is made to undergo a turn, then the electric field must necessarily be stronger at the inner wall. This is illustrated in Fig. 1. Since the electric field is tangential to the insulating walls, the equipotential surfaces AA′ and BB′ are orthogonal to the channel centerline. This implies that the same potential drop takes place over a shorter distance, AB, at the inner edge of the channel; thus, the electric field is stronger. Therefore, molecules in the band move faster at the inner edge and, furthermore, have a shorter distance to travel in order to clear the bend (the so-called "racetrack" effect). Thus, the band is sheared as it clears the bend as shown in Fig. 1. Once such a distortion is introduced, cross-stream molecular diffusion will act on it in accordance with the Taylor-Aris mechanism discussed later, leading to enhanced axial dispersion. It is possible to minimize such "geometric dispersion" by carefully designing the channel shape and cross section, and this is an interesting example of the usefulness of mathematical modeling in microfluidic technology [26]. We conclude this discussion with some formulas for estimating the extent of geometric dispersion in channels of uniform width.

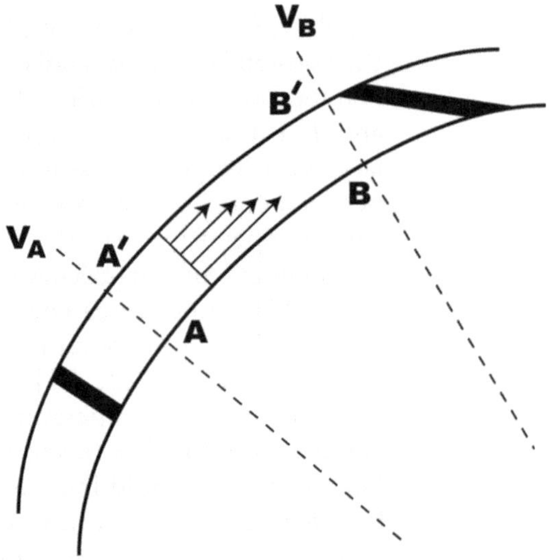

Fig. 1 Illustrating the mechanism of dispersion caused by channel curvature. Reproduced with permission from [25]

In essence, the analysis depends on the relative magnitudes of advection and diffusion as measured by two time scales: the diffusion time across the capillary (of width w), $t_d \sim w^2/D$, and the characteristic residence time in the curved region, $t_a \sim R\theta/u_e$ (R being the radius of curvature, θ the turn angle, and u_e a characteristic migration speed). Thus, the ratio of the advection to diffusion time is

$$\frac{t_a}{t_d} = Pe^{-1}\frac{R}{w}\theta \tag{14}$$

where $Pe = wu_e/D$ is a dimensionless number called the "Peclet number." The expression for axial dispersion depends on this time scale ratio. Griffiths and Nilson [27] proposed an empirical formula for the added peak variance in a planar channel upon traversing a turn that reduces to the correct limiting forms when $t_a/t_d \ll 1$ or $t_a/t_d \gg 1$:

$$\left(\frac{\sigma}{w}\right)^2 = \frac{\theta^2 Pe}{15R\theta w + 3Pe} + \frac{2R\theta}{wPe} \tag{15}$$

In the diffusion-dominated limit, $t_a/t_d \gg 1$, Eq. 15 reduces to $\sigma^2 \sim 2Dt_a$, which is just the effect of axial spreading due to molecular diffusion over the turn time. In the advection-dominated limit, $t_a/t_d \ll 1, \sigma^2 \sim \theta^2/3$. The distortion in this case is purely geometric and depends solely on the angle of the turn. Both limits are relevant in microfluidic applications. For example, if we take $w \sim 100$ μm, $R \sim 1$ cm, and $\theta = \pi$, then $R\theta/w \sim 300$. Since $Pe \sim 10^1 - 10^2$ for small molecules but $Pe \sim 10^3 - 10^4$ for macromolecules, it is clear from Eq. 14 that the "high," the "low," as well as the "intermediate" Peclet number regimes are all relevant for microfluidic applications.

Equation (15) suggests two possible strategies for minimizing dispersion: (a) increase R and (b) decrease Pe. Each of these possibilities has led to design ideas. Culbertson and others [28, 29] have pursued the idea of etching the CE channel in the form of a spiral with the inlet reservoir at the center. This approach relies on ensuring that the spreading peak encounters a progressively increasing radius of curvature. An alternate design [30] uses straight channel sections that undergo 180° turns at the edge of the chip. However, the channel is "pinched" at the turn so that it becomes very narrow in the curved sections, thereby effectively reducing Pe locally by decreasing w. A third possibility is to redesign the channel geometry at the bend so as to compensate for both the higher electric field and the "racetrack" effect. Effectively this is equivalent to attempting to reduce the prefactor multiplying the Pe in the first term of Eq. 15 by altering the velocity distribution. These approaches have been used to come up with optimal bend geometries using computational algorithms [26, 31–33]. Others have attempted to combine several of these ideas, for example [34], spirals with inner walls that are wavy to compensate for the shorter path, thereby both increasing R and reducing the geometric prefactor in Eq. 15 or by attempting to modify the wall ζ-potential through laser ablation, which will also change the pattern of EOF in the channel [35]. A similar effect may be achieved using a "faceted" design [36, 37] in which the spiral is replaced by an approximating polygonal shape or even "pleated channels" [27] where some of the turn-induced dispersion is "undone" at the one that follows in the opposite direction.

3.2.3 Electromigration Dispersion

Electromigration dispersion (EMD) is caused by perturbations in the local electrical conductivity since the solute composition in the vicinity of the peak is altered [38]. The effect becomes significant when sample concentration becomes comparable to the concentration of background ions, either due to sample overloading or when buffers of low conductivity are employed. Since local electroneutrality in solution requires conservation of current, and the current density in solution is the product of the electric field and conductivity (Ohm's law), perturbations of conductivity imply corresponding changes in the local axial electric field. These field perturbations lead to changes in the ion migration speed which, in turn, affects the concentration distribution. In practice, the presence of EMD is often identifiable by a characteristic highly asymmetric "wedge-shaped" peak profile [39]. We have seen that in order to minimize Joule heat effects, a low buffer conductivity is advantageous. A high sample concentration is also desirable in CE, because detector sensitivity over the extremely short optical path length across the microcapillary is often a limiting factor. However, any attempt to improve device performance by increasing sample concentration or lowering buffer conductivity runs into EMD, which is therefore a very significant anomalous dispersion mechanism. The characteristic triangular peak shape is reminiscent of one-dimensional nonlinear waves familiar in various other contexts such as breaking water waves, shock waves in gas dynamics, or even traffic flow on highways. This is no accident, as it can be shown that under certain simplifying assumptions, the sample concentration in EMD is described by the same nonlinear equation as these other physical processes [40].

The relation between EMD and nonlinear waves was recognized by early investigators who constructed one-dimensional mathematical models and deduced the triangular peak shape [38, 41–44]. These models however all neglected molecular diffusivity of the sample. As a result, they only describe the limit of strong nonlinearity and do not reduce to the ordinary one-dimensional advection-diffusion equation when EMD effects are weak. Here we describe an alternate mathematical model where we assume a three-component (analyte, anion in the background electrolyte, cation in the background electrolyte) equi-diffusive, and fully dissociated system of ions (*see* **Note 2**). EOF is neglected (*see* **Note 3**). It may be shown [40] that for such an idealized three-component system, the analyte concentration c obeys a single nonlinear partial differential equation. This equation may be conveniently expressed in terms of the dimensionless quantity $\phi = c/c_n^{(\infty)}$ where $c_n^{(\infty)}$ is the concentration of the anions in the background electrolyte:

$$\frac{\partial \phi}{\partial t} + u_{\text{ep}} \frac{\partial}{\partial x}\left(\frac{\phi}{1 - \alpha\phi}\right) = D\frac{\partial^2 \phi}{\partial x^2} \tag{16}$$

Here u_{ep} is the electrophoretic velocity of an isolated sample ion, and α is a dimensionless parameter defined as

$$\alpha = \frac{(\mu - \mu_{\text{n}})\left(\mu - \mu_{\text{p}}\right)}{\mu_{\text{n}}\left(\mu_{\text{p}} - \mu_{\text{n}}\right)} \tag{17}$$

in terms of the electrophoretic mobilities $\mu_{\text{p}} > 0$ (cation), $\mu_{\text{n}} < 0$ (anion), and μ (analyte). It may be easily shown that α is positive if μ is in the interval $(\mu_{\text{n}}, \mu_{\text{p}})$ and is negative otherwise. The concentrations of the anion and cation in the background electrolyte may be expressed in terms of ϕ using the conservation laws. If $\phi \ll 1$ (a "weakly nonlinear" system), we may make the approximation $(1 - \alpha\phi)^{-1} \approx 1 + \alpha\phi$, so that Eq. 16 reduces to

$$\frac{\partial \phi}{\partial t} + 2\alpha u_{\text{ep}}\phi\frac{\partial \phi}{\partial x} = D\frac{\partial^2 \phi}{\partial x^2} \tag{18}$$

where we have transformed away the $u_{\text{ep}}(\partial\phi/\partial x)$ term by shifting to a reference frame moving at a uniform velocity u_{ep} relative to the laboratory frame. Equation 18 is Burgers' equation that describes nonlinear waves in various contexts. It is exactly solvable for any initial condition $\phi(x, 0)$. For our application, a special case of interest is

$$\phi(x, 0) = \Gamma\delta(x) \tag{19}$$

where Γ is a positive constant and $\delta(x)$ denotes the Dirac delta function. This corresponds to a peak that initially has no spread. Since $\int_{-\infty}^{+\infty} c(x, 0) = c_n^{(\infty)} \int_{-\infty}^{+\infty} \phi(x, 0)dx = \Gamma c_n^{(\infty)}$, Γ clearly has units of length. It is actually the length along the capillary that contains the same number of anions of the background electrolyte as there are sample ions in solution. For this choice of initial conditions, the integrals appearing in the formal solution of Burgers' equation can be evaluated exactly. Thus, we obtain the following explicit expression for the sample distributions at any time

$$\phi(x,t) = \frac{1}{2\alpha u_{\mathrm{ep}}}\left(\frac{D}{\pi t}\right)^{1/2} F\left(\frac{x - u_{\mathrm{ep}}t}{\sqrt{4Dt}}\right) \tag{20}$$

Here

$$F(x) = \frac{Ae^{-x^2}}{1 + (A/2)\,\mathrm{erfc}(x)} \tag{21}$$

where $\mathrm{erfc}(x) = (2/\sqrt{\pi})\int_x^\infty \exp(-t^2)\,dt$ is the complementary error function and $A = e^{\alpha P} - 1$ (*see* **Note 4**). The dimensionless parameter $P = \Gamma u_{\mathrm{ep}}/D$ is a "Peclet number" based on the characteristic length scale Γ, and it measures the strength of the nonlinearity or extent of sample overloading in the system. Indeed, if αP is small, $A \approx 0$, and in this case the denominator in Eq. 21 may be approximated by one. In this limit, Eq. 20 reduces to a spreading Gaussian profile as expected in the absence of EMD. If αP~1, A is non-negligible in magnitude and has the same sign as α. In this case, the quantity A in the denominator of Eq. 21 creates an asymmetry in the profile. If $\alpha > 0$, $A > 0$ and the Gaussian profile steepens in the direction of peak propagation. Conversely, if $\alpha < 0$, $A < 0$ and the Gaussian profile steepens behind the peak (*see* Fig. 2). For $\alpha > 0$, if μ lies in the window (μ_{n}, μ_{p}), the peak steepens at its leading edge if the analyte ions are of intermediate mobility, a well-known experimental fact. The dimensionless quantity $|\alpha|P$ that measures the strength of the nonlinearity may also be viewed as a ratio of time scales. If a is a characteristic capillary width, then $\tau_{\mathrm{d}} = a^2/D$ is a diffusive time in the cross-stream direction. A corresponding advection time is $\tau_{\mathrm{a}} = a/u_{\mathrm{ep}}$. The new length scale Γ in our problem allows us to introduce yet another time scale, $\tau_{\mathrm{e}} = |\alpha|\Gamma/u_{\mathrm{ep}}$, which may be interpreted as a time scale associated with sample loading. Then, $|\alpha|P = \tau_{\mathrm{e}}\tau_{\mathrm{d}}/\tau_{\mathrm{a}}^2$, that is, $|\alpha|P$ is the square of the ratio of the geometric mean $\sqrt{\tau_{\mathrm{e}}\tau_{\mathrm{d}}}$ and the advection time scale τ_{a}.

A comparison of the analytical solution of Eq. 20, a numerical integration of Eq. 16, and the analytical solution of the nonlinear wave equation (i.e., with $D = 0$) are shown in Fig. 2 for a set of characteristic parameters. It is seen that Eq. 20 agrees with the full numerical solution both for strong and weak sample loading, whereas the nondiffusive nonlinear wave solution is able to reproduce the correct behavior only in the strongly nonlinear case. This is to be expected, since the limit $D \to 0$ implies $P = u_{\mathrm{ep}}\Gamma/D \to \infty$, which is the limit of strong nonlinearity.

The EMD theory presented above has been extended [45] to a four-component system consisting of analyte species, a weak acid HX (neutral), and its dissociation products H^+ and X^-. All species are assumed to have the same diffusivity (and therefore, mobility), but they have different valences so that the electrophoretic mobilities are different. The idealization that is invoked in this model is

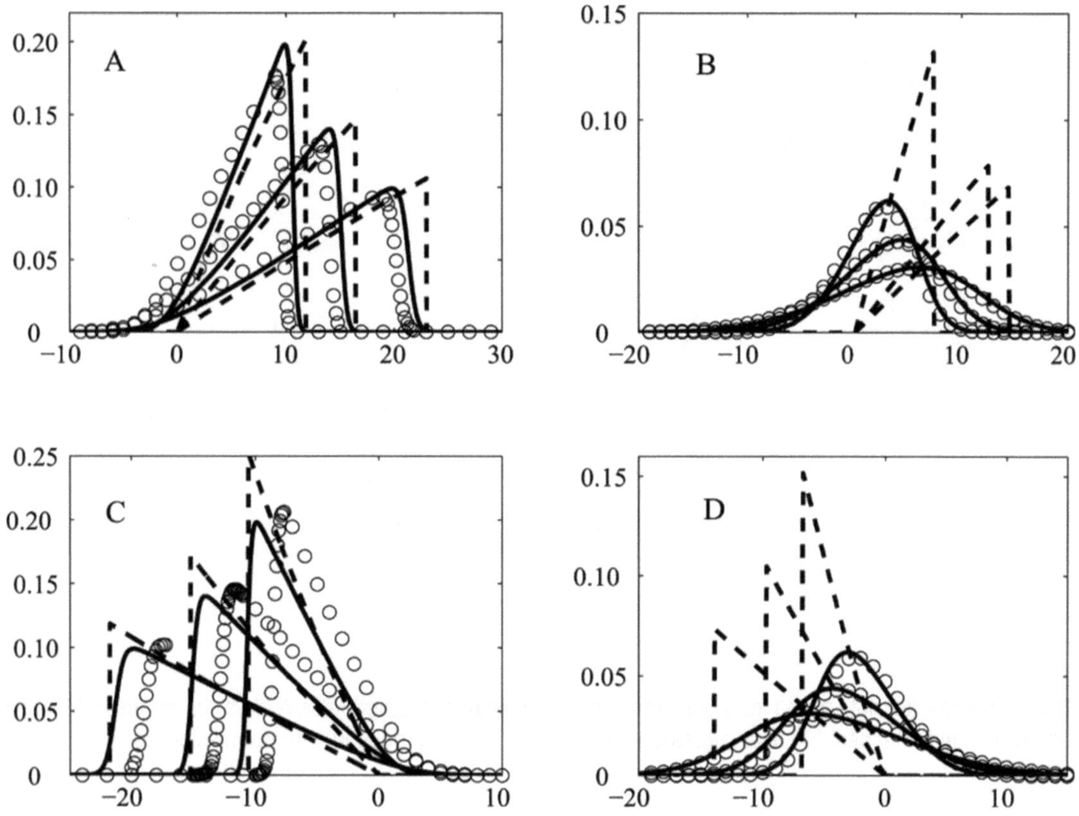

Fig. 2 Comparison of the normalized concentration ϕ (y-axis) as a function of $(x - u_{ep}t)/\sigma_0$ (x-axis) obtained from Eq. 20 (solid line) compared to the numerical solution of Eq. 16 (symbols) at fixed instants of time $u_{ep}t/\sigma_0 = 50, 100, 200$. The dashed lines are solutions of the nonlinear wave equation, Eq. 18, with $D = 0$. The parameters are $P = 62.7$: $\alpha = 0.5$ (panel **a**), $\alpha = -0.5$ (panel **c**), and $P = 5$: $\alpha = 0.5$ (panel **b**), $\alpha = -0.5$ (panel **d**). The initial peak shape was a Gaussian of standard deviation σ_0 centered on $x = 0$. Reproduced with permission from [40]

that of a "perfect buffer," that is, the hydrogen ion concentration (pH) remains exactly constant throughout the solution. It may then be shown that Eq. 16 is once again recovered for the dimensionless analyte concentration ϕ, provided that α in Eq. 17 is replaced by a different expression. It may be shown that Eq. 20 implies a linear dependence of N^{-1} with the concentration c_* of sample in the analyte plug. This relation is verified in Fig. 3 using experimental data presented by Lukacs and Jorgenson [46]. The best fit parameters for the linear regression may be shown to be consistent with known or estimated values of physical parameters.

3.2.4 Flow-Induced Dispersion

In CE, electrophoretic migration of analytes is usually accompanied by EOF, as the capillary walls have a significant zeta potential. The EOF is sometimes suppressed by chemical coatings (e.g., if the sample contains cationic proteins that have strong nonspecific

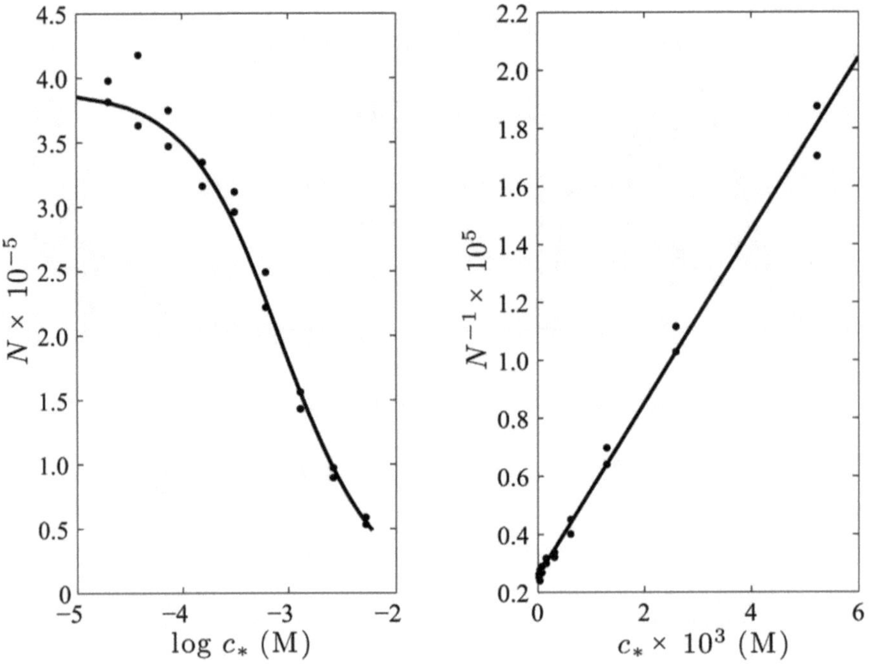

Fig. 3 Symbols are data from Fig. 3 of Lukacs and Jorgenson [46] replotted as N^{-1} vs. c_* on the right panel. The solid line is the best fit linear regression $N^{-1} = a + bc_*$. Reproduced with permission from [45]

adsorption to the walls), but in other applications, EOF is actually desirable. A strong EOF reduces analysis times and ensures that cationic and anionic components elute at the same end of the capillary. If the capillary is wide in comparison with the Debye length (typically ~ 1 nm) and material properties are uniform in the axial direction, the EOF is uniform over the cross section, and its strength is given by Eq. 4. Such a uniform flow in the capillary has no effect on band broadening; in fact, it reduces the dispersion observed at the detector by shortening the time between injection and detection. There are however situations when the EOF is disturbed and it is no longer uniform. This results in anomalous dispersion that we now discuss.

1. *Taylor-Aris dispersion*: The basic principle underlying dispersion due to flow perturbations relates to a mechanism first understood by Taylor [47] and developed further by Aris [48]. These ideas, known as "Taylor-Aris dispersion," were further extended by Brenner and Edwards [49] into a general formalism for the analysis of dispersion in periodic and random media. We summarize here a version of the theory in a form most suited to our application. Consider an infinitely long capillary of arbitrary cross-sectional shape. The capillary contains a spatially varying steady flow field with axial velocity u. All axial variations are "slow," in the following sense: if L is a

characteristic axial distance over which geometric and flow properties vary, then, $L \gg w$, where w is a characteristic capillary width. The characteristic cross-channel diffusion time is then $t_{\mathrm{d}} \sim w^2/D$, where D is the molecular diffusivity of the solute. After a time $t \gg t_{\mathrm{d}}$ has passed since sample injection, the cross-sectionally averaged (*see* **Note 5**) concentration field $\bar{c}(x, t)$ may be described by [50, 51] the following advection-diffusion equation:

$$A\frac{\partial \bar{c}}{\partial t} + \frac{\partial}{\partial x}(A\bar{u}\bar{c}) = \frac{\partial}{\partial x}\left(AD_{\mathrm{eff}}\frac{\partial \bar{c}}{\partial x}\right) \tag{22}$$

where D_{eff} is the effective diffusivity defined as

$$D_{\mathrm{eff}} = D + \gamma\bar{u}^2/D \tag{23}$$

Here $\bar{u}$ is the cross-sectionally averaged velocity (volume flux per unit cross-sectional area), and A is the cross-sectional area that can change slowly along the channel. The quantity γ is defined as

$$\gamma = -\overline{gu}/\bar{u} \tag{24}$$

where g is a function defined over the two-dimensional domain Ω representing the cross-sectional shape by the boundary value problem

$$\nabla^2 g = u\bar{u} - 1 \tag{25}$$

with boundary condition

$$\vec{\nabla} g \cdot \hat{n} = 0 \tag{26}$$

Here $\hat{n}$ is the unit normal at the bounding line of the planar 2D domain Ω. In order to define g uniquely, we also assume an auxiliary condition

$$\bar{g} = 0 \tag{27}$$

The physical basis of the Taylor-Aris theory is that for narrow channels, $t_{\mathrm{d}} = w^2/D$ is quite short, so that the concentration field homogenizes over the cross section relatively quickly compared to the transit time along the capillary. As an example, if we take $w \sim 100$ μm and $D \sim 10^{-5}$ cm^2/s, then $t_{\mathrm{d}} \sim 10$ s which is usually short compared to typical CE run times $\sim 10 - 30$ min. The departure from uniformity of the concentration field over the cross section, though small, is nevertheless finite and may be calculated from the formula

$$c = \bar{c} + \frac{\bar{u}g}{D}\frac{\partial \bar{c}}{\partial x} \tag{28}$$

Thus, the concentration c is "slaved" to the averaged concentration $\bar{c}$, which evolves according to the one-dimensional equation, Eq. 22.

2. *Debye layer overlap*: It is clear that in order to apply Eq. 22, one needs to know the velocity distribution u. In the case of a pressure-driven flow, the velocity profile is parabolic, and γ can be evaluated exactly. The corresponding expressions for D_{eff} are well known [47]. In flow through a channel that is much wider than the Debye length and uniform in the axial direction, the flow is uniform over the cross section ("plug" flow) as discussed earlier. In this case, the right-hand side of Eq. 25 vanishes, and we have $g = 0$ and therefore $D_{\text{eff}} = D$. There is no anomalous dispersion in plug flow. If the channel width is not very large compared to the Debye length, then the flow is no longer a plug flow, and Taylor-Aris dispersion is present. Finite Debye length effects could be important in several situations: (1) if very narrow capillaries (e.g., $w \sim 100$ nm) are used, this has an advantage as Joule heating is minimized allowing higher voltages and thereby greater resolution, (2) a low conductivity buffer is used where the Debye length may be relatively large, and (3) the capillary is filled with beads and the interstitial spaces between the beads are comparable to the Debye length. The axial flow speed for uniform capillaries at fixed wall zeta potential (ζ) may be written [25] in general as

$$u = \frac{\epsilon E}{\eta}\left(\phi^{\text{EDL}} - \zeta\right) \tag{29}$$

where ϕ^{EDL} is the equilibrium potential distribution in the Debye layer. An explicit analytical form for ϕ^{EDL} can be provided in some special geometries within the Debye-Hückel limit. For example, for a planar 2D channel of width w

$$\phi^{\text{EDL}} = \zeta \frac{\cosh(\kappa y)}{\cosh(\kappa w/2)} \tag{30}$$

where y is the coordinate in the cross-channel direction with origin on the centerline. For a circular capillary of radius R

$$\phi^{\text{EDL}} = \zeta \frac{I_0(\kappa r)}{I_0(\kappa R)} \tag{31}$$

where r is the radial distance from the axis and I_0 is the modified Bessel function of the first kind of order zero. If Eq. 29 is substituted in Eq. 24 and the integral evaluated, D_{eff} can be calculated. In some cases an analytical evaluation of the integral is possible [50, 52, 53]. For a symmetric binary electrolyte, the equivalent of Eq. 30 could be written down even for large zeta potentials where Debye-Hückel theory is no longer applicable [1].

3. *Lubrication theory*: We have seen that Eq. 23 provides a way for calculating the effective diffusivity and therefore band broadening, provided that the perturbed flow velocity u is known. A

common situation in which the flow deviates from the uniform profile is when channel properties, such as cross section and zeta potential, vary in the axial direction. A general expression for the axial flow velocity u may be given for straight channels of arbitrary geometric shape, provided that the channel width is much larger than the Debye length and the channel properties are slowly varying in the axial direction [54]:

$$u = -\frac{u_p}{\eta}\frac{dp}{dx} + \frac{\epsilon I}{\sigma \eta}\frac{\psi}{A} \tag{32}$$

$$Q = -\frac{A\bar{u}_p}{\eta}\frac{dp}{dx} + \frac{\epsilon I \bar{\psi}}{\sigma \eta} \tag{33}$$

$$E = \frac{I}{\sigma A} \tag{34}$$

Here I and Q are constants representing, respectively, the current and volume flow rate through any cross section, and σ is the (constant) buffer conductivity. We denote by Ω_x the 2D domain defined by the intersection of the channel interior with the plane orthogonal to the x-axis at location "x." The bounding curve of this domain will be indicated by $\partial\Omega_x$. The area of this domain is $A(x)$, the fluid pressure and axial electric field at this cross section are $p(x)$ and $E(x)$, and $u_p(x, y, z)$ and $\psi_p(x, y, z)$ are functions defined on Ω_x through the following boundary value problems:

$$\nabla^2 u_p = -1; \qquad u_p\big|_{\partial\Omega_x} = 0 \tag{35}$$

and

$$\nabla^2 \psi_p = 0; \qquad \psi_p\Big|_{\partial\Omega_x} = -\zeta \tag{36}$$

where $\zeta(x, y, z)$ is the zeta potential distribution on $\partial\Omega_x$ and $\nabla^2 = \partial_{yy} + \partial_{zz}$ is the two-dimensional Laplacian operator. Physically, u_p and ψ_p represent, respectively, the axial velocity distribution in a uniform channel of cross-sectional Ω_x in response to a unit pressure gradient and the potential distribution in a domain Ω_x when the potential at the boundary is specified (Dirichlet problem). Equations 32–34 determine simultaneously the fields u, p, and E, with Q and I playing the role of integration constants. If, instead of Q and I, the applied drop in pressure (Δp) and voltage (ΔV) between inlet and outlet are known, Q and I may be determined by means of the following equations:

$$I = \frac{\sigma}{L}\frac{\Delta V}{\langle A^{-1}\rangle} \tag{37}$$

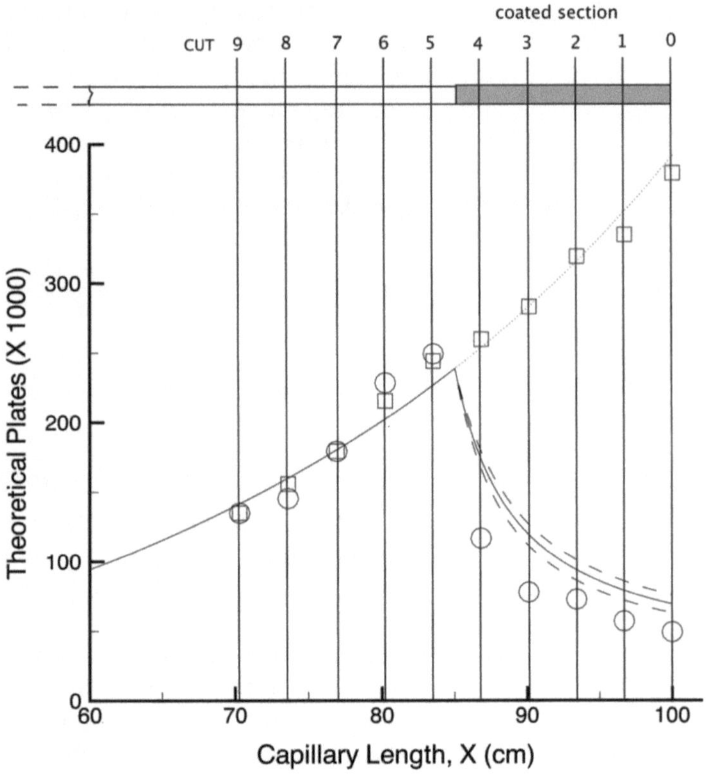

Fig. 4 Dispersion caused by a step change in the ζ-potential. Experimental data replotted from Towns and Regnier [55]; lines represent theoretical calculation (*see* text). Reproduced with permission from [25]

$$Q = \frac{\Delta p}{\eta L \left\langle \bar{u}_{\mathrm{p}}^{-1} A^{-1} \right\rangle} + \frac{\epsilon \Delta V}{\eta L} \frac{\left\langle A^{-1} \bar{u}_{\mathrm{p}}^{-1} \overline{\Psi}_{\mathrm{p}} \right\rangle}{\left\langle A^{-1} \right\rangle \left\langle \bar{u}_{\mathrm{p}}^{-1} A^{-1} \right\rangle} \tag{38}$$

Once the axial velocity u is known, the effective diffusivity may be evaluated from Eq. 23. To illustrate this formalism, we calculate the number of theoretical plates N for a model experiment [55] illustrated in Fig. 4. In the experiment, the first 15 cm of a 100 cm-long CE microcapillary was coated with a polymeric material, thereby creating a step change in the zeta potential. Such a nonuniform capillary would have an EOF that deviates from uniformity and therefore show anomalous dispersion due to the Taylor-Aris effect. A neutral marker which does not adsorb to the wall and has no electrophoretic velocity is introduced at the inlet, and N is calculated from the detector output. The experiment is then repeated after excising a 3 cm length of capillary from the inlet section (indicated by "cut 1" in Fig. 4) while simultaneously adjusting the potential to keep the electric field strength at the same level. The process is continued, with the removal of 3 cm of the capillary length at each step. Clearly, after cut 5, one is left with a uniform capillary that should not show any anomalous dispersion. The experimental data (circles) is shown in Fig. 4 together with the results of an identical control experiment (squares) where the capillary is not coated. As expected, one sees a monotonic increase in N with capillary length (*see* **Note 6**) for a uniform capillary but a sharp drop in N for a capillary with a nonuniform zeta potential distribution. The solid line shows the theoretically predicted [56] result based on Eq. 23 and the lubrication theory (*see* **Note 7**) model for the axial velocity field (Eqs. 32–38). The dashed lines show the effect of uncertainties due to the spread in the molecular diffusivity of the solute that was obtained from the literature. The comparison shown in Fig. 4 is based on measured values; no fitting parameters were used.

4. *Wall adsorption kinetics*: In the controlled experiment on a model system described above, a variable zeta potential was specified a priori. A more common situation arises when the zeta potential changes dynamically in response to the adsorption and desorption of analyte species to the capillary walls. This is the case, for example, in CEC where differential adsorption to walls is the basis of separation. Another situation involves CE of cationic components such as proteins that show nonspecific adsorption to capillary walls. The CE channel must always be coated in such cases, but coatings are not perfect, they may greatly reduce but not eliminate adsorption. Such situations can be handled by means of asymptotic theories based on the premise of slow axial variations that combine chromatographic principles with ideas from lubrication theory and Taylor-Aris dispersion. In the case of a single analyte species of concentration $c(x, y, z, t)$ in solution and absorbed surface concentration $s(x, y, z, t)$ on the wall, Eqs. 22–28 are modified as follows [51, 57]:

$$\frac{\partial \bar{c}}{\partial t} + \bar{u}\frac{\partial \bar{c}}{\partial x} = \frac{\partial}{\partial x}\left(D_{\text{eff}}\frac{\partial \bar{c}}{\partial x}\right) - \bar{p}\frac{\partial \bar{s}}{\partial t} - \frac{\bar{p}\bar{u}}{D}\frac{\partial}{\partial x}\left(\bar{g}_{\text{w}}\frac{\partial \bar{s}}{\partial t}\right) \quad (39)$$

$$D_{\text{eff}} = D + \gamma \bar{u}^2/D \quad (40)$$

$$\frac{\partial s}{\partial t} = f(c_{\text{w}}, s) \quad (41)$$

$$c = \bar{c} + \frac{\bar{u}g}{D}\frac{\partial \bar{c}}{\partial x} - \frac{\bar{p}G}{D}\frac{\partial \bar{s}}{\partial t} \quad (42)$$

where G is defined by the boundary value problem on the 2D domain Ω defining the capillary cross-sectional shape

$$\nabla^2 G = 1; \qquad \vec{\nabla} G \cdot \hat{n} = 1\bar{p} \quad (43)$$

In Eq. 39, $\bar{p}$ is the ratio of the perimeter of the channel to its cross-sectional area (a purely geometric quantity of dimension inverse length), and the suffix w on any variable indicates its value at the wall. Here we have assumed that the channel cross section does not change along the capillary to keep matters simple (*see* **Note 8**). Following our usual notation, a bar over a scalar field indicates its average over the cross section as defined in **Note 5**. In the case of s, since this variable is defined only on the wall of the channel, the overbar indicates its average over the line $\partial\Omega_x$. The function $f(c_w, s)$ indicates the law of adsorption/desorption at the wall of which a commonly used model is that of Langmuir:

$$f(c_{\mathrm{w}}, s) = k_{\mathrm{a}} c_{\mathrm{w}}(s_{\mathrm{m}} - s) - k_{\mathrm{d}} s \tag{44}$$

where k_a and k_d are adsorption and desorption rates and $s = s_m$ is the saturation concentration for adsorption. Equations 39–42 are coupled to the fluid flow equations (Eqs. 32–38) through the dependence of the zeta potential on the adsorbed solute concentration. Any such functional dependence may be used, but the simplest such model would be a linear dependence

$$\zeta = \zeta_0 - As \tag{45}$$

where A is a constant coefficient. Figure 5 shows the result of a numerical integration of the one-dimensional coupled system

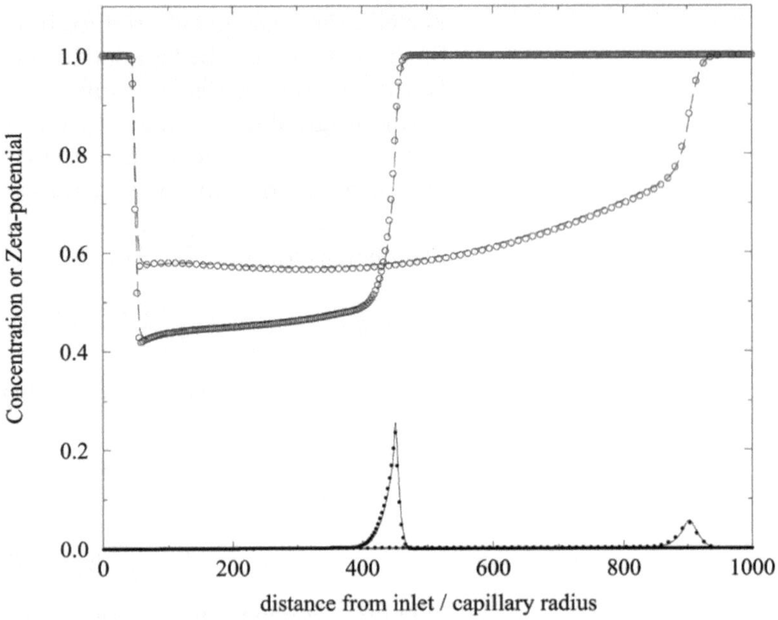

Fig. 5 Comparison of asymptotic theory (symbols) with numerical simulation (lines) of the cross-sectionally averaged analyte concentration (lower curves) and ζ-potential (upper curves) at two time instants. Reproduced with permission from [58]

described above for some typical values of the parameters. Also shown (symbols) are the result of a direct numerical integration of the primitive equations requiring far greater computational effort. The results are essentially indistinguishable from the reduced one-dimensional model presented here. The distortion of peak shape and loss in amplitude of the signal are clearly seen. In this example, desorption is weak so that once the zeta potential is reduced by the adsorbed analyte, it recovers very slowly following passage of the analyte band. Notice that the "tailing" of the analyte peak seen here is quite different from the characteristic triangular shape seen in Fig. 2 in the case of electromigration dispersion.

3.3 Concluding Remarks

A set of methods for calculating the amount of band broadening in FSCZE is presented here. These ideas could be generalized to other CE modes. In some instances, explicit formulas are given for calculating the variance due to anomalous dispersion. When this is not possible, one-dimensional partial differential equations are presented for calculating band broadening and peak shapes. These equations can be easily implemented on platforms such as MATLAB or even on web-based applications such as Java and provide a much more efficient alternative to the solution of the full three-dimensional primitive equations for transport. Thus, they may serve as a helpful tool in designing better microfluidic systems or improved methods of separation on existing systems.

4 Notes

1. Should not be confused with the "mobility" which is the velocity per unit applied *force*.
2. Here we depart in our description from a general buffer to one of a specific composition. A three-component system is a "minimal model" for the description of CZE that must have a sample species and at least one co-ion and counterion. Since the diffusivities of all of these ions are presumed equal, by the stokes-Einstein relation, they must all have the same mobility. However, the charges on the three ionic species may be different, so that their electrophoretic mobilities are in general unequal. We also assume here that all ionic species are fully dissociated.
3. This restriction can be removed [59].
4. Equation 20 may be used in place of the empirical "Haarhoff-van der Linde (HVL) function" sometimes used to fit experimental data on asymmetric peaks [60, 61].
5. Throughout this chapter, an overbar indicates an average over the area (A) of the capillary cross section

$$\bar{f} = \frac{1}{A} \int f(x, y, z) dy\, dz$$

and angular brackets indicate an axial average over the length (L) of the capillary

$$\langle f \rangle = \frac{1}{L} \int f(x, y, z)\, dx$$

6. There is no contradiction here with Eq. 8 which shows that N is independent of L at a fixed *voltage*. Here the electric field is held fixed; hence, N would be expected to increase linearly with L. The slight departure from linearity is due to a nonzero initial variance in the injected plug.
7. The slowly varying assumption of lubrication theory should not apply to a discontinuous zeta potential; however, it may be shown that corrections to N due to departure from lubrication theory are small [56].
8. This assumption can be relaxed [51].

References

1. Russel WB, Saville DA, Schowalter WR (1989) Colloidal dispersions. Cambridge University Press, Cambridge
2. Saville DA (1977) Electrokinetic effects with small particles. Annu Rev Fluid Mech 9:321–337
3. Anderson JL (1989) Colloid transport by interfacial forces. Annu Rev Fluid Mech 21:61–99
4. Morrison FA Jr (1970) Electrophoresis of a particle of arbitrary shape. J Colloid Interface Sci 34(210–214):45–54
5. Ghosal S (2007) Effect of salt concentration on the electrophoretic speed of a polyelectrolyte through a nanopore. Phys Rev Lett 98 (23):238104
6. Vreeland WN, Desruisseaux C, Karger AE, Drouin G, Slater GW, Barron AE (2001) Molar mass profiling of synthetic polymers by free-solution capillary electrophoresis of DNA-polymer conjugates. Anal Chem 73:1795–1803
7. Ren H, Karger AE, Oaks F, Menchen S, Slater GW, Drouin G (1999) Separating DNA sequencing fragments without a sieving matrix. Electrophoresis 20:2501–2509
8. Sherwood JD, Stone HA (1995) Electrophoresis of a thin charged disk. Phys Fluids 7 (4):697–705
9. Weinberger R (2000) Practical capillary electrophoresis. Academic Press, San Diego
10. Landers JP (ed) (1996) Introduction to capillary electrophoresis. CRC Press, Boca Raton, FL
11. Camilleri P (ed) (1998) Capillary electrophoresis, theory & practice. CRC Press, Boca Raton, FL
12. Everaerts FM, Verheggen ThPEM (1987) Capillary isotachophoresis, chapter 14. American Chemical Society, Washington, DC
13. Mikkers FEP, Everaerts FM, Peek JAF (1979) Isotachophoresis: the concepts of resolution, load capacity and separation efficiency. J Chromatogr 168:293–315
14. Probstein R (1994) Physicochemical hydrodynamics. John Wiley and Sons, Inc., New York
15. Terabe S, Otsuka K, Ando T (1985) Electrokinetic chromatography with micellar solution and open-tubular capillary. Anal Chem 57:834
16. Terabe S, Otsuka K, Ichikawa K, Tsuchiya A, Ando T (1984) Electrokinetic separations with micellar solutions and open-tubular capillaries. Anal Chem 56:111
17. Giddings JC (1969) Generation of variance, "theoretical plates," resolution, and peak capacity in electrophoresis and sedimentation. J Sep Sci 4(3):181–189
18. Jorgenson JW (1987) Capillary zone electrophoresis. In: Jorgenson JW, Phillips M (eds) New directions in electrophoretic methods. American Chemical Society, Washington, DC

19. Knox JH (1988) Thermal effects and band spreading in capillary electro-separation. Chromatographia 26:329–337
20. Burgi DS, Salomon K, Chien RL (1991) Methods for calculating the internal temperature of capillary columns during capillary electrophoresis. J Liq Chromatogr 14(5):847–867
21. Davis JM (1990) Influence of thermal variation of diffusion coefficient on non-equilibrium plate height in capillary zone electrophoresis. J Chromatogr 517:521–547
22. Grushka E, McCormick RM, Kirkland JJ (1989) Effect of temperature gradients on the efficiency of capillary zone electrophoresis separations. Anal Chem 61:241–246
23. Andreev VP, Lisin EE (1992) Investigation of the electroosmotic flow effect on the efficiency of capillary electrophoresis. Electrophoresis 13:832–837
24. Jones AE, Grushka E (1989) Nature of temperature gradients in capillary zone electrophoresis. J Chromatogr 466:219–225
25. Ghosal S (2006) Electrokinetic flow and dispersion in capillary electrophoresis. Annu Rev Fluid Mech 38(38):309–338
26. Zubritsky E (2000) Taming turns in microchannels. Anal Chem 72(1):687A–690A
27. Griffiths SK, Nilson RH (2002) Design and analysis of folded channels for chip-based separations. Anal Chem 74(13):2960–2967
28. Culbertson CT, Jacobson SC, Ramsey JM (2000) Microchip devices for high-efficiency separations. Anal Chem 72:5814–5819
29. Gottschlich N, Jacobson SC, Culbertson CT, Ramsey JM (2001) Two-dimensional electrochromatography/capillary electrophoresis on a microchip. Anal Chem 73(11):2669–2674
30. Paegel BM, Hutt LD, Simpson PC, Mathies RA (2000) Turn geometry for minimizing band broadening in microfabricated capillary electrophoresis channels. Anal Chem 72 (14):3030–3037
31. Mohammadi B, Molho JI, Santiago JG (2000) Design of minimum dispersion fluidic channels in a cad-free framework. In: Moin P, Reynolds WC, Mansour NN (eds) Studying turbulence using numerical simulation databases – VIII, proceedings of the 2000 summer program, Nov 2000. Center for Turbulence Research, Stanford University, Stanford, CA, pp 49–62
32. Molho JI, Herr AE, Mosier BP, Santiago JG, Kenny TW (2001) Optimization of turn geometries for microchip electrophoresis. Anal Chem 73(6):1350–1360
33. Griffiths SK, Nilson RH (2001) Low-dispersion turns and junctions for microchannel systems. Anal Chem 73(2):272–278
34. Dutta D, Leighton DT (2002) A low dispersion geometry for microchip separation devices. Anal Chem 74:1007–1016
35. Johnson TJ, Ross D, Gaitan M, Locascio LE (2001) Laser modification of preformed polymer microchannels: application to reduce band broadening around turns subject to electrokinetic flow. Anal Chem 73(15):3656–3661
36. Miyahara ZY, Miura J, Watanabe Y, Miyagi H, Sato K (1990) Design of an open-tubular column liquid chromatograph using silicon chip technology. Sens Actuators B1:249–255
37. Fiechtner GJ, Cummings EB (2003) Faceted design of channels for low-dispersion electrokinetic flows in microfluidic systems. Anal Chem 75(18):4747–4755
38. Mikkers FEP, Everaerts FM, ThPEM V (1979) Concentration distributions in free zone electrophoresis. J Chromatogr 169:1–10
39. Boušková E, Presutti C, Gebauer P, Fanali S, Beckers JL, Boček P (2004) Experimental assessment of electromigration properties of background electrolytes in capillary zone electrophoresis. Electrophoresis 25:355–359
40. Ghosal S, Chen Z (2010) Nonlinear waves in capillary electrophoresis. Bull Math Biol 72 (8):2047–2066
41. Mikkers FEP (1999) Concentration distributions in capillary electrophoresis: Cze in a spreadsheet. Anal Chem 71:522–533
42. Babskii VG, Zhukov MY, Yudovich VI (1989) Mathematical theory of electrophoresis. Consultants Bureau, Plenum Publishing, New York
43. Găs B (2009) Theory of electrophoresis: fate of one equation. Electrophoresis 30:S7–S15
44. Thormann W, Caslavska J, Breadmore MC, Mosher RA (2009) Dynamic computer simulations of electrophoresis: three decades of active research. Electrophoresis 30:S16–S26
45. Chen Z, Ghosal S (2011) Electromigration dispersion in capillary electrophoresis. Bull Math Biol 74(2):346–355
46. Lukacs KD, Jorgenson JW (1985) Capillary zone electrophoresis: effect of physical parameters on separation efficiency and quantitation. J High Resolut Chromatogr 8 (8):407–411
47. Taylor GI (1953) Dispersion of soluble matter in solvent flowing slowly through a tube. Proc R Soc Lond A 219:186–203
48. Aris R (1956) On the dispersion of a solute in a fluid flowing through a tube. Proc R Soc Lond A 235:67–77
49. Brenner H, Edwards DA (1993) Macrotransport processes. Butterworth-Heinemann, Stoneham, MA

50. Datta R, Kotamarthi VR (1990) Electrokinetic dispersion in capillary electrophoresis. AICHE J 36:916–926
51. Datta S, Ghosal S (2008) Dispersion due to wall interactions in microfluidic separation systems. Phys Fluids 20:012103
52. Griffiths SK, Nilson RH (1999) Hydrodynamic dispersion of a neutral non-reacting solute in electroosmotic flow. Anal Chem 71 (24):5522–5529
53. McEldoon JP, Datta R (1992) Analytical solution for dispersion in capillary liquid chromatography with electroosmotic flow. Anal Chem 64:227–230
54. Ghosal S (2002) Lubrication theory for electroosmotic flow in a microfluidic channel of slowly varying cross-section and wall charge. J Fluid Mech 459:103–128
55. Towns JK, Regnier FE (1992) Impact of polycation adsorption on efficiency and electroosmotically driven transport in capillary electrophoresis. Anal Chem 64:2473–2478
56. Ghosal S (2002) Band broadening in a microcapillary with a stepwise change in the ζ-potential. Anal Chem 74(16):4198–4203
57. Datta S, Ghosal S (2009) Characterizing dispersion in microfluidic channels. Lab Chip 9:2537–2550
58. Ghosal S (2004) Fluid mechanics of electroosmotic flow and its effect on band broadening in capillary electrophoresis. Electrophoresis 25:214–228
59. Ghosal S, Chen Z (2012) Electromigration dispersion in a capillary in the presence of electroosmotic flow. J Fluid Mech 697:436–454
60. Erny GL, Bergstrŏm ET, Goodall DM (2001) Predicting peak shape in capillary electrophoresis: a generic approach to parametrizing peaks using the Haarhoff-van der Linde (hvl) function. Anal Chem 73:4862–4872
61. Erny GL, Bergstrŏm ET, Goodall DM (2002) Electromigration dispersion in capillary zone electrophoresis experimental validation of use of the Haarhoff-van der Linde function. J Chromatogr A 959:229–239

Chapter 12

Estimating Stream Broadening in Free-Flow Electrophoretic Systems Based on the Method-of-Moments Formulation

Debashis Dutta

Abstract

The resolving power of free-flow electrophoretic assays is often limited by the broadening of analyte streams as they migrate through the separation chamber. While molecular diffusion and flow hydrodynamics inherently contribute to such dispersion, non-idealities such as Joule heating and unwanted pressure-driven cross-flows among others can also significantly modify these contributions. In this chapter, we describe a theoretical approach to estimating stream broadening in free-flow electrophoretic systems under steady-state conditions based on the method-of-moments formulation. This methodology allows the determination of the spatial moments of solute concentration, thereby yielding measures for the mean position and spatial variance of the analyte stream.

Key words Electrodynamic dispersion, Free-flow electrophoresis, Free-flow isoelectric focusing, Hydrodynamic dispersion, Method-of-moments

1 Introduction

Continuous separation of a target sample fraction from a complex mixture is desirable in several applications that either require an enrichment of the analyte molecules in the specimen for its further downstream processing or a reduction in interference from the matrix species during analyte detection. Free-flow electrophoretic (FFE) techniques offer a useful approach in this regard enabling the extraction of analytes based on their electrophoretic mobilities or isoelectric points and are applicable to a wide range of chemical and biological samples [1]. Moreover, the use of relatively gentle operating conditions in these techniques combined with their high throughput makes them particularly valuable tools for simplifying the sample feed to an electrophoretic or chromatographic system. Free-flow electrophoretic assays typically rely on the continuous pressure-driven flow of a sample stream through a separation chamber in the presence of an electric field applied perpendicular to this flow direction. In this situation, analyte

Debashis Dutta (ed.), *Microfluidic Electrophoresis: Methods and Protocols*, Methods in Molecular Biology, vol. 1906,
https://doi.org/10.1007/978-1-4939-8964-5_12,

molecules are deflected from their original path based on their electrophoretic mobilities causing them to exit the separation chamber at different lateral positions. The resolution of the separation process consequently is determined by the differences in the electrophoretic transport rates or isoelectric points of the solute species as well as the amount of band dispersion they undergo during their transit through the analysis column [2].

Stream broadening in free-flow electrophoretic systems primarily occurs due to molecular diffusion orthogonal to the flow direction as well as a variation in the transit time for the analyte molecules when traveling along the different streamlines in the system. While the first contribution to stream broadening is relatively simple to estimate, the latter one is determined by a complex interplay between the local streamline velocity and analyte diffusion across these streamlines [3]. It has been established that the latter effect is particularly prominent at moderate to large transverse electric fields and can originate for a variety of reasons. In this chapter, we present a general approach to estimating the mean position and spatial variance of analyte streams as they migrate through a free-flow electrophoresis column. This approach is based on the method-of-moments formulation and can assess the resolving power of FFE methods under ideal conditions over a wide range of time and length scales. In addition, the noted approach allows the determination of leading order effects of common non-idealities such as Joule heating and unwanted pressure-driven cross-flows among others on the resolution of FFE separations, thereby providing insights for developing strategies to minimize them.

2 Materials

This chapter discusses theoretical results on stream broadening in free-flow electrophoretic systems which are applicable to a broad range of separation conditions, e.g., buffer compositions, electrophoretic mobilities, electric fields, pressure-driven flow rates, channel dimensions, etc. However, our theoretical models require an estimate for the magnitudes of these quantities as well as their dependence on various parameters such as temperature, buffer pH, etc., in order to formulate the governing mathematical equations. In addition, we often need to assume certain properties about the separation buffer (e.g., Newtonian fluid) and analyte species (e.g., isotropic diffusion) in order to simplify our mathematical treatment of the problem which will be noted at appropriate points in this chapter. While many of these assumptions/approximations may not be strictly valid for a specific experimental condition, the results included here can be still useful for broadly assessing the separation performance of FFE assays, thereby enabling the preliminary optimization of their operating conditions.

3 Methods

3.1 Mathematical Formulation

To evaluate stream dispersion in an FFE system, we consider the flow of an analyte stream between two parallel plates (*see* **Note 1**) separated by a distance d (*see* Fig. 1) under the influence of a pressure-driven flow in the axial direction (along the z-coordinate) and an electric field (E) applied across the width of the separation chamber (along the x-coordinate). In order to simplify our mathematical analysis, we assume the locations of the parallel plates to be $y = \pm\, d/2$ yielding a pressure-driven velocity profile $u_z = \left(3\bar{U}_z/2\right)\left(1 - 4y^2/d^2\right)$ with $\bar{U}_z$ being the spatially averaged value of u_z.

The advection-diffusion equation governing the concentration of the analyte species (C) under steady state in this situation may be written as (*see* **Note 2**)

$$\frac{\partial(u_x C)}{\partial x} + u_z\frac{\partial C}{\partial z} = D\left(\frac{\partial^2 C}{\partial x^2} + \frac{\partial^2 C}{\partial z^2}\right) + \frac{\partial}{\partial y}\left(D\frac{\partial C}{\partial y}\right) \quad (1)$$

where u_x and D refer to the analyte migration velocity along the x-coordinate and analyte diffusion coefficient in the system, respectively. Notice that the quantity u_x in this formulation may vary with the x- and y-coordinates as in a free-flow isoelectric focusing assay with an unwanted pressure-driven cross-flow while the diffusion coefficient D may only be a function of y say due to a temperature variation across the channel depth arising from Joule heating effects. Upon normalizing all length scales with respect to d, i.e., $x^*, y^*, z^* = x/d, y/d, z/d$, the sample concentration by its inlet value (C_0), i.e., $C^* = C/C_0$, and the diffusion coefficient by its magnitude at a reference point, e.g., channel walls ($y^* = \pm\, 1/2$), Eq. 1 may be reduced to the dimensionless form

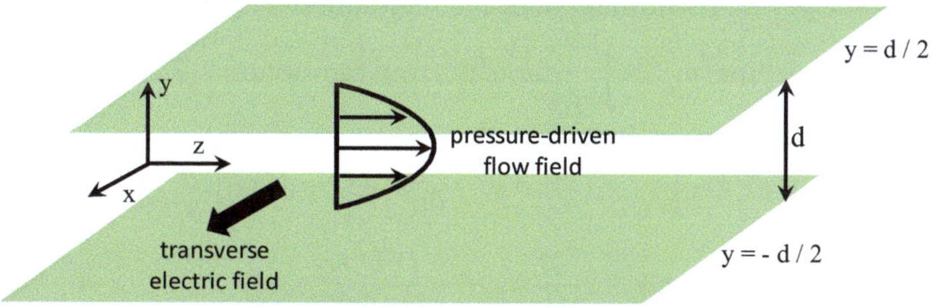

Fig. 1 Schematic of the free-flow electrophoretic separation process between two parallel plates as described in this chapter. Reproduced with permission from Ref. 3

$$Pe_x \frac{\partial (u_x^* C^*)}{\partial x^*} + \frac{3}{2} Pe_z (1 - 4y^{*2}) \frac{\partial C^*}{\partial z^*} = \frac{D}{D_0} \left(\frac{\partial^2 C^*}{\partial x^{*2}} + \frac{\partial^2 C^*}{\partial z^{*2}} \right) + \frac{\partial}{\partial y^*} \left(\frac{D}{D_0} \frac{\partial C^*}{\partial y^*} \right) \quad (2)$$

where $u_x^* = u_x / \bar{U}_x$. The quantities $Pe_x = \bar{U}_x d / D_0$ and $Pe_z = \bar{U}_z d / D_0$ here denote the Péclet numbers in the x- and z-directions, respectively, yielding measures of advective transport relative to diffusion along the width and length of the separation compartment. The analyte concentration in this system is subjected to the boundary conditions $\partial C^* / \partial y^* = 0$ at $y^* = \pm 1/2$, $C^*, \partial C^* / \partial x^* = 0$ as $x^* \to \pm \infty$ and $C^* = 1$ at $z^* = 0$, $-b/2d \leq x^* \leq b/2d$ where b denotes the width of the sample stream at the inlet location ($z^* = 0$). In addition to these constraints, the amount of analyte flowing per unit time through any $x-y$ plane is a constant M in our analysis which can be equated to the integral $(3C_0 \bar{U}_z d^2/2) \int_{-1/2}^{1/2} \int_{-\infty}^{\infty} (1 - 4y^{*2}) C^* dx^* dy^*$ to satisfy the material balance in the system. Now multiplying Eq. 2 with x^{*p} followed by integrating it along the x^*-coordinate from $-\infty$ to ∞, it is possible to show that [4]

$$\frac{3}{2} Pe_z (1 - 4y^{*2}) \frac{\partial \phi_p}{\partial z^*} - p Pe_x \int_{-\infty}^{\infty} x^{*p-1} u_x^* C^* dx^*$$
$$= p(p-1) \frac{D}{D_0} \phi_{p-2} + \frac{\partial}{\partial y^*} \left(\frac{D}{D_0} \frac{\partial \phi_p}{\partial y^*} \right) + \frac{D}{D_0} \frac{\partial^2 \phi_p}{\partial z^{*2}}$$

$$\text{Boundary conditions}: \left. \frac{\partial \phi_p}{\partial y^*} \right|_{y^* = \pm 1/2} = 0; \int_{-1/2}^{1/2} (1 - 4y^{*2}) \phi_0 \, dy^* = \frac{2M}{3C_0 \bar{U}_z d^2} = \frac{2\delta}{3} \quad (3)$$

where $\phi_p = \int_{-\infty}^{\infty} x^{*p} C^* dx^*$. Further integrating Eq. 3 along the y^*-coordinate over the region between the parallel plates and defining $m_p = \int_{-1/2}^{1/2} \phi_p \, dy^*$, one can obtain

$$\frac{3}{2} Pe_z \int_{-1/2}^{1/2} \left[(1 - 4y^{*2}) \frac{\partial \phi_p}{\partial z^*} \right] dy^* - p Pe_x \int_{-1/2}^{1/2} \int_{-\infty}^{\infty} x^{*p-1} u_x^* C^* dx^* dy^*$$
$$= p(p-1) \frac{D}{D_0} m_{p-2} + \frac{D}{D_0} \frac{\partial^2 m_p}{\partial z^{*2}} \quad (4)$$

with $m_0|_{z^*=0} = \delta$, $m_1|_{z^*=0} = 0$ and $m_2|_{z^*=0} = \delta^3/12$ (*see* **Notes 3** and **4**). Notice that the quantity m_p in this formulation represents the pth moment of C^* in any $x-y$ plane with m_1 representing the normalized x^*-position of the center of mass for the analyte stream and the quantity $m_2/m_0 - m_1^2/m_0^2$ equaling its normalized spatial variance along the x-axis.

3.2 Hydrodynamic Dispersion in Free-Flow Zone Electrophoresis

In an ideal free-flow zone electrophoresis (FFZE) assay, the transverse analyte velocity can be simply written as $u_x = \mu E$ where μ refers to the net electrokinetic mobility (algebraic sum of the electrophoretic and electroosmotic mobilities) of the analyte species under the operating conditions. In addition, the diffusion coefficient D may be assumed to be uniform at all locations due to the isothermal nature of the system, i.e., $D = D_0$. In this situation, the equations governing ϕ_p and m_p reduce to the forms

$$\frac{3}{2}Pe_z\left(1-4y^{*2}\right)\frac{\partial\phi_p}{\partial z^*} - pPe_x\phi_{p-1} = p(p-1)\phi_{p-2} + \frac{\partial^2\phi_p}{\partial y^{*2}} + \frac{\partial^2\phi_p}{\partial z^{*2}} \tag{5}$$

$$\frac{3}{2}Pe_z\int_{-1/2}^{1/2}\left[\left(1-4y^{*2}\right)\frac{\partial\phi_p}{\partial z^*}\right]dy^* - pPe_x m_{p-1} = p(p-1)m_{p-2} + \frac{d^2 m_p}{dz^{*2}} \tag{6}$$

Moreover, the integral boundary condition conserving the amount of analyte flowing per unit time through any $x - y$ plane suggests that ϕ_0, and hence m_0, cannot be a function of z^*, yielding the solutions $\phi_0 = m_0 = \delta$. Now upon substituting these values into the equation governing ϕ_1, i.e., Eq. 5 with $p = 1$, it is possible to show that

$$\phi_1 = \frac{Pe_x\delta}{Pe_z}z^* + \frac{Pe_x\delta}{4}\left(y^{*2} - 2y^{*4}\right) - \frac{7Pe_x\delta}{480} \quad \text{and} \quad m_1 = \frac{Pe_x\delta}{Pe_z}z^* \tag{7}$$

Further substituting these expressions in Eq. 5 for $p = 2$, one can obtain

$$\phi_2 = \frac{Pe_x^2\delta}{Pe_z^2}z^{*2} + \left[\frac{Pe_x^2\delta}{2Pe_z}\left(y^{*2} - 2y^{*4}\right) + \frac{2\delta}{Pe_z}\left(1 + \frac{Pe_x^2}{Pe_z^2}\right) - \frac{11Pe_x\delta}{560Pe_z}\right]z^* + g(y^*)$$

$$\text{where } g(y^*) = \frac{Pe_x^2\delta}{1612800}\left(172800y^{*8} - 188160y^{*6} + 49440y^{*4} - 240y^{*2} - 253\right) - \frac{\delta}{240}\left(1 + \frac{Pe_x^2}{Pe_z^2}\right)\left(240y^{*4} - 120y^{*2} + 7\right) + \frac{\delta^3}{12} \tag{8}$$

Integrating Eq. 8 across the separation gap between the parallel plates, we get

$$m_2 = \frac{Pe_x^2 \delta}{Pe_z^2} z^{*2} + \left[\frac{2\delta}{Pe_z} \left(1 + \frac{Pe_x^2}{Pe_z^2} \right) + \frac{Pe_x^2}{105 Pe_z} \right] z^* + \frac{\delta^3}{12} \tag{9}$$

Finally, the spatial variance of the sample stream (σ^2) in an FFZE system may be evaluated as

$$\frac{\sigma^2}{d^2} = \frac{m_2}{m_0} - \frac{m_1^2}{m_0^2} = \left[\frac{2}{Pe_z} \left(1 + \frac{Pe_x^2}{Pe_z^2} \right) + \frac{Pe_x^2}{105 Pe_z} \right] z^* + \frac{\delta^2}{12} \tag{10}$$

which shows that this quantity grows linearly with the axial position (z-coordinate) in the separation chamber. The factor multiplying z^* in this equation is a measure of the effective dispersion coefficient for the analyte zone, and the term $\delta^2/12$ represents the contribution from the injection width of the sample stream. Because the normalized axial position z^* scales linearly with the residence time (t) of the analyte molecules in the FFZE compartment, it is then possible to express $\sigma^2 = 2Kt + b^2/12$ yielding a Taylor-Aris dispersion coefficient (K) [3–5]:

$$K = D \left[1 + \left(\frac{Pe_x}{Pe_z} \right)^2 + \frac{1}{210} Pe_x^2 \right] \tag{11}$$

Equation 11 shows that the overall broadening of a sample stream during its transit through an FFZE chamber occurs as a result of molecular diffusion of the analyte species orthogonal to their flow direction as well as hydrodynamic dispersion stemming from a variation in the pressure-driven flow velocity across the gap between the parallel plates. The first contribution to this broadening process is captured by the $D[1 + (Pe_x/Pe_z)^2]$ term in which the $D(Pe_x/Pe_z)^2$ component (orientation factor) arises due to the fact that σ is measured along the x-axis while the sample stream makes an angle $\theta = \cot^{-1}(Pe_x/Pe_z)$ with respect to this line. The $D(Pe_x^2/210)$ term in Eq. 11 quantitates the hydrodynamic dispersion component and interestingly has the same numerical coefficient, i.e., $1/210$, as expected during the advection of analyte bands by an axial pressure gradient. However, it must be noted that while the Péclet number multiplying this coefficient is evaluated based on the lateral electrophoretic velocity of the analyte species in an FFZE assay, this dimensionless group in the other case depends on the longitudinal pressure-driven flow speed in the system. It is important to point out that the expression derived for quantitating the hydrodynamic dispersion component in the current work is identical to that empirically proposed by Fonslow and Bowser in a recent publication based on their experimental observations [6]. In this situation, the mathematical analysis presented here offers a rigorous justification for the agreement between Fonslow and Bowser's proposed theory and experimental results.

Although the normalized expression in Eq. 11 describes stream dispersion in FFZE separations in terms of the smallest set of variables, it does not highlight the dependence of σ^2 on the different operating parameters very well. To bring out these relationships, the stream variance along the x-coordinate at the channel exit may be rewritten in its dimensional form as

$$\sigma^2 = \underbrace{\frac{2D_0L}{\bar{U}_z}}_{\text{diffusive broadening}} + \underbrace{\frac{2\mu^2E^2D_0L}{\bar{U}}{}^{3}_{z}}_{\text{orientation factor}} + \underbrace{\frac{1}{105}\frac{\mu^2E^2d^2L}{\bar{U}_z}D_0}_{\text{hydrodynamic broadening}} + \underbrace{\frac{b^2}{12}}_{\text{injection variance}} \tag{12}$$

where L denotes the length of the FFZE channel. In order to emphasize the relative importance of the various contributions to stream broadening in Eq. 12, we have plotted them individually in Fig. 2a as a function of the pressure-driven velocity, $\bar{U}_z$. The values for the other operating conditions in this figure were chosen to be $E = 500$ V/cm, $d = 20$ μm, $D_0 = 8 \times 10^{-6}$ cm^2/s, $\mu = 5 \times 10^{-4}$ cm^2/Vs, $L = 2.5$ cm, and $b = 100$ μm that are commonly employed in microfluidic FFZE fractionations. As may be seen from the figure, the dominant contribution to σ^2 under the chosen conditions comes from hydrodynamic dispersion of the analyte stream for values of $\bar{U}_z$ in the range of 0.05–5 cm/s. What is also interesting to note is that this component diminishes with an increase in the value of $\bar{U}_z$ following a reverse trend to that observed in pressure-driven band separations. This result arises from the fact

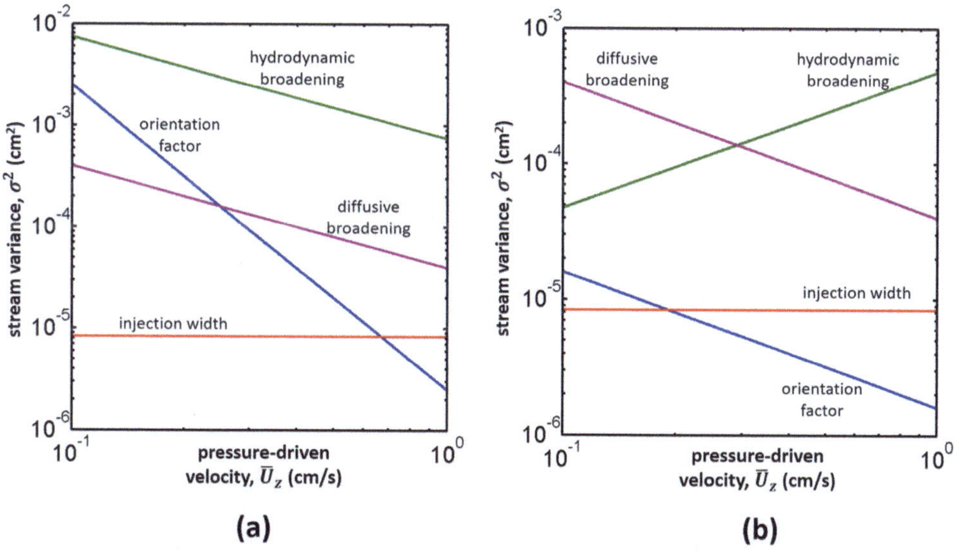

Fig. 2 Contributions to stream variance by the various factors in a FFZE system (**a**) for a constant lateral electric field ($E = 500$ V/cm) and (**b**) for a constant lateral distance migrated by the analyte molecules before exiting the separation chamber ($S = 5$ mm). The values for the other operating parameters in both these figures were chosen to be $d = 20$ μm, $D = 8 \times 10^{-6}$ cm^2/s, $\mu = 5 \times 10^{-4}$ cm^2/Vs, $L = 2.5$ cm, and $b = 100$ μm. Reproduced with permission from Ref. 3

that the residence time of the analyte molecules in the separation channel is reduced with an increase in the pressure-driven velocity for a fixed value of E. Alternatively, if one compares the different contributions to σ^2 for a fixed migration distance of the analyte zone in the lateral direction ($S = \mu EL/\bar{U}_z$), Eq. 12 translates to

$$\begin{aligned}\sigma^2 &= \frac{2D_0L}{\bar{U}_z} + \frac{2S^2D_0}{\bar{U}_zL} + \frac{1}{105}\frac{S^2d^2\bar{U}_z}{D_0L} + \frac{b^2}{12} \\ &= \underbrace{\frac{b^2}{12}}_{\text{A}} + \underbrace{\left(2D_0L + \frac{2S^2D_0}{L}\right)}_{\text{B}}\frac{1}{\bar{U}_z} + \underbrace{\frac{1}{105}\frac{S^2d^2}{D_0L}}_{\text{C}}\bar{U}_z\end{aligned} \tag{13}$$

yielding an expression similar to the Van Deemter equation [7] commonly used for describing sample dispersion in band separations. In Fig. 2b, we have compared the different additive terms in Eq. 13 for the same set of operating parameters used in Fig. 2a except for replacing the condition $E = 500$ V/cm with $S = 5$ mm. In this comparison, the hydrodynamic dispersion component is seen to dominate only for $\bar{U} \gg 3$ mm/s with the diffusive broadening contribution dictating σ^2 in the opposite limit. The A, B, and C terms defined in Eq. 13 were further evaluated in our work by fitting experimental measurements of the stream variance reported by Fonslow and Bowser (data taken from Fig. 6a in Ref. 6) to a curve of the form $\sigma^2 = A + B/\bar{U}_z + C\bar{U}_z$ (*see* Fig. 3a). In Fig. 3b,

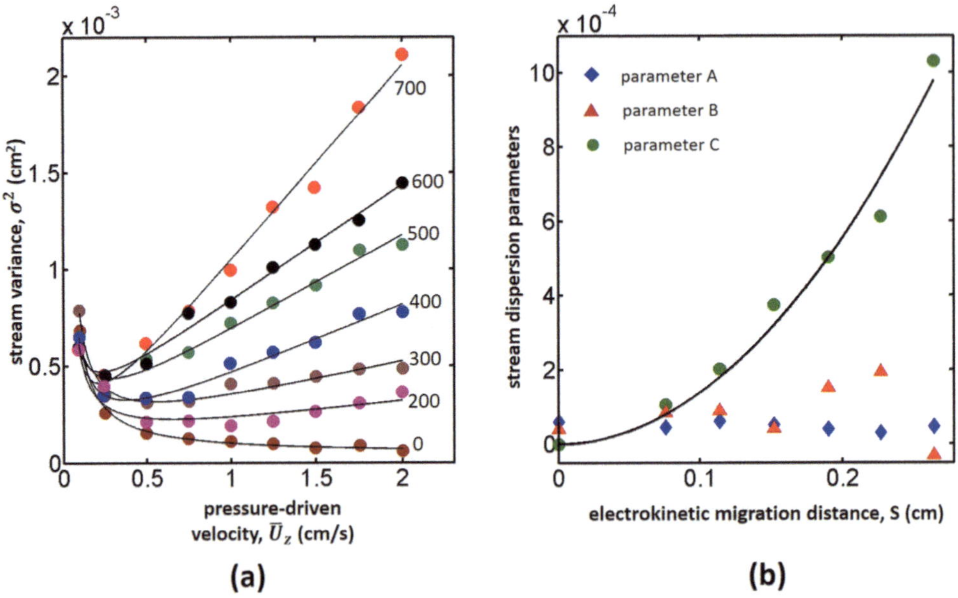

Fig. 3 (**a**) Fitting of the experimentally measured stream variance reported by Fonslow and Bowser [6] to an equation of the form $\sigma^2 = A + B/\bar{U}_z + C\bar{U}_z$. (**b**) Estimated values of A, B, and C from the regression analysis presented in (**a**) and fitting of the parameter C to an equation of the form $C = kS^2$. Reproduced with permission from Ref. 3

we have then plotted the estimates of these quantities as a function of the parameter S in the system. The figure shows the C term in this mathematical description to only vary with S indicating the contribution to stream variance from the orientation factor to be negligible in Fonslow and Bowser's work. Subsequently, the estimates for the C term were fitted to an equation of the form $C = kS^2$ to yield $A = 4.9 \times 10^{-5}$ cm^2, $B = 8.3 \times 10^{-5}$ cm^3/s, and $k = 1.4 \times 10^{-2}$ s/cm. These values correspond to an injection stream width (b) and analyte diffusion coefficient (D_0) of 242 μm and 1.67×10^{-5} cm^2/s, respectively. The estimate for the quantity k on the other hand yielded a factor 1/7, instead of 1/105, for the numerical coefficient in the hydrodynamic dispersion term suggesting that this contribution exceeded our theoretical prediction by a factor of 15 in Fonslow and Bowser's device likely due to non-idealities in the system.

3.3 Electrodynamic Dispersion in Free-Flow Zone Electrophoresis

Electrodynamic dispersion of sample streams in FFZE assays can originate due to partial or complete blockage of electroosmotic flow (EOF) across the channel width by the sidewalls of the conduit. Such blockage of EOF then generates a pressure-driven back-flow in the transverse direction for maintaining a flow balance in the system leading to additional stream broadening. In this situation, the analyte velocity in the transverse direction may be expressed as $u_x = \mu E[1 - (3\alpha/2)(1 - 4y^2/d^2)]$ in regions far away from the channel sidewalls [8]. Notice that the spatially averaged value of u_x in this description equals $\mu E(1 - \alpha)$ which determines the lateral migration distance for the analyte stream at the channel exit. Moreover, the diffusion coefficient D may again be assumed to be the uniform at all locations as the system is isothermal, i.e., $D = D_0$. These considerations yield equations governing ϕ_p and m_p given by

$$\frac{3}{2}Pe_z\left(1 - 4y^{*2}\right)\frac{\partial \phi_p}{\partial z^*} - \frac{pPe_x}{(1-\alpha)}\left[1 - \frac{3}{2}\alpha\left(1 - 4y^{*2}\right)\right]\phi_{p-1} = p(p-1)\phi_{p-2} + \frac{\partial^2 \phi_p}{\partial y^{*2}} + \frac{\partial^2 \phi_p}{\partial z^{*2}} \quad (14)$$

$$\frac{3}{2}Pe_z\int_{-1/2}^{1/2}\left[\left(1 - 4y^{*2}\right)\frac{\partial \phi_p}{\partial z^*}\right]dy^* - \frac{pPe_x}{(1-\alpha)}\int_{-1/2}^{1/2}\left[1 - \frac{3}{2}\alpha\left(1 - 4y^{*2}\right)\right]\phi_{p-1}dy^* = p(p-1)m_{p-2} + \frac{d^2 m_p}{dz^{*2}} \quad (15)$$

The integral boundary condition conserving the amount of analyte flowing per unit time through any $x-y$ plane again suggests that ϕ_0, and hence m_0, cannot be a function of z^*, yielding the solutions $\phi_0 = m_0 = \delta$. Moreover, upon substituting these values into the equation governing ϕ_1, i.e., Eq. 14 with $p = 1$, it is possible to show that

$$\phi_1 = \frac{Pe_x\delta}{Pe_z} z^* + \frac{Pe_x\delta}{480(1-\alpha)} \left(120y^{*2} - 240y^{*4} - 7\right) \quad \text{and} \quad m_1 = \frac{Pe_x\delta}{Pe_z} z^* \tag{16}$$

Further substituting these expressions in Eq. 14 for $p = 2$, one can obtain

$$\phi_2 = \frac{Pe_x^2\delta}{Pe_z^2} z^{*2} + \left[\frac{Pe_x^2\delta}{2Pe_z(1-\alpha)}\left\{\left(y^{*2} - 2y^{*4}\right) - \frac{(33-49\alpha)}{840(1-\alpha)}\right\} + \frac{2\delta}{Pe_z}\left(1 + \frac{Pe_x^2}{Pe_z^2}\right)\right] z^* + g(y^*)$$

$$\text{where } g(y^*) = \frac{Pe_x^2\delta}{1612800(1-\alpha)^2}\left(172800y^{*8} - 188160y^{*6} + 49440y^{*4} - 240y^{*2} - 253\right) - \frac{\delta}{240}\left(1 + \frac{Pe_x^2}{Pe_z^2}\right)\left(240y^{*4} - 120y^{*2} + 7\right) + \frac{\delta^3}{12} \tag{17}$$

Integrating Eq. 17 across the separation gap between the parallel plates, we get

$$m_2 = \frac{Pe_x^2\delta}{Pe_z^2} z^{*2} + \left[\frac{2\delta}{Pe_z}\left(1 + \frac{Pe_x^2}{Pe_z^2}\right) + \frac{Pe_x^2\delta}{105Pe_z(1-\alpha)^2}\right] z^* + \frac{\delta^3}{12} \tag{18}$$

Finally, the spatial variance of the sample stream (σ^2) in an FFZE system may be evaluated as

$$\frac{\sigma^2}{d^2} = \frac{m_2}{m_0} - \frac{m_1^2}{m_0^2} = \left[\frac{2}{Pe_z}\left(1 + \frac{Pe_x^2}{Pe_z^2}\right) + \frac{Pe_x^2}{105Pe_z(1-\alpha)^2}\right] z^* + \frac{\delta^2}{12} \tag{19}$$

which shows that this quantity grows linearly with the axial position (z^*-coordinate) in the separation chamber. The factor multiplying z^* in this equation is a measure of the effective dispersion coefficient for the analyte zone, while the term $\delta^2/12$ represents the contribution from the injection width to the overall sample variance. Because the normalized axial position z^* scales linearly with the residence time ($t = z/\bar{U}_z$) of the analyte molecules in the FFZE compartment, it is then possible to express $\sigma^2 = 2Kt + b^2/12$ yielding a Taylor-Aris dispersion coefficient (K):

$$K = D\left[1 + \left(\frac{Pe_x}{Pe_z}\right)^2 + \frac{1}{210(1-\alpha)^2}\frac{Pe_x^2}{}\right] \tag{20}$$

Equation 20 shows that the overall broadening of a sample stream during its transit through an FFZE chamber occurs as a result of molecular diffusion of the analyte species orthogonal to its flow direction as well as hydrodynamic dispersion stemming from a variation in the analyte velocity in the *x*- and *z*-directions across the gap between the parallel plates. The first contribution to this broadening process is captured by the $D_0[1 + (Pe_x/Pe_z)^2]$ term in which the $D_0(Pe_x/Pe_z)^2$ component (orientation factor) simply arises due to the fact that σ is measured along the *x*-axis, while the sample stream makes an angle $\theta = \cot^{-1}(Pe_x/Pe_z)$ with respect to this line. Notice that for a given transverse electric field (E) and axial pressure-driven flow speed $\left(\bar{U}_z\right)$, the orientation factor varies as $D_0[\mu E(1 - \alpha)/\bar{U}_z]^2$, whereas for a given lateral migration distance $S = \mu E(1 - \alpha)L/\bar{U}_z$ of the analyte stream, this factor equals $D_0(S/L)^2$ and is independent of α. The other term in Eq. 20, i.e., $D_0\left[(1/210)\left\{Pe_x^2/(1 - \alpha)^2\right\}\right]$, quantitates the hydrodynamic dispersion component in the system and differs by a factor of $1/(1 - \alpha)^2$ from the case of no transverse pressure-driven flow. This comparison, however, is only valid for a fixed Pe_x or lateral migration distance (S) of the sample zone. For a given transverse electric field in the FFZE assay, the hydrodynamic dispersion component reduces to $D_0(\mu Ed/D_0)^2/210$ which is identical to the case of no transverse pressure-driven flow. This is a very interesting result which says that although the pressure-driven cross-flow modifies the lateral migration distance for an analyte zone, it does not alter its spatial variance for a given transverse electric field in the assay. In other words, the pressure-driven cross-flow does not lead to any additional dispersion in the system. In fact, if one solves the transport equations relevant to our problem in the absence of a transverse electric field but with a non-zero pressure-driven cross-flow, the Taylor-Aris dispersion coefficient can be shown to be given by $K = D_0\left[1 + \left(\bar{U}_x/\bar{U}_z\right)^2\right]$ where $\bar{U}_x$ and $\bar{U}_z$ correspond to the spatially averaged pressure-driven velocities in the *x*- and *z*-directions, respectively. This expression again does not have any contribution from hydrodynamic dispersion, consistent with our analysis of an FFZE process with a lateral pressure-driven backflow. The lack of hydrodynamic dispersion in the presence of both axial and transverse pressure-gradients is a consequence of the fact that although solute molecules traveling along the channel center ($y^* = 0$) in these systems experience a higher lateral velocity, their transit time in the channel is shorter as they migrate along the faster moving streamlines in the axial direction. Consequently, the lateral distance drifted by these entities is identical to that traveled by molecules residing close to the channel walls at $y^* = \pm 1/2$.

For a given lateral migration distance in an FFZE device, however, pressure-driven cross-flows can significantly influence the performance of the assay irrespective of whether they arise due to blockage of EOF by the channel sidewalls or as a result of unwanted pressure gradients across the width of the separation chamber. If one analyzes the expression describing the variance of a sample stream under these conditions by rewriting Eq. 19 as

$$\sigma^2 = \underbrace{\frac{b^2}{12}}_{A} + \underbrace{\left(2D_0L + \frac{2S^2D_0}{L}\right)}_{B}\frac{1}{\bar{U}_z} + \underbrace{\frac{1}{105}\frac{S^2d^2}{(1-\alpha)^2D_0L}}_{C}\bar{U}_z \quad (21)$$

the C term in it can be significantly underestimated when any pressure-driven backflow is ignored. The discrepancy noticed in the coefficient multiplying the ratio S^2d^2/D_0L as estimated from experiments and theory in Fonslow and Bowser's work [6], for example, may at least in part be explained by the presence of a transverse pressure gradient in their system. Moreover, the greater dispersion caused by such a backflow for a fixed magnitude of S as predicted by Eq. 21 shifts the optimum axial flow velocity $(\bar{U}_{z,\ \mathrm{opt}})$ at which the stream variance is minimized to lower values as

$$\bar{U}_{z,\ \mathrm{opt}} = \sqrt{105\left(2D_0L + \frac{2S^2D_0}{L}\right)\frac{D_0L(1-\alpha)^2}{S^2d^2}} \quad (22)$$

The minimum stream variance realized under these conditions is also seen to increase with α as

$$\sigma^2_{\mathrm{min}} = \frac{b^2}{12} + \sqrt{\frac{4}{105}\left(2D_0L + \frac{2S^2D_0}{L}\right)\left\{\frac{S^2d^2}{(1-\alpha)^2D_0L}\right\}} \quad (23)$$

Notice that for pressure-driven backflows originating due to blockage of transverse EOF in an FFZE chamber, α varies between -1 and 1 based on our definition for this parameter. In our formulation, α assumes a negative value whenever the electrophoretic and electroosmotic analyte velocities oppose each other with the former exceeding the latter. In this situation, Eq. 22 suggests that $\bar{U}_{z,\mathrm{opt}}$ can be over- and underestimated by factors between $1-\infty$ and $1-2$ for positive and negative values of α, respectively, if the pressure-driven backflow in the system is ignored. Interestingly, the present analysis is also applicable to FFZE assays with a pressure-driven cross-flow that does not originate due to blockage of transverse EOF. In such a scenario, α can vary between $-\infty$ and ∞ causing both the over- and underestimation factors range from 1 to ∞.

3.4 Electrodynamic Dispersion in Free-Flow Isoelectric Focusing Assays

Pressure-driven cross-flows can arise in free-flow isoelectric focusing (FFIEF) assays due to a nonuniform electroosmotic flow velocity along the channel width induced by the pH gradient in this direction. In addition, variations in the channel cross section as well as unwanted differences in hydrostatic heads at the buffer/sample inlet ports can also lead to such pressure gradients which besides altering the equilibrium position of the sample zones have a tendency to substantially broaden their widths deteriorating the separations. As a result, the analyte velocity in the x-direction (u_x) in FFEIF systems with a pressure-driven cross-flow can be assumed to comprise two additive contributions from the electrokinetic and pressure-driven cross-flows as given by $u_x = u_{EK} + u_c$ [9]. Moreover, the electrokinetic contribution (u_{EK}) can be assumed to originate both due to electrophoresis and electroosmosis of the analyte molecules in the x-direction with the gradient $\partial u_{EK}/\partial x$ dominated by the electrophoretic component. Such a situation would be desirable in an FFIEF device as gradients in electroosmotic flow automatically produce unwanted pressure-driven cross-flows in the system leading to additional stream broadening. The transverse electrokinetic velocity under these conditions may be expressed as $u_{EK} = -k_{EP}x + U_{EOF}$ around the analyte's equilibrium position with U_{EOF} being the local electroosmotic flow component. An appropriate choice for the constant k_{EP} governing the electrophoretic component in an FFIEF chamber is $\Delta\mu E\omega/W$ where $\Delta\mu$ refers to absolute change in the electrophoretic mobility of the solute molecules across the width of the analysis chamber (W). The parameter ω here denotes the local dimensionless gradient in the analyte's electrophoretic mobility (μ_{EP}) with respect to the x-coordinate around its equilibrium position and assumes a value of 1 for a uniformly linear variation in μ_{EP} across the channel width. Notice that the minus sign associated with the electrophoretic cross-flow velocity in our analysis arises from the fact that the electric field in an FFIEF system is always applied in the direction of decreasing electrophoretic mobility for the solute molecules in order to drive them toward their equilibrium position. The gradient in the local pressure-driven cross-flow velocity $\partial u_c/\partial x$ around the analyte's equilibrium position in this situation may be neglected and the quantity u_c expressed as $(3\bar{U}_c/2)(1 - 4y^2/d^2)$ with $\bar{U}_c$ being its spatially averaged value with respect to the y-coordinate (*see* **Note 5**). Also, recognizing the fact that the analyte stream in an FFIEF device flows along the z-direction under equilibrium conditions due to a net zero transverse velocity further allows one to drop all concentration gradients with respect to the z-coordinate in the system. Finally, the diffusion coefficient D may again be assumed to be uniform at all locations, i.e., $D = D_0$, due to a uniform temperature in the system leading to equations governing ϕ_p and m_p given by

$$pPe_{EP}\phi_p - p\phi_{p-1} - \frac{3}{2}pPe_c(1 - 4y^{*2})\phi_{p-1} = p(p-1)\phi_{p-2} + \frac{d^2\phi_p}{dy^{*2}} \tag{24}$$

$$pPe_{\mathrm{EP}}m_p - pm_{p-1} - \frac{3}{2}pPe_{\mathrm{c}}\int_{-1/2}^{1/2}(1-4y^{*2})\phi_{p-1}dy^* = p(p-1)m_{p-2} \tag{25}$$

The quantities $Pe_{\mathrm{EP}} = \Delta\mu E\omega d^2/WD_0$, $Pe_{\mathrm{EOF}} = U_{\mathrm{EOF}}d/D_0$, and $Pe_{\mathrm{c}} = \bar{U}_{\mathrm{c}}d/D_0$ in the above equations denote the Péclet numbers based on the electrophoretic, electroosmotic, and pressure-driven cross-flow velocities, respectively. Notice that because the analyte velocity at the stream equilibrium position is zero in an FFIEF system, it is more appropriate to capture the effect of analyte migration in the x-direction on the separation performance using the Péclet numbers defined above rather than Pe_x as noted in Eq. 2. With these considerations in mind, it can now be shown that Eq. 24 reduces to $d^2\phi_0/dy^{*2} = 0$ for $p = 0$ yielding a solution $\phi_0 = m_0 = \delta$ upon application of the relevant boundary conditions. The corresponding equation for $p = 1$ may be written as

$$\frac{d^2\phi_1}{dy^{*2}} = Pe_{\mathrm{EP}}\phi_1 - Pe_{\mathrm{EOF}}\delta - \frac{3}{2}Pe_{\mathrm{c}}(1-4y^{*2})\delta \tag{26}$$

yielding a solution

$$\begin{aligned}\phi_1 &= \frac{6Pe_{\mathrm{c}}\delta\cosh\left(\sqrt{Pe_{\mathrm{EP}}}\,y^*\right)}{Pe_{\mathrm{EP}}^{3/2}\sinh\left(\sqrt{Pe_{\mathrm{EP}}}/2\right)} - \frac{6Pe_{\mathrm{c}}\delta}{Pe_{\mathrm{EP}}}y^{*2}\\ &+\frac{3}{2}\frac{Pe_{\mathrm{c}}\delta}{Pe_{\mathrm{EP}}} + \frac{Pe_{\mathrm{EOF}}\delta}{Pe_{\mathrm{EP}}} - 12\frac{Pe_{\mathrm{c}}\delta}{Pe_{\mathrm{EP}}^2}\end{aligned} \tag{27}$$

The normalized lateral equilibrium position for the analyte stream in this situation is given by

$$\frac{m_1}{m_0} = \frac{Pe_{\mathrm{EOF}} + Pe_{\mathrm{c}}}{Pe_{\mathrm{EP}}} \tag{28}$$

which represents the x^*-position where the transverse analyte velocity averaged over the y^*-coordinate vanishes, i.e., $\int_{-1/2}^{1/2} u_x dy^* = 0$. Proceeding further, the governing differential equation for $p = 2$ can be expressed as

$$\frac{d^2\phi_2}{dy^{*2}} = 2Pe_{\mathrm{EP}}\phi_2 - 2Pe_{\mathrm{EOF}}\phi_1 - 3Pe_{\mathrm{c}}(1-4y^{*2})\phi_1 - 2\delta \tag{29}$$

which upon integration from $y^* = -1/2$ to $y^* = 1/2$ and subsequent rearrangement yields

$$\begin{aligned}m_2 &= \frac{Pe_{\mathrm{EOF}}(Pe_{\mathrm{c}} + Pe_{\mathrm{EOF}})\delta}{Pe_{\mathrm{EP}}^2}\\ &+ \frac{\delta}{Pe_{\mathrm{EP}}} + \frac{3}{2}Pe_{\mathrm{c}}\int_{-1/2}^{1/2}(1-4y^{*2})\phi_1 dy^*\end{aligned} \tag{30}$$

Finally, the spatial variance of a sample stream (σ^2) in an FFIEF system with a pressure-driven cross-flow may be shown to be given by

$$\frac{\sigma^2}{d^2} = \frac{m_2}{m_0} - \frac{m_1^2}{m_0^2} = \frac{1}{Pe_{\mathrm{EP}}} + \frac{Pe_{\mathrm{c}}^2}{Pe_{\mathrm{EP}}^2}\left[72\frac{\coth\left(\sqrt{Pe_{\mathrm{EP}}}/2\right)}{Pe_{\mathrm{EP}}^{3/2}} - \frac{144}{Pe_{\mathrm{EP}}^2} - \frac{12}{Pe_{\mathrm{EP}}} + \frac{1}{5}\right] \tag{31}$$

under equilibrium conditions. In order to understand the expression for σ^2 presented above, it is important to realize that this variance arises from the interplay between three different factors, i.e., spatial gradient in the transverse electrophoretic velocity, molecular diffusion, and transverse pressure-driven velocity. While the gradient in the transverse electrophoretic velocity (first factor) in an FFIEF device tends to focus the solute molecules at their equilibrium position, molecular diffusion (second factor) counters this effect producing a steady-state concentration distribution. These two factors lead to a spatial variance represented by the first term in Eq. 31, i.e., $d^2/Pe_{\mathrm{EP}} = DW/(\Delta\mu E\theta)$, which is identical to the expression derived by Svensson [10] given that the local gradient in the electrophoretic mobility of the solute particles with respect to the x-coordinate is $\Delta\mu\theta/W$ in our analysis. The normalized spatial variance of the analyte stream (σ^2/d^2) in this situation is governed by a single dimensionless parameter Pe_{EP} which represents the ratio of the characteristic electrophoretic to diffusive migration rate of the sample molecules in the system. Introduction of a pressure-driven cross-flow (third factor) under these conditions perturbs the noted steady state in two ways. Firstly, it shifts the lateral position where the analyte molecules experience a net zero transverse velocity modifying the value of m_1 in the device. Moreover, because such pressure gradients also induce a variation in the streamline velocity with the y-coordinate, this equilibrium position changes across the depth of the assay chamber. The latter effect produces additional broadening of the analyte stream that is captured by the terms additive to $1/Pe_{\mathrm{EP}}$ in Eq. 31. The noted equation also predicts that the Péclet number based on the electroosmotic cross-flow velocity in an FFIEF device does not influence the broadening of the analyte stream under equilibrium conditions. This result is a consequence of the uniformity in the electroosmotic flow profile across the channel depth which modifies the equilibrium positions of all solute molecules to an equal extent. The magnitude and direction of this cross-flow, however, alter the equilibrium position of the solute particles and therefore enter the expression for m_1 in our analysis.

As may be seen from Eq. 31, the contribution to stream variance arising from the pressure-driven cross-flow velocity in an FFIEF device scales with the square of the Péclet number based on the parameter $\bar{U}_{\mathrm{c}}$ but again diminishes with increasing values of

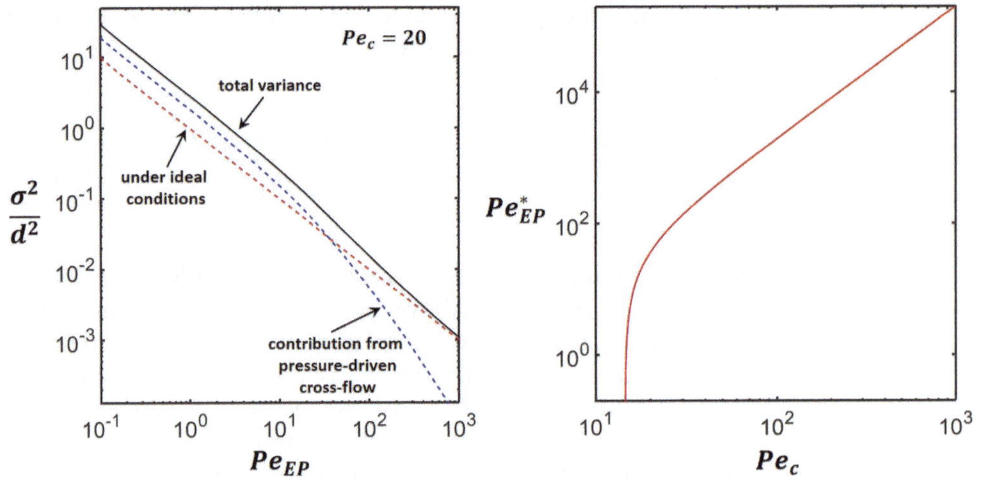

Fig. 4 (**a**) Dependence of the equilibrium stream variance on the quantity Pe_{EP} as predicted by Eq. 31 in the presence of a steady pressure-driven cross-flow in an FFIEF system. "Ideal conditions" here correspond to the situation when $Pe_{\mathrm{c}} = 0$. (**b**) Variation in the value of Pe^*_{EP} as function of Pe_{c} in an FFIEF assay with a steady pressure-driven cross-flow. The quantity Pe^*_{EP} here denotes the value of Pe_{EP} at which the additional spatial variance from the pressure-driven cross-flow equals that expected under ideal conditions, i.e., d^2/Pe_{EP}. Reproduced with permission from Ref. 9

Pe_{EP}. The noted trends arise due to the dispersal effect of fluid shear and the focusing effect of the gradient in electrophoretic velocity across the channel width, which are measured in terms of the parameters Pe_{c} and Pe_{EP}, respectively, in our analysis. In Fig. 4a, the different contributions to σ^2 as well as the total stream variance in an FFIEF system have been plotted as a function of these two relevant Péclet numbers. The figure shows several interesting trends both for small and large values of Pe_{c} and Pe_{EP}. In the limit of $Pe_{\mathrm{EP}} \ll 1$, for example, the contribution to σ^2 from the pressure-driven cross-flow is seen to scale with Pe_{EP} in an identical way as that under ideal FFIEF conditions, i.e., ~$1/Pe_{\mathrm{EP}}$. This occurs as the term

$$72\frac{\coth\left(\sqrt{Pe_{\mathrm{EP}}}/2\right)}{Pe_{\mathrm{EP}}^{3/2}} - \frac{144}{Pe_{\mathrm{EP}}^2} - \frac{12}{Pe_{\mathrm{EP}}} + \frac{1}{5} \quad \longrightarrow \quad \frac{Pe_{\mathrm{EP}}}{210} \tag{32}$$

in this regime yielding an expression for the stream variance given by

$$\frac{\sigma^2}{d^2} \approx \frac{1}{Pe_{\mathrm{EP}}}\left(1 + \frac{Pe_{\mathrm{c}}^2}{210}\right) \tag{33}$$

As may be noted, Eq. 33 predicts the pressure-driven cross-flow to increase the spatial variance of the analyte stream by a factor $1 + Pe_{\mathrm{c}}^2/210$, identical to that known for regular pressure-driven flow systems. In this limit of small electrophoretic migration rates, cross streamline diffusion interacts with the parabolic profile of the pressure-driven cross-flow in the same manner as was described by Taylor and Aris [4, 5]. Consequently, the curves depicting the ideal stream variance and the additional contribution from transverse pressure-gradients in Fig. 4a are seen to line up parallel to each other for $Pe_{\mathrm{EP}} \ll 1$. The value of Pe_{c} dictates their relative magnitude in this regime with the pressure-driven cross-flow contribution exceeding the ideal component when $Pe_{\mathrm{c}} > \sqrt{210}$. In the opposite limit of $Pe_{\mathrm{EP}} \gg 1$, the term

$$72\frac{\coth\left(\sqrt{Pe_{\mathrm{EP}}}/2\right)}{Pe_{\mathrm{EP}}^{3/2}} - \frac{144}{Pe_{\mathrm{EP}}^{2}} - \frac{12}{Pe_{\mathrm{EP}}} + \frac{1}{5} \longrightarrow \frac{1}{5} \tag{34}$$

yielding an expression for the stream variance given by

$$\frac{\sigma^2}{d^2} \approx \frac{1}{Pe_{\mathrm{EP}}}\left(1 + \frac{1}{5}\frac{Pe_{\mathrm{c}}^2}{Pe_{\mathrm{EP}}}\right) \tag{35}$$

Under these conditions, the pressure-driven cross-flow contribution decays with increasing values of Pe_{EP} much more rapidly than its $1/Pe_{\mathrm{EP}}$ counterpart eventually having the overall stream variance approach its ideal limit. Physically this occurs, as the pressure-driven cross-flow now becomes negligible relative to the electrophoretic migration rate in the system causing the additional stream broadening induced by the transverse pressure-gradient to become vanishingly small. Figure 4a also shows the noted trends lead to a cross-over of the ideal and pressure-driven cross-flow components to σ^2 for operating conditions with $Pe_{\mathrm{c}} > \sqrt{210}$. The value of Pe_{EP} where this cross-over occurs, denoted as Pe_{EP}^*, has been plotted in Fig. 4b as a function of the parameter, Pe_{c}, in the system. The figure shows Pe_{EP}^* to approach zero when $Pe_{\mathrm{c}} \to \sqrt{210}$ as expected and increase for larger values Pe_{c}. In the limit of $Pe_{\mathrm{c}} \gg \sqrt{210}$, the quantity $Pe_{\mathrm{EP}}^* \approx Pe_{\mathrm{c}}^2/5$ as predicted by Eq. 35.

3.5 Stream Dispersion in Free-Flow Zone Electrophoresis with Joule Heating Effects

The use of an electric field in free-flow zone electrophoresis (FFZE) automatically leads to Joule heating yielding a higher temperature at the center of the separation chamber relative to that around the channel walls. For small amounts of heat generated, this thermal effect introduces a variation in the equilibrium position of the analyte molecules due to the dependence of liquid viscosity and analyte diffusivity on temperature leading to a modification in the position of the analyte stream as well as the zone width [11]. The Joule heating of the background electrolyte under these conditions leads to a fully developed temperature (T) profile across the channel gap determined by the following expression based on Fourier's law

$$\frac{d}{dy}\left(k\frac{dT}{dy}\right) = -\lambda E^2 \quad \text{boundary conditions}: \quad T = T_0 \quad \text{at } y = \pm\frac{d}{2} \tag{36}$$

where T_0 refers to the absolute temperature at the parallel plates and the quantities k and λ denote the ionic and thermal conductivities of the background electrolyte, respectively. It is important to point out that although the conductivity parameters k and λ in Eq. 36 are both functions of temperature, their dependence on T for aqueous solutions are significantly different from each other. For example, while the thermal conductivity of water is known to increase by less than 1% for small variations in T ($<10^{\circ}$C) around room temperature (25°C), the corresponding change in λ can be as large as 25%. In this situation, the leading order effect of Joule heating may be captured by assuming k to be spatially constant (0.6 W/m/K for water at $T_0 = 25^{\circ}$C) [11] and the ionic conductivity to vary as $\lambda = \lambda_0 I_0[1 + \alpha(T - T_0)]$ where $\lambda_0 = 0.14$ m^2 S/mol (for aqueous KCl solutions), $\alpha = 0.025$/K (for water), and I_0 equals the ion concentration in the background electrolyte in mol/m^3 [12].

Notice that the current analysis ignores any axial temperature gradients in the FFZE chamber due to an incompletely developed thermal profile, which, for microchip-based capillary zone electrophoretic devices, have been experimentally shown to exist only around the channel terminals extending about a few millimeters into the duct [13]. The results presented here are therefore strictly applicable to the middle section of the assay compartment with lengths about a centimeter or more, where the end effects from the channel terminals may be neglected. Fortunately, most microchip-based FFZE chambers are several centimeters long in which case the overall stream dispersion in them is likely to be dominated by zone broadening mechanisms applicable to the central region of the chamber where the thermal profile is completely developed. In addition, a temperature gradient across the channel width (x-direction) of an FFZE chamber can arise either due to heat loss through its sidewalls or when the electrical and thermal properties of the flow buffer used is significantly different from that of the sample stream. Interestingly, the effect of heat loss through the channel sidewalls on sample transport may be neglected for large aspect ratio FFZE chambers (channel width to depth ratio $\gg 1$) especially when the sample stream is conveyed far away from these edges. Again, this situation is often applicable to microchip-based FFZE systems and therefore has been considered in the present analysis. Moreover, the current work also ignores any significant differences in the electrical and thermal properties of the sample stream relative to that of the flow buffer unlike in isoelectric focusing systems, for example. As a result, the thermal profile relevant to the stream dispersion process in FFZE assays is mostly dictated by energy transport across the narrower dimension in the separation chamber (y-direction) as has been considered in Eq. 36.

Proceeding further, Eq. 2 may now be normalized to the form

$$\frac{d^2\theta}{dy^{*2}} = -(1 + \alpha\, \Delta T\, \theta) \quad \text{boundary conditions : } \theta = 0 \quad \text{at } y^* = \pm\frac{1}{2} \tag{37}$$

after accounting for the variation in the ionic conductivity with temperature where $\theta = (T - T_0)/\Delta T$, $\Delta T = \lambda_0 I_0 E^2 d^2/k$, and $y^* = y/d$ (*see* **Notes 6** and 7). It is well known that Eq. 37 admits a closed-form solution which for small values of ΔT predicts θ to vary quadratically with y^* as noted below:

$$\theta = \frac{1}{\alpha\Delta T}\left[\frac{\cos\left(\sqrt{\alpha\Delta T}\, y^*\right)}{\cos\left(\sqrt{\alpha\Delta T}/2\right)} - 1\right] = \frac{1 - 4y^{*2}}{8\cos\left(\sqrt{\alpha\Delta T}/2\right)} + O\left(\Delta T^2\right) \tag{38}$$

Equation 38 shows that Joule heating between two parallel plates leads to a parabolic thermal profile with the temperature differential between the channel center and the walls scaling with ΔT for small values of this parameter, i.e., $\Delta T < 25\,^\circ\mathrm{C}$. Moreover, the noted temperature differential assumes values less than $0.14\,\Delta T$, i.e., $<3.5\,^\circ\mathrm{C}$, under these conditions which are particularly relevant to microfluidic separation systems with channel depths in the range of 10–100 μm. To highlight this point, the dependence of ΔT on the electric field (E) in an FFZE device has been plotted in Fig. 5a for different channel depths (d) and ion concentrations in the background electrolyte (I_0) commonly employed in microfluidic assays. The figure shows ΔT to assume values mostly under $25\,^\circ\mathrm{C}$ for d and I_0 in the range of 10–25 μm and 10–25 mM, respectively.

Having determined the temperature profile between the parallel plates, sample transport in this system may be analyzed by considering the analyte velocity along the x-direction (u_x) to comprise two additive contributions from electroosmosis and electrophoresis, i.e., $u_x = \mu_{\mathrm{EP}}E + \mu_{\mathrm{EOF}}E$. Interestingly, however, while the nonuniformity in liquid temperature described by Eq. 38 produces a variation in the electrophoretic component across the channel gap, the electroosmotic counterpart remains spatially uniform provided the thickness of the Debye layer around the channel walls can be neglected relative to d. In this situation, the spatial variation in the quantity u_t may be expressed as $u_x = \mu_{\mathrm{EP}}^0 E\eta_0/\eta(y^*) + \mu_{\mathrm{EOF}}^0 E$ where μ_{EP}^0 and μ_{EOF}^0 refer to the values of μ_{EP} and μ_{EOF} at the wall temperature T_0, respectively. Now approximating the properties of the background electrolyte by those of water, the spatial variation in η may be captured using the correlation

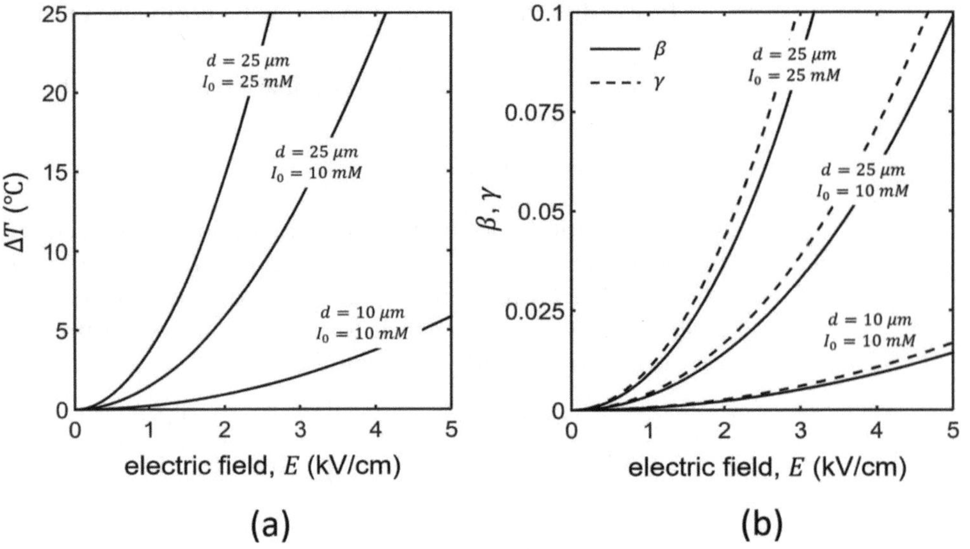

Fig. 5 (**a**) Estimated characteristic temperature rise as a function of the applied electric field in an FFZE device. (**b**) Estimated values of the dimensionless parameters (β and γ) that determine the effect of Joule on analyte transport in an FFZE system. Reproduced with permission from Ref. 11

$$\eta = A\exp\left(\frac{B}{T}\right) = \eta_0\left(1 - \frac{B\Delta T\theta}{T_0^2}\right) = \eta_0\left[1 - \frac{B\Delta T(1-4y^{*2})}{8T_0^2\cos\left(\sqrt{\alpha\Delta T}/2\right)}\right] + O(\Delta T^2) \quad (39)$$

where $A = 2.76 \times 10^{-6}$ Pa s, $B = 1713$ K, and $\eta_0 = A\exp(B/T_0)$.

The spatial variation in u_z may be determined by solving the Navier-Stokes equation while accounting for the temperature dependence of liquid viscosity yielding $u_z = \bar{U}_z^0\left[3(1-4y^{*2})/2 + 3\beta(1-4y^{*2})^2/4\right]$ to the leading order in ΔT where $\bar{U}_z^0$ denotes the spatially averaged pressure-driven flow velocity at the wall temperature T_0. To capture the spatial variation in D on the other hand, the Stokes-Einstein relationship for solute diffusivity, i.e., $D = k_B T/(6\pi\eta R)$, has been used in the present analysis where the symbols k_B and R correspond to the Boltzmann constant and the effective radius of the analyte molecule, respectively. In this situation, the equations governing ϕ_p and m_p may be written as

$$-pPe_{\mathrm{EP}}f\phi_{p-1} - pPe_{\mathrm{EOF}}\phi_{p-1} + Pe_z g\frac{\partial\phi_p}{\partial z^*} = p(p-1)h\phi_{p-2} + \frac{\partial}{\partial y^*}\left(h\frac{\partial\phi_p}{\partial y^*}\right) + h\frac{\partial^2\phi_p}{\partial z^{*2}} \quad (40)$$

$$-pPe_{\mathrm{EP}}\int_{-1/2}^{1/2} f\phi_{p-1}\,dy^* - pPe_{\mathrm{EOF}}m_{p-1} + Pe_z\int_{-1/2}^{1/2}\left(g\frac{\partial\phi_p}{\partial z^*}\right)dy^*$$

$$= p(p-1)\int_{-1/2}^{1/2} h\phi_{p-2}\,dy^* + \int_{-1/2}^{1/2}\left(h\frac{\partial^2\phi_p}{\partial z^{*2}}\right)dy^* \tag{41}$$

where the functions $f = \eta_0/\eta = 1 + \beta(1 - 4y^{*2})$, $g = u_P/\bar{U}_z^0 = 3(1-4y^{*2})/2 + 3\beta(1-4y^{*2})^2/4$, and $h = D/D_0 = 1 + \gamma(1 - 4y^{*2})$ describe the spatial variation in the analyte electrophoretic velocity, liquid pressure-driven flow speed, and analyte diffusivity, respectively, to the leading order in ΔT. The various parameters relevant to these functions are defined as $\beta = B\Delta T/\left\{8T_0^2\cos\left(\sqrt{\alpha\Delta T}/2\right)\right\}$, $\gamma = \Delta T(1 + B/T_0)/\{8T_0\cos(\sqrt{\alpha\Delta T}/2)\}$, $\bar{U}_z^0 = \Delta P d^2/(12\eta_0 L)$, and $D_0 = k_B T_0/(6\pi\eta_0 R)$ with ΔP being the pressure drop applied across the axial length of the FFZE chamber (L) and D_0 referring to the analyte diffusivity at the wall temperature T_0. The three Péclet numbers included in Eq. 40 here offer a measure of solutal advection via electrophoresis, electroosmosis, and pressure-driven flow relative to their diffusive migration in the system and have been evaluated as $Pe_{\mathrm{EP}} = \mu_{\mathrm{EP}}^0 Ed/D_0$, $Pe_{\mathrm{EOF}} = \mu_{\mathrm{EOF}}^0 Ed/D_0$, and $Pe_z = \bar{U}_z^0 d/D_0$, respectively. For simplicity, the above expressions assume the radius of the analyte molecules (R) as well as the electrical charge carried by them to be unaffected by Joule heating which is appropriate for assessing sample transport in electrophoretic systems with a small temperature differential across the channel gap. Notice that because the analyte velocity across the channel width in this analysis has an electrophoretic contribution that varies spatially and an electroosmotic one that does not, it is more appropriate to capture the effect of analyte migration in the x-direction on the separation performance using two separate parameters Pe_{EP} and Pe_{EOF} as defined above rather than a single one, e.g., Pe_x, as noted in Eq. 2. As may be noted, Eq. 40 suggests that the leading order effects of Joule heating on the separation performance of FFZE assays are captured by the dimensionless parameters β and/or γ which are both proportional to the quantity ΔT. In this situation, it is important to have a sense of the magnitude of these quantities which have been included in Fig. 5b. The figure shows that for aqueous electrolytes (with $B = 1713$ K) and a wall temperature of $T_0 = 298$ K, the parameter γ always exceeds β by about 17% ($=T_0 \times 100/B$) with both quantities assuming values less than 0.1 under operating conditions commonly applicable to microfluidic FFZE assays.

The integral boundary condition in Eq. 3 implies that ϕ_0, and hence m_0, cannot be functions of z^*. Upon incorporating this result into the differential equation governing ϕ_0, it can then be shown $d(hd\phi_0/dy^*)/dy^* = 0$ yielding the solutions $\phi_0 = m_0 = \delta$. The corresponding equation for ϕ_1 in this situation becomes

$$Pe_z g \frac{\partial \phi_1}{\partial z^*} - h \frac{\partial^2 \phi_1}{\partial z^{*2}} - \frac{\partial}{\partial y^*}\left(h \frac{\partial \phi_1}{\partial y^*}\right) = Pe_{\text{EP}} f \delta + Pe_{\text{EOF}} \delta \quad (42)$$

which is solved by assuming ϕ_1 to vary linearly with z^*, i.e., $\phi_1 = \psi_{11} z^* + \psi_{10}$. Upon substituting this functional form of ϕ_1 in Eq. 42 and later matching the coefficients of z^*, it is possible to show that $d(hd\psi_{11}/dy^*)/dy^* = 0$ which admits the following solution to the leading order in ΔT (*see* **Note 8**):

$$\psi_{11} = \frac{(Pe_{\text{EP}} + Pe_{\text{EOF}})\delta}{Pe_z} + \frac{2(2Pe_{\text{EP}} - 3Pe_{\text{EOF}})\beta\delta}{15Pe_z} \quad (43)$$

It must be noted that the parameters β and γ both scale with the quantity ΔT in the present analysis and are therefore involved in terms that capture the leading order effect of Joule heating on sample transport. Proceeding further, the terms independent of z^* in Eq. 42 are equated to obtain $d(hd\psi_{10}/dy^*)/dy^* = Pe_z g \psi_{11} - Pe_{\text{EP}} f \delta - Pe_{\text{EOF}} \delta$ which is solved to the leading order in ΔT yielding

$$\psi_{10} = (Pe_{\text{EP}} + Pe_{\text{EOF}})\delta \left[\frac{(120y^{*2} - 240y^{*4} - 7)}{480} + \left(\frac{\beta}{5600} - \frac{\gamma}{1680} \right)(420y^{*2} - 1680y^{*4} + 2240y^{*6} - 19) \right] \quad (44)$$

The normalized lateral equilibrium position of the analyte stream in this situation can be evaluated as

$$\frac{m_1}{m_0} = \left[\frac{(Pe_{\text{EP}} + Pe_{\text{EOF}})}{Pe_z} + \frac{2\beta(2Pe_{\text{EP}} - 3Pe_{\text{EOF}})}{15Pe_z} \right] z^* \quad (45)$$

which shows Joule heating to shift this quantity by an additional term proportional to β. Moreover, Eq. 45 predicts that the noted term can assume both positive and negative values depending on the amount of electroosmosis relative to electrophoresis in the system, allowing the analyte stream to drift either way of the equilibrium position realized in the absence of Joule heating. In the limit of diminishing electrophoresis $(Pe_{\text{EP}}/Pe_{\text{EOF}} = \mu^0_{\text{EP}}/\mu^0_{\text{EOF}} \rightarrow 0)$, for example, the analytic theory predicts the stream position to drift by an amount $-2\beta Pe_{\text{EOF}}/5$, implying a smaller deflection of the analyte stream by the electric field in the presence of Joule heating. This occurs as the lateral migration of the solute molecules under these conditions is completely dominated by

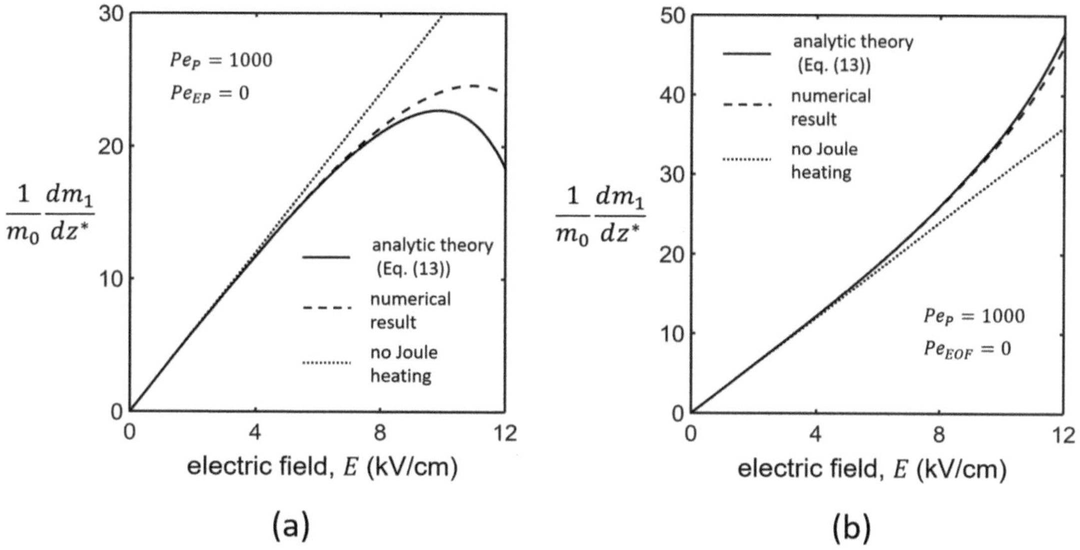

Fig. 6 Effect of electric field on the drift in the lateral position of the analyte stream due to Joule heating in the absence of (**a**) electrophoresis and (**b**) electroosmosis. Reproduced with permission from Ref. 11

electroosmosis which is unaffected by the temperature rise at the center of the channel. Nevertheless, for a given pressure drop (ΔP) across the channel length, the lower liquid viscosity in the device increases the sample velocity along the z-coordinate by a factor of $1 + 2\beta/5$ to the leading order reducing the residence time of the analyte stream by a factor $1 - 2\beta/5$ (*see* Fig. 6a). In the opposite scenario when the magnitude of $\mu^0_{\mathrm{EOF}}/\mu^0_{\mathrm{EP}}$ is small, i.e., $\mu^0_{\mathrm{EOF}}/\mu^0_{\mathrm{EP}} \rightarrow 0$, the lateral position of the analyte molecules is completely determined by their electrophoretic migration along the x-coordinate which speeds up by a factor of $1 + 2\beta/3$ to the leading order due to the reduced viscosity in the system. In this limit, the combined effects of the increased pressure-driven and electrophoretic flow speeds lead to a change in the quantity m_1/m_0 by a factor $1 + 4\beta/15$ as shown in Fig. 6b. It must be noted that while the solid curves in Fig. 6a/b correspond to the predictions of the analytic theory, the dashed ones represent simulation results obtained by numerically solving Eq. 42 without linearizing the expressions for the liquid viscosity, the analyte diffusivity, and the temperature profile across the channel gap with respect to ΔT. As expected, the deviation between the two curves is seen to grow for larger values of ΔT with the analytic theory underpredicting the numerical simulations for the case of no electrophoresis. Nevertheless, this mismatch is still below 0.2% for $\beta \leq 0.1$ and the operating conditions considered in Fig. 6a. Interestingly, the analytic theory overpredicts the simulation results in the opposite scenario of zero electroosmosis, but the deviation between the two outcomes is even smaller for the conditions considered in Fig. 6b.

Now for estimating the second moment of C^*, Eq. 40 is again referred back to arrive at the differential equation governing ϕ_2:

$$Pe_z g \frac{\partial \phi_2}{\partial z^*} - h \frac{\partial^2 \phi_2}{\partial z^{*2}} - \frac{\partial}{\partial y^*}\left(h \frac{\partial \phi_2}{\partial y^*}\right) = 2Pe_{\mathrm{EP}} f \phi_1 + 2Pe_{\mathrm{EOF}} \phi_1 + 2h\delta \tag{46}$$

which is solved by assuming the functional form $\phi_2 = \psi_{22} z^{*2} + \psi_{21} z^* + \psi_{20}$. Upon substituting this quadratic polynomial in z^* into Eq. 46 and later matching the coefficients for z^{*2}, it can be shown that $d(h d\psi_{22}/dy^*)/dy^* = 0$ which admits the following solution to the leading order in ΔT:

$$\psi_{22} = \frac{(Pe_{\mathrm{EP}} + Pe_{\mathrm{EOF}})^2 \delta}{Pe_z^2} + \frac{4(Pe_{\mathrm{EP}} + Pe_{\mathrm{EOF}})(2Pe_{\mathrm{EP}} - 3Pe_{\mathrm{EOF}})\beta\delta}{15 Pe_z^2} \tag{47}$$

Later matching the coefficients for z^* in Eq. 46 yields $d(h d\psi_{21}/dy^*)/dy^* = 2Pe_z g \psi_{22} - 2Pe_{\mathrm{EP}} f \psi_{11} - 2Pe_{\mathrm{EOF}} \psi_{11}$ which is solved to the leading order in ΔT to obtain

$$\begin{aligned}
\psi_{21} = {} & \frac{(Pe_{\mathrm{EP}} + Pe_{\mathrm{EOF}})^2 \delta (280 y^{*2} - 560 y^{*4} - 11)}{560 Pe_z} + \frac{2\delta}{Pe_z}\left[1 + \frac{(Pe_{\mathrm{EP}} + Pe_{\mathrm{EOF}})^2}{Pe_z^2}\right] \\
& + \frac{(Pe_{\mathrm{EP}} + Pe_{\mathrm{EOF}})\delta\beta}{25200 Pe_z}\left[Pe_{\mathrm{EP}}(20160 y^{*6} - 21840 y^{*4} + 7140 y^{*2} - 367\right. \\
& \left. + Pe_{\mathrm{EOF}}(20160 y^{*6}) - 5040 y^{*4} - 1260 y^{*2} + 123)\right] \\
& + \frac{(Pe_{\mathrm{EP}} + Pe_{\mathrm{EOF}})^2 \delta\gamma (5040 y^{*4} - 6720 y^{*6} - 1260 y^{*2} + 41)}{2520 Pe_z} \\
& - \frac{4\delta\beta}{5 Pe_z} + \frac{4\delta\gamma}{3 Pe_z}\left[1 + \frac{(Pe_{\mathrm{EP}} + Pe_{\mathrm{EOF}})^2}{Pe_z^2}\right] + \frac{4(Pe_{\mathrm{EP}} + Pe_{\mathrm{EOF}})\delta\beta(Pe_{\mathrm{EP}} - 9 Pe_{\mathrm{EOF}})}{15 Pe_z^3}
\end{aligned} \tag{48}$$

Similarly, it is possible to determine the spatial dependence of ψ_{20} on y^* by equating the terms independent of z^* after substituting the functional form of ψ_2 in Eq. 46. However, it is important to realize that the second moment of C^*, i.e., m_2, only depends on the spatially averaged value of ψ_{20}, i.e., $\int_{-1/2}^{1/2} \psi_{20}\, dy^*$, which interestingly is simply equal to $\delta^3/12$ as the analyte molecules are assumed to be uniformly distributed over a normalized width δ at the channel inlet ($z^* = 0$). Combining these results with the integrals of Eqs. 47 and 48 over the channel gap, the second moment of C^* in any x–y plane may be estimated as

$$m_2 = \frac{(Pe_{\mathrm{EP}} + Pe_{\mathrm{EOF}})^2 \delta}{Pe_z^2} \left[1 + \frac{4(2Pe_{\mathrm{EP}} - 3Pe_{\mathrm{EOF}})\beta}{15(Pe_{\mathrm{EP}} + Pe_{\mathrm{EOF}})}\right] z^{*2}$$
$$+ \left[\frac{2\delta}{Pe_z}\left\{1 + \frac{(Pe_{\mathrm{EP}} + Pe_{\mathrm{EOF}})^2}{Pe_z^2} + \frac{(Pe_{\mathrm{EP}} + Pe_{\mathrm{EOF}})^2}{210}\right\}\right.$$
$$+ + \frac{4\delta\gamma}{3Pe_z}\left\{1 + \frac{(Pe_{\mathrm{EP}} + Pe_{\mathrm{EOF}})^2}{Pe_z^2} - \frac{(Pe_{\mathrm{EP}} + Pe_{\mathrm{EOF}})^2}{210}\right\} \tag{49}$$
$$\left. + \frac{4\delta\beta(Pe_{\mathrm{EP}} + Pe_{\mathrm{EOF}})(Pe_{\mathrm{EP}} - 9Pe_{\mathrm{EOF}})}{15Pe_z^3} - \frac{4\delta\beta}{5Pe_z}\right] z^* + \frac{\delta^3}{12}$$

The normalized spatial variance of the sample stream (σ^2/d^2) evaluated as $m_2/m_0 - m_1^2/m_0^2$ in this situation turns out to be

$$\sigma^{*2} = \frac{\sigma^2}{d^2} = \left[\frac{2}{Pe_z}\left\{1 + \frac{(Pe_{\mathrm{EP}} + Pe_{\mathrm{EOF}})^2}{Pe_z^2} + \frac{(Pe_{\mathrm{EP}} + Pe_{\mathrm{EOF}})^2}{210}\right\}\right.$$
$$+ \frac{4\beta}{5Pe_z}\left\{\frac{(Pe_{\mathrm{EP}} + Pe_{\mathrm{EOF}})(Pe_{\mathrm{EP}} - 9Pe_{\mathrm{EOF}})}{3Pe_z^2} - 1\right\}$$
$$\left. + \frac{4\gamma}{3Pe_z}\left\{1 + \frac{(Pe_{\mathrm{EP}} + Pe_{\mathrm{EOF}})^2}{Pe_z^2} - \frac{(Pe_{\mathrm{EP}} + Pe_{\mathrm{EOF}})^2}{210}\right\}\right] z^* + \frac{\delta^2}{12} \tag{50}$$

which shows this quantity to increase linearly with the axial position (z-coordinate) in the separation chamber as is the case in the absence of Joule heating. The factor multiplying z^* in this equation is a measure of the effective dispersion coefficient for the analyte zone with the term $\delta^2/12$ representing the contribution from the injection width. Equation 50 also shows that the leading order effect of Joule heating on the stream variance in FFZE assays is captured by two terms proportional to β and γ which both scale with ΔT. While the first of these two terms quantitates the effect of a modified electrophoretic and pressure-driven flow velocity due to a lower liquid viscosity around the channel center, the second one captures the influence of an increased analyte diffusivity in the system. Figure 7 shows that Joule heating tends to reduce the overall stream broadening in FFZE assays in the limit when this quantity is dominated by the hydrodynamic dispersion component, i.e., Pe_z, $Pe_{\mathrm{EP}} + Pe_{\mathrm{EOF}} \gg \sqrt{210}$. This result may be best analyzed in terms of the Taylor-Aris dispersion coefficient (K) in the system which relates to the stream variance as $\sigma^2 = 2Kt + b^2/12$ where $t = z^* d(1 - 2\beta/5)/\bar{U}_P^0$ denotes the residence time of the analyte stream in the FFZE chamber. Upon incorporating this information in Eq. 50 and making the relevant rearrangement, it is now possible to express

$$K = D_0 \left[1 + \left(\frac{Pe_{EP} + Pe_{EOF}}{Pe_z} \right)^2 + \frac{(Pe_{EP} + Pe_{EOF})^2}{210} + \frac{2\beta}{5} \left\{ \frac{2(Pe_{EP} + Pe_{EOF})(2Pe_{EP} - 3Pe_{EOF})}{3Pe_z^2} + \frac{(Pe_{EP} + Pe_{EOF})^2}{210} \right\} + \frac{2\gamma}{3} \left\{ 1 + \frac{(Pe_{EP} + Pe_{EOF})^2}{Pe_z^2} - \frac{(Pe_{EP} + Pe_{EOF})^2}{210} \right\} \right] \quad (51)$$

Equation 51 shows that the modified electrophoretic and pressure-driven flow profile in the presence of Joule heating alters both the orientation factor and the hydrodynamic dispersion contributions in the system. In fact, the term $4\beta D_0(Pe_{EP} + Pe_{EOF})(2Pe_{EP} - 3Pe_{EOF})/(15Pe_P^2)$ exactly accounts for the change in the orientation factor to the leading order in ΔT based on the drift in the lateral equilibrium stream position predicted by Eq. 45. The other term scaling with β in Eq. 51, i.e., $\beta D_0(Pe_{EP} + Pe_{EOF})^2/525$, captures the effect of Joule heating on the hydrodynamic dispersion component and always assumes a positive value due to an increased shear in the system. The terms proportional to the parameter γ in Eq. 51 quantitate the influence of an increased analyte diffusivity under the operating conditions. In particular, this expression shows that faster diffusion along the *x*-coordinate broadens the sample zone by an additional term given by $2\gamma D_0\left\{1 + (Pe_{EP} + Pe_{EOF})^2/Pe_z^2\right\}/3$. Whereas the increased diffusivity along the *y*-coordinate forces a reduction in the hydrodynamic dispersion component by an amount equal to $\gamma D_0(Pe_{EP} + Pe_{EOF})^2/315$, such a reduction occurs as analyte molecules are now able to equilibrate at a faster rate across the different flow streamlines compared to the situation when Joule heating is absent. Overall, however, the subtractive hydrodynamic dispersion component arising from faster analyte diffusion offsets the corresponding additive contribution resulting from larger fluid shear as γ always exceeds β based on their definitions (*see* Fig. 5b). Figure 7 also shows that Joule heating can be expected to reduce stream broadening by about 20% in the absence of electrophoresis under conditions when $Pe_z = 1000$ and $Pe_{EOF} = 2000$ with the analytic theory underpredicting the numerical simulation results by around 5%. However when electoosmosis is eliminated from the system and $Pe_{EP} = 2000$, the reduction in zone dispersion due to Joule heating is as much as 36% for the same value of Pe_z with the analytic theory now overpredicting the simulation results by almost 20%. It must be pointed out that the Péclet

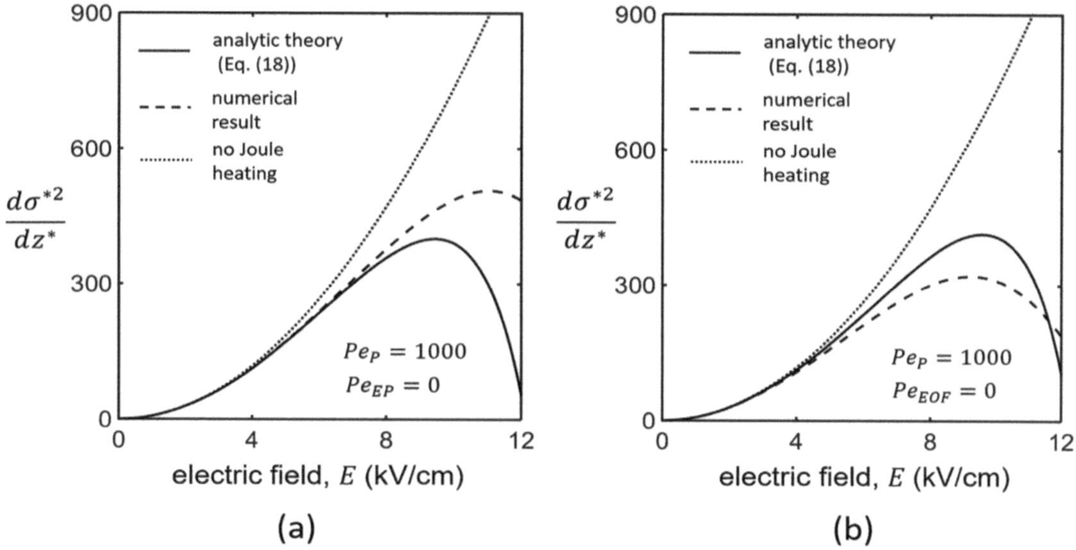

Fig. 7 Effect of electric field on the spatial variance of the analyte stream in the absence of (**a**) electrophoresis and (**b**) electroosmosis. In the above figures, the quantity $\sigma^* = \sigma/d$. Reproduced with permission from Ref. 11

numbers noted above may be realized in a 25 μm deep FFZE channel for a choice of $\bar{U}_P^0 = 1$ cm/s, $E = 8$ kV/cm, and μ_{EP}^0 or $\mu_{\mathrm{EOF}}^0 = 3 \times 10^{-4}$ cm^2/Vs.

3.6 Concluding Remarks

While the mathematical formulation described in this chapter has been applied to evaluating the mean position and spatial variance of a sample stream in free-flow electrophoretic assays, it can also be extended to calculating the higher-order moments of the analyte concentration profile, e.g., skewness, kurtosis, etc. However, the present analysis is only valid for assays in which the transit time for the analyte molecules in the separation chamber is much larger compared to their characteristic diffusion time across the gap between the parallel plates, i.e., $L/\bar{U}_z \gg d^2/D_0$. Moreover, our mathematical model ignores the effect of the channel sidewalls on the pressure-driven flow profile and hence on the hydrodynamic dispersion component of the sample stream. Furthermore, it does not account for any additional band broadening introduced by the abrupt changes in the channel cross section at the entrance and exit of the separation chamber. Fortunately, these effects tend to be insignificant in most free-flow electrophoretic systems rendering the current mathematical model highly useful in assessing the performance of this fractionation method.

4 Notes

1. The parallel-plate geometry considered here ignores the effect of channel sidewalls on fluid and analyte transport. In this situation, the present analysis is only valid for large aspect ratio separation chambers, i.e., $d/W \ll 1$, and in regions significantly away from the sidewalls.
2. The analyte concentration and liquid flow profiles have been assumed to be decoupled in our mathematical model. This decoupling of the mass and momentum conservation equations is strictly valid for dilute samples for which the solute concentration profile does not influence any of the material properties in the system.
3. The choice of the condition $m_1|_{z^*=0} = 0$ is not essential but simplifies the mathematical expressions in our analysis by introducing the sample stream at the origin of the *x*-coordinate.
4. The condition $m_2|_{z^*=0} = \delta^3/12$ assumes that analyte concentration (C) is uniform across the *x*- and *y*-coordinates at the channel entrance $z = 0$.
5. For a channel of uniform cross section, if the pressure-driven cross-flow (u_c) originates due to a variation in EOF along the *x*-coordinate induced by the pH gradient in the system, the quantity $\partial u_c/\partial x$ may be ignored provided $\partial U_{EOF}/\partial x \ll \partial u_{EK}/\partial x$.
6. The thermal profile in a straight microchamber may be calculated independently of the velocity field provided the temperature gradient along the *x*- and *z*-coordinates and the advective velocity along the *y*-coordinate may be neglected.
7. The transverse advective velocity component along the *y*-coordinate may be neglected for small Reynolds number flows in the absence of any natural convection in the channel (small Grashof number).
8. The higher-order terms in ΔT may be estimated following a similar method by retaining the higher-order terms in the expressions for η and D in our analysis.

References

1. Pamme N (2007) Continuous flow separations in microfluidic devices. Lab Chip 7:1644–1659
2. Turgeon RT, Bowser MT (2009) Micro free-flow electrophoresis: theory and applications. Anal Bioanal Chem 394:187–198
3. Dutta D (2014) A method-of-moments formulation for describing hydrodynamic dispersion in free-flow zone electrophoresis. J Chromatogr A 1340:134–138
4. Aris R (1956) On the dispersion of a solute in a fluid flowing through a tube. Proc R Soc Lond A 235:67–77
5. Taylor GI (1953) Dispersion of soluble matter in solvent flowing slowly through a tube. Proc R Soc Lond A 219:186–203

6. Fonslow BR, Bowser MT (2006) Optimizing band width and resolution in micro-free flow electrophoresis. Anal Chem 78:8236–8244
7. van Deemter JJ, Zuiderweg FJ, Klinkenberg A (1956) Chem Eng Sci 5:271–289
8. Dutta D (2015) An analytic description of electrodynamic dispersion in free-flow zone electrophoresis. J Chromatogr A 1404:124–130
9. Dutta D (2017) Broadening of analyte streams due to a transverse pressure-gradient in free-flow isoelectric focusing. J Chromatogr A 1484:85–92
10. Svensson H (1961) Isoelectric fractionation, analysis and characterization of ampholytes in neutral pH gradients. I. The differential equation of solute concentrations at a steady state and its solution for simple cases. Acta Chem Scand 15:325–341
11. Dutta D (2018) Joule heating induced stream broadening in free-flow zone electrophoresis. Electrophoresis. https://doi.org/10.1002/elps.201700307
12. Incropera FP, Dewitt DP (1990) Fundamentals of heat and mass transfer. Wiley, New York
13. Xuan XC, Sinton D, Li DQ (2004) Thermal end effects on electroosmotic flow in a capillary. Int J Heat Mass Transf 47:3145–3157

Chapter 13

Microchip Electrophoresis Tools for the Analysis of Small Molecules

Federico J. V. Gomez and María Fernanda Silva

Abstract

Microchip electrophoresis (ME) results from miniaturization of capillary electrophoresis (CE) to a microfabricated separation device. Both techniques have common characteristics, but in some aspects, the microfluidic separation device has unique features resulting from its planar miniaturized format. Here we describe the process to transfer of CE to ME and the benefits and drawbacks of the chip with respect to the capillary. A practical guide for method development on the microchip for small ionizable molecules such as phenolic compounds, amino acids, or alkaloids is also presented.

Key words Lab-on-a-chip separations, Capillary zone electrophoresis, Phenolic compounds, Indolamines

1 Introduction

Micro total analytical systems (μTAS), also known as lab-on-a-chip (LOC), were introduced about three decades ago with the revolutionary idea of generating an analysis system using microfluidics, preferably with all components of the analysis including sample preparation, sample enrichment, separation of analytes, and detection integrated onto the same microchip [1]. Professor Andreas Manz is regarded as a founding father of the exploration of this area, and his group has contributed inspiring fundamental and applied research discoveries [1]. Born as the blend of microelectronics technology and capillary electrophoresis (CE), microchip electrophoresis (ME) opens new possibilities for the miniaturization of conventional chemical and biochemical analysis. Some of the advantages of the ME over conventional benchtop systems include tailor-made design, marked reduced consumption of reagents and sample, lower waste generation, and increased analysis speed and portability [2]. In ME, the basic CE configuration is miniaturized and transferred to a chip format. With ME, it is possible to integrate injection, separation, and detection on a single

Debashis Dutta (ed.), *Microfluidic Electrophoresis: Methods and Protocols*, Methods in Molecular Biology, vol. 1906,
https://doi.org/10.1007/978-1-4939-8964-5_13,

microchip with typical channel lengths of 5–15 cm. The separation mechanism for ME is similar to that of CE in that differential migration of analytes is governed by their effective ionic mobilities and the mobility of the electroosmotic flow (EOF) under an electric field. Furthermore, in ME the EOF can be generated on different materials [3].

Initially, the most used detection method in microanalytical devices was laser-induced fluorescence (LIF). Nevertheless, the cost of optical instrumentation, the need for analyte derivatization, and the limited portability of LIF have directed to electrochemical detection (ECD), an attractive alternative for microchip devices [2]. The main advantage of electrochemical detection over other detection methods is the ability to miniaturize both the detector and control instrumentation. Four different detection modes can be distinguished in the electrochemical detection: amperometry, voltammetry, conductometry, and potentiometry. Recently, mass spectrometry detectors have also been successfully coupled with MCE. In contrast to CE, UV/Vis absorption is not widely used for on-chip detection because of its low detection sensitivity. Microchip electrophoresis is an excellent separation method for the analysis of biologically important molecules. Separations employing microchip electrophoresis are fast (seconds), highly efficient, and require low sample volumes (pL-to-nL). These qualities make microchip electrophoresis ideal for the analysis of precious biological samples involving clinical, environmental, and food safety studies [4].

2 Materials

2.1 Chip Material and Layout

ME has been performed on many different substrates like glass and polymers, the last mostly using polydimethylsiloxane (PDMS) and poly(methyl methacrylate) (PMMA). This is because both, glass and polymers, exhibit excellent electroosmotic and electrokinetic properties [5, 6]. Desirable characteristics of the substrates that are essential to achieve reasonable separation include (1) stable EOF, (2) good optical clarity at the monitoring wavelengths (in cases of optical-based detectors), (3) accessibility of the investigator to different microfabrication technologies, (4) compatibility with the background electrolyte (BGE), and (5) good thermal and electrical properties (in order to minimize Joule's heating) [7]. Glass is the most commonly used as a substrate in ME since it provides excellent optical clarity and reproducible EOF and exhibits similar surface chemistry to that of fused silica capillaries in conventional CE [3, 8].

In the literature both commercial and lab-made devices have been used, the latter being preferred. In this sense, a variety of strategies have been employed in the production of devices for

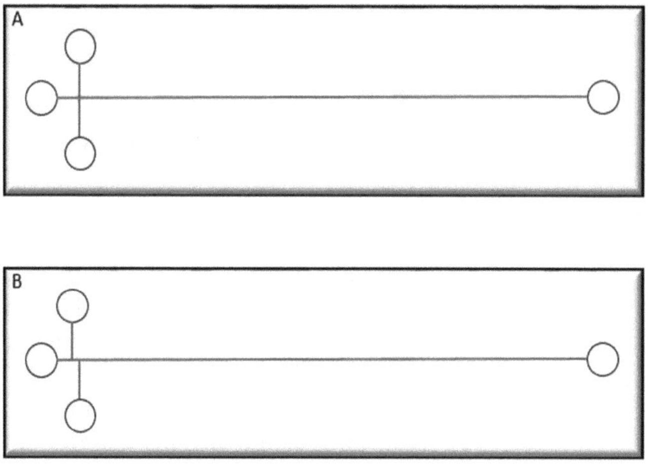

Fig. 1 Most common MC layouts. (**a**) Simple cross and (**b**) double tee

ME ranging from standard photolithography [9] to rapid prototyping [10–12] and assembly [13, 14]. Several microfluidic shapes and designs can be achieved and are up to the analyst's imagination. Nevertheless, the typical microchip design consists of microchannels arranged into a cross shape (or double tee) with one elongated limb (so-called separation channel), where the separation process takes place (Fig. 1). The typical channel depths are 15–50 μm, widths are 50–200 μm, and the lengths of the separation channel are 1–10 cm. Four fluid reservoirs are positioned at the ends of the channels, two for introduction of sample and background electrolyte (buffer solution) and the other two serving as waste reservoirs. Electrodes present in all reservoirs provide connection to a high voltage power supply (*see* **Notes 1–3**) [15].

2.2 Detection

Many modes of detection have been employed to monitor separations on microanalytical devices. The detection on the microchip is usually made at the opposite end of the separation channel, most commonly by laser-induced fluorescence (LIF) and electrochemical detection (ECD). Recently, also mass spectrometry detectors have been successfully coupled with ME. In contrast to CE, UV/Vis absorption is not widely used for on-chip detection because of its low detection sensitivity [15]. However, the cost of optical instrumentation, the need for analyte derivatization, and the limited portability of LIF has led to the investigation of electrochemical detection (ECD), an attractive alternative for microchip devices [16, 17]. Since many compounds are electrochemically active, many applications can be found [2].

The selection of the detection mode should be made taking into account the following:

1. Disponibility.
2. Chemical nature of the target analytes.
3. Desired sensibility.
4. Coupling mode to the system.

2.3 Rinse Solutions

It is common practice in ME to perform some sort of micro channel washing step(s) every day before use and between separations. Typically, a standard washing procedure is used involving a few minutes wash with H_2O followed by a few minutes wash with 0.1 M NaOH and finally by a few minutes reconditioning with the run buffer [18]. The microchip could be rinsed manually using a conventional syringe pump adapted to the system. Prepare all solutions using ultrapure water (prepared by purifying deionized water, to attain a sensitivity of 18 MΩ-cm at 25 °C) and analytical grade reagents. Prepare and store all reagents at indicated temperature. All rinsing solutions should be filtrated before use. In case of microchannels obstructions, several rinses should be made (*see* **Notes 2–4**).

2.4 Background Electrolyte

A wide range of buffers has been employed in ME systems. The most common are phosphate, borate, citrate, acetate, and tris (trishydroxymethylamino methane). However, the selection of a buffer for ME should be based on several factors, including:

1. pH value desired,
2. Charge of the buffer relative to the analytes and the micro channel wall.
3. Effects on detection.

The significance of the micro channel wall in controlling the process of separation in ME cannot be overstated. The separation, however, takes place in the separation buffer. It is here that the conditions are such that the differences in mobility can exist. Buffers can either be made or purchased. Buffer quality is independent of the instrument system used. However, even the best instrument system will not perform properly with a poorly prepared buffer.

Once the buffer composition has been selected, it is essential to prepare it properly. In order to make a buffer consistent, it is critical to make the buffer the same way every time. In addition, in ME it is extremely important to properly control pH since it affects analyte charge, electroosmotic flow, and current. Thus, small changes in pH tend to have greater impact in ME than CE.

Preparation steps:

1. Decide on the kind and the concentration of the buffer. Most buffers are used at a concentration between 0.1 and 10 M. if you make up a concentrated buffer solution, you can dilute it as needed.
2. Decide on the pH for your buffer. This pH should be within one pH unit from the pK_a of the acid/conjugate base.
3. Prepare carefully the solution in a little less than the volume of the flask.
4. Check the pH using a pH meter and adjust the pH as necessary.
5. Once you have reached the desired pH, add water to bring the total volume of the flask.
6. If you prepared this buffer as a stock solution, you can dilute it to make up buffers at other concentrations.

2.5 Standard Solutions

In ME, like in a large number of analytical methods, the calibration of the instrument is required. This typically involves the preparation of a set of standards containing a known amount of the analyte of interest, measuring the instrument response for each standard and establishing the relationship between the instrument response and analyte concentration.

The issues an analyst needs to consider when planning a calibration study are as follows:

1. The number of calibration standards.
2. The concentration of each of the calibration standards.
3. The number of replicates at each concentration.
4. Preparation of the calibration standards.

For an initial assessment of the calibration function, as part of method validation, for example, standards with at least seven different concentrations (including a blank) should be prepared. The standard concentrations should cover, at least, the range of concentrations encountered during the analysis of test samples and be evenly spaced across the range. Ideally, the calibration range should be established so that the majority of the test sample concentrations fall toward the center of the range.

Typically, calibration solutions are prepared from a pure substance with a known purity value or a solution of a substance with a known concentration. The uncertainty associated with the property value (i.e., the purity or the concentration) needs to be considered to ensure that it is fit for purpose (*see* **Note 5**).

Standard solutions are recommended to be prepared daily, and if this is not possible, you must strictly comply with the storage protocols depending on the species you are working with. For example, there are some molecules that are degraded quickly with certain types of light and in these cases are recommendable work with dim light.

The resolution (R) between peaks could be calculated using the following equation:

$$R = \frac{1.18\,(t_2 - t_1)}{w_1 + w_2}$$

where t_1 and t_2 are migration times of corresponding analytes and w_1 and w_2 are the corresponding peak width at the half height of two adjacent analyte bands.

2.6 Samples

In contrast to CE, the injection on the microchip is realized through the intersection of the microchannels, which serves as an integrated injector. The intersection has a form of a simple cross or, alternatively, the form of so-called double tee (*see* Fig. 1), where the two arms of the sample channel are offset to form a larger injector region. The sample is usually injected electrokinetically, by applying an electric field across the sample channel. The portion of the sample present in the intersection represents the injection plug, which is subjected to separation when the electric field is applied across the separation channel. The injection and separation steps are often controlled by independent high voltage power supplies [15].

The accurate extraction of target analytes from the sample (most cases complex matrices) is of utmost importance. It is only possible to conduct further separation, identification, and characterization of target analytes applying an appropriate extraction process. Without sample pretreatment, interferences often will occur when analyzing complex samples. In this sense, one of the most important reasons for performing sample preparation in ME is to concentrate the analyte(s) of interest. This is often necessary because of the limited concentration sensitivityof the most commonly used detection methods. To achieve sample pre-concentration, an extraction step is usually performed, followed by evaporation of the extraction solvent, and reconstitution in a volume smaller than that of the original sample. A potential bonus may be the removal of other components in the sample [18]. The most common extraction procedures for ME are the same than for other techniques like CE and HPLC, and these are solid-phase extraction (SPE), liquid-liquid extraction (LLE), and dispersive liquid-liquid microextraction (DLLME). Miniaturized approaches are ideal for sample prep prior to ME.

3 Methods

In order to transfer capillary electrophoresis to planar microfabricated separation devices, the following considerations should be considered.

3.1 Benefits Also Drawbacks for the Transition from CE to ME

1. Reduced analysis time (seconds) and lower consumption of reagents (microliters) and samples (picoliters). In addition, bear in mind that sample throughput can be further drastically increased using microchip array.
2. Rapid development of a separation method. Microchip, as well as capillary, offers different separation modes for separation of a wide range of analytes. However, in some aspects the microfluidic separation device has distinctive characteristics resulting from its planar miniaturized format.
3. Limited separation efficiency. On the microchip, the injection plug length frequently exceeds the diffusion-controlled zone length and thus becomes the dominant source of dispersion. Consequently, the separation efficiency on the microchip is very often much lower. Therefore, consider that lower electrophoretic efficiencies can be overcome with selective detection systems such as electrochemical approaches.
4. Imprecise injection. In contrast to CE, the injection on the microchip (usually electrokinetic) is performed through the intersection of the microchannels, which serves as an integrated injector (simple cross or double tee). The portion of the sample present in the intersection represents the injection plug, which is subjected to separation when the electric field is applied across the separation channel. So, if possible, work with shorter injection plugs by reducing the width of the injected zone by reducing the dimensions of the cross section of the microchip. Thus, it could be necessary to work with more concentrated samples or including an online pre-concentration step to achieve the same detection sensitivity.

3.2 Optimization Conditions for the Separation of Small Molecular Weight Compounds

The statistics-based optimization of variables using response surface methodology (RSM) can be advantageous over the classical one-variable-at-a-time (OVAT) approach because it allows the evaluation of interacting effects between variables and variable optimization in overall scope from fewer experiments.

3.2.1 Buffer Selection

Capillary zone electrophoresis (CZE) is by far the separation mode most applied in ME. Analytes' acid-base properties should be considered in order to select the appropriate separation conditions. In CZE, the pH of the running buffer should be strategically established according to the pK_a values so all the analytes become ionized. Also, the use of additives such as organic solvents could enhance the resolution of the analytes in electrophoresis. This is owing to the effect of decreasing both the conductivity of the buffer and EOF in the presence of organic solvents. As the conditions of the measurement can be changed promptly on the chip, modifiers have to be tested with the aim to achieve the ME separation of all selected compounds (*see* **Note 6**). Enhancement in resolution may result from a combination of the decreased EOF, decreased thermal diffusion, and improved analyte solubility.

3.2.2 Effect of Voltage

Typical separation voltages for ME are within +500 to +3000 V. Separation voltage affects signal-to-noise (S/N) characteristics; higher voltages produce increases of electrophoretic velocities with the subsequent decrease of the diffusion effect and the narrowing of the peak widths. Also, pronounced decreases at migration times are observed. On the other hand, high voltages produce noise caused by Joule effect. Consequently, a compromise between resolution and separation time should be found in order to find the optimal separation voltage.

3.2.3 Detection Potential

If amperometric detection is applied, hydrodynamic voltammograms for the oxidation of the analytes have to be performed in order to find the optimal detection potential giving the best S/N characteristics (*see* **Note 7**).

3.2.4 Calibration Curve and Analytical Figures of Merit

A study of the effect of each experimental variable on matrix effects should be performed. Thus, create calibration curves from spiked matrix and spiked pure solvent samples after the whole procedure, and calculate matrix to solvent slope ratios for each of the studied analytes. Matrix to solvent ratios higher that 20% indicate severe matrix effect, so standard addition or matrix-matched calibration protocols should be performed depending on the case.

Then, calculate precision, robustness, linearity, and limits of detection and limits of quantification. Additionally, evaluate electrode reproducibility using different electrodes and by evaluating their electroanalytical response.

3.2.5 Sample Analysis

Finally, apply the optimized approach for the analysis of real samples, and compare the results with a certified material, standard procedure (if available), or a previously reported method.

4 Notes

1. Safety considerations: The high voltage supply should be handled with extreme care in order to avoid electrical shock.
2. It is a useful idea to have an optical microscope to observe the microchannels looking for obstructions.
3. All rinse solution should be filtrated previous it use, preferably using filters with pore sizes of 0.1 or 0.22 μm.

4. In case of microchannel obstructions, several rinses should be made with H_2O. Nevertheless, if the obstruction persists, it is useful to use methanol or some surfactant like SDS.
5. It is recommendable the use of some mathematical software (like origin) for the manipulation and analysis of the obtained data.
6. Methanol and acetonitrile are commonly used for this purpose in concentrations about 5–20%.
7. Hydrodynamic voltammograms are obtained by plotting S/N vs. separation voltage (V).

Acknowledgments

This work was supported by funding from the Agencia Nacional de Promoción Científica y Tecnológica, University of Cuyo, and Consejo Nacional de Investigaciones Científicas y Tecnológicas (CONICET).

References

1. Haddad PR (2015) Editorial on "Present state of microchip electrophoresis: state of the art and routine applications" by Eric R. Castro and Andreas Manz. J Chromatogr A 1382:65
2. García CD, Henry CS (2003) Direct determination of carbohydrates, amino acids, and antibiotics by microchip electrophoresis with pulsed amperometric detection. Anal Chem 75(18):4778–4783
3. Nuchtavorn N, Suntornsuk W, Lunte SM, Suntornsuk L (2015) Recent applications of microchip electrophoresis to biomedical analysis. J Pharm Biomed Anal 113:72–96
4. Saylor RA, Reid EA, Lunte SM (2015) Microchip electrophoresis with electrochemical detection for the determination of analytes in the dopamine metabolic pathway. Electrophoresis 36(16):1912–1919
5. Gomez FJV, Silva MF (2016) Microchip electrophoresis for wine analysis. Anal Bioanal Chem 408(30):8643–8653
6. Escarpa A, González MC, Crevillén AG, Blasco AJ (2007) CE microchips: an opened gate to food analysis. Electrophoresis 28 (6):1002–1011
7. Shadpour H, Musyimi H, Chen J, Soper SA (2006) Physiochemical properties of various polymer substrates and their effects on microchip electrophoresis performance. J Chromatogr A 1111(2):238–251
8. Manz A, Harrison DJ, Verpoorte EMJ, Fettinger JC, Paulus A, Lüdi H, Widmer HM (1992) Planar chips technology for miniaturization and integration of separation techniques into monitoring systems. J Chromatogr A 593 (1):253–258
9. Segato TP, Coltro WKT, de Jesus Almeida AL, de Oliveira Piazetta MH, Gobbi AL, Mazo LH, Carrilho E (2010) A rapid and reliable bonding process for microchip electrophoresis fabricated in glass substrates. Electrophoresis 31 (15):2526–2533
10. Gabriel EFM, Duarte Junior GF, Garcia PT, de Jesus DP, Coltro WKT (2012) Polyester-toner electrophoresis microchips with improved analytical performance and extended lifetime. Electrophoresis 33(17):2660–2667
11. Lucio do Lago C, Torres da Silva HD, Neves CA, Alves Brito-Neto JG, Fracassi da Silva JA (2003) A dry process for production of microfluidic devices based on the lamination of laser-printed polyester films. Anal Chem 75 (15):3853–3858
12. Waldbaur A, Rapp H, Lange K, Rapp BE (2011) Let there be chip-towards rapid prototyping of microfluidic devices: one-step manufacturing processes. Anal Methods 3 (12):2681–2716
13. Gabriel EFM, Coltro WKT, Garcia CD (2014) Fast and versatile fabrication of PMMA

microchip electrophoretic devices by laser engraving. Electrophoresis 35(16):2325–2332

14. Segato TP, Bhakta SA, Gordon MT, Carrilho E, Willis PA, Jiao H, Garcia CD (2013) Microfabless microfluidic capillary electrophoresis devices. Anal Methods 5(7):1652–1657
15. Vlcková M (2008) Microchip electrophoresis bioanalytical applications. University of Basel
16. Wang J (2002) Electrochemical detection for microscale analytical systems: a review. Talanta 56(2):223–231
17. Verpoorte E (2002) Microfluidic chips for clinical and forensic analysis. Electrophoresis 23(5):677–712
18. Lloyd DK (1996) Chapter 7: Sample preparation for capillary electrophoresis. Prog Pharm Biomed Anal 2:309–326

Chapter 14

Integrated Microfluidic System for Rapid DNA Fingerprint Analysis: A Miniaturized Integrated DNA Analysis System (MiDAS)—Swab Sample-In to DNA Profile-Out

Jianing Yang, Cedric Hurth, Alan Nordquist, Stan Smith, and Frederic Zenhausern

Abstract

A fully automated rapid DNA analysis system requires integrating several operational elements performing multiple steps onto one single microfluidic platform. The functions to include on the microfluidic platform consist of substrate lysis, lysate DNA extraction, single or multiplexed PCR amplification, amplicon separation, and product readout. Here we describe a fully automated integrated system for forensic short tandem repeat (STR) analysis of reference samples, achieving buccal swab-in and DNA profile-out.

Key words Microfluidics, Integration, STR analysis, DNA extraction, Multiplexed PCR, Capillary electrophoresis

1 Introduction

Since the introduction of the micro total analysis system (μTAS) [1], many integrated microfluidic systems have been developed for DNA/RNA analyses by incorporating different functional modules of cell lysis, nucleic acid extraction or pre-concentration, microfluidic polymerase chain reaction (PCR), and detection with microchip capillary electrophoresis (CE) or microarrays [2, 3]. The integrated microfluidic platform offers many advantages over the benchtop processes. Advantages include reducing sample input, reducing reagent consumption, expediting total analysis time, eliminating the sample transfer steps, and minimizing contamination. All of this is of interest in applications for pathogen detection, drug discovery, and clinical point of analysis (POA) as well as in forensic sciences [4–8].

Rapid genetic analysis used for human identification in a miniaturized platform allows for automation and integration of multistep processes combining DNA extraction, PCR amplification, and

Debashis Dutta (ed.), *Microfluidic Electrophoresis: Methods and Protocols*, Methods in Molecular Biology, vol. 1906,
https://doi.org/10.1007/978-1-4939-8964-5_14,

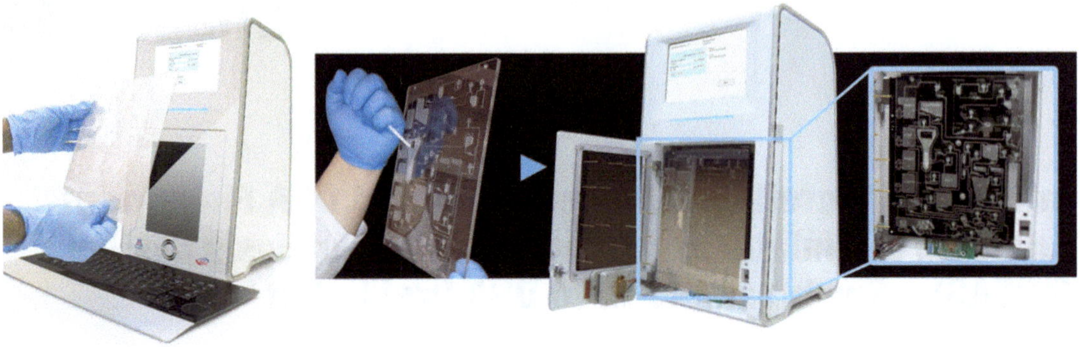

Fig. 1 Illustration of a miniaturized microfluidic-based integrated DNA analysis system—MiDAS. This rapid DNA analysis system is integrated, portable, and mobile. The sample preparation cartridge is plastic, self-contained, and disposable

electrophoretic separation and detection into one single microfluidic device [5, 6]. With a plastic, multichannel microdevice interfaced with a single instrument, four buccal swab samples can be analyzed simultaneously [9]. Using a unique enzymatic liquid for DNA template preparation and microliter non-contact PCR, a high-resolution separation was achieved, matching the profiles obtained from conventional process within 2 h. The DNAscan/ANDE™ rapid DNA analysis platform is a benchtop instrument [10]. It combines both merits of ANDE [11] (Accelerated Nuclear DNA Equipment) and DNAscan [12] and can reliably process five reference swab samples in parallel with a fully integrated STR allele calling expert analysis system. Another integrated STR analysis system, namely, RapidHIT™, developed by IntegenX for forensic reference samples [13, 14], could automatically process five buccal swab samples simultaneously and generate the complete profiles in less than 2 h. The machine is commercially available and deployed at different locations including forensics laboratories.

Initially co-developed by the Forensic Science Service (FSS), we designed a miniaturized integrated DNA analysis system (MiDAS) for forensic reference sample STR analysis [15] (Fig. 1). A modular approach was implemented which allows a customized assay platform and provides a wide range of nucleic acid analyses, not only for human identification, pathogen sample preparation (bacterial and viral DNAs), and detection from liquid biopsy samples but also biomarker detection (RNA and mRNA) from stroke or sepsis patient blood samples (unpublished data), toward a point-of-care (POC) device.

In MiDAS, the integrated microfluidic cartridge workflow consists of five interdependent modules—swab lysis [16], DNA extraction [15], polymerase chain reaction (PCR) [17, 18], PCR product transfer, and CE separation/optical readout [19]. The MiDAS instrument and sample processing cartridge have gone through

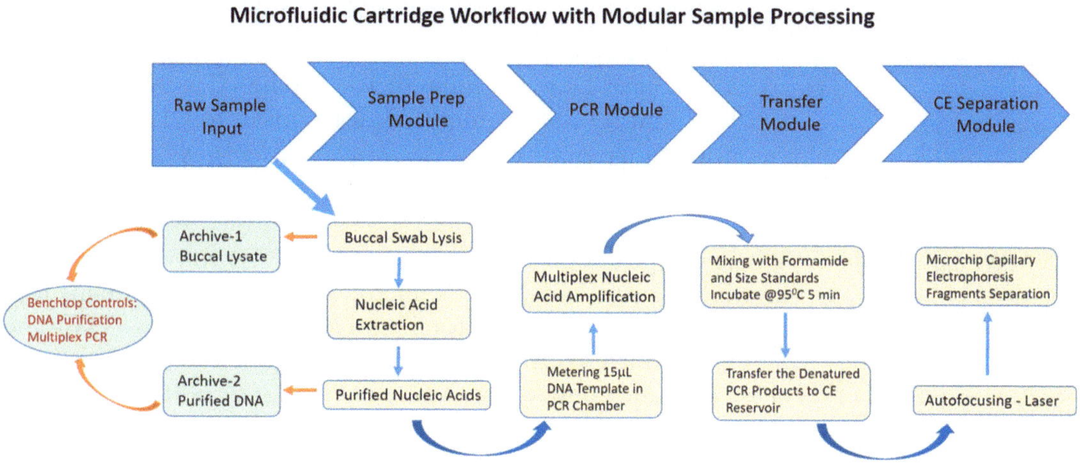

Fig. 2 A schematic presentation of microfluidic cartridge workflow with modular sample processing—sample preparation module, PCR module, transfer module, and CE separation module, completing from the raw swab sample-in to the DNA profile-out

several iterations and modifications [20]. Unlike other integrated STR analysis systems with all reagents onboard, we manually preload the reagents needed onto the cartridge and conserve the versatility of the instrument for a variety of applications. In this chapter, we describe the workflow and process for an example of STR analysis from a buccal swab input to a CE profile output. The system can be easily adapted to other rapid nucleic acid analysis processes (Fig. 2).

2 Materials

2.1 Cartridge Plastics and Elements

1. Substrate: 254 mm × 254 mm × 3-mm-thick polycarbonate (PC) sheet (Bayer Makrolon® or equivalent).
2. Capping layer: 254 mm × 254 mm × 0.5-mm-thick polycarbonate sheet (Bayer Makrolon® or equivalent).
3. Pressure-sensitive adhesive (PSA) layer: ARcare 90106 (Adhesives Research). Two-sided hybrid acrylic adhesive system.
4. Thermally activated single-shot polymer valves (open and close) (Sigma-Aldrich#411663-1Kg 65 °C phase change temperature).
5. Electrodes for the electrochemical (EC) pumps (Al wire, 1 mm, Alfa Aesar 43427-G5). Cut to length, UV-glued (Dymax 1180-MT UV Epoxy) in location.
6. Input points are sealed with laser-cut single-sided PSA (Adhesives Research 7815 or biocompatible 90697 for the PCR chamber).

7. 19 mm of the PEEK tubing (IDEX 1568 1/32″ OD × 0.015 ID) for CE transfer.
8. 200 mm of Dow Corning Silastic tubing (Cat. No. 508-006, 1.47 mm ID × 1.96-mm OD) for transfer to CE.
9. Ethanol (Fisher Scientific, Cat. No. 2701) to clean the cartridge.

2.2 Cartridge Design

The cartridge design is created using AutoCAD software into a 2D drawing format. The MiDAS microfluidic cartridge contains four electrochemical pumps, a total of 19 opening valves (OV) and 29 closing valves (CV), on-cartridge mixing elements, functional chambers, and connecting channels (Fig. 3) (*see* **Note 1**).

2.3 Electronics and Printed Circuit Boards

2.3.1 Manufactured Components

1. Main Board—The main board provides servo controls and interfaces between the PC and the circuit boards.
2. Sample Preparation Board—The sample preparation board provides discreet circuits for heating paraffin valves, chamber heaters, and controlling the electrochemical pump.
3. Thermoelectric Cooler (TEC) Controller and Peltiers (2 ea)—The thermoelectric cooler (TEC) controller and Peltiers (2 ea) controls the cooling fan, drives the Peltiers, and monitors temperatures in a closed-loop system.
4. High-Voltage Power Supply Board—The high-voltage power supply (HVPS) provides ground and/or voltage from 50 V up to 4000 V to four different points in the chip for capillary electrophoresis (CE). Additionally, it controls a chip heater board to keep the CE chip at a constant and uniform 50 °C. The HVPS also stores data during a CE run for voltage, current, and temperature of all HVPS-controlled components.

2.3.2 Commercial Components

An integrated, Windows-based computer runs the rapid DNA system. The PC has custom software to provide a user interface to the system. Connections to the laser and the charge-coupled device (CCD) camera for detection and the main board are required to control the system. There is a single power supply for power distribution.

2.4 Nucleic Acid Extraction Reagents

1. Buccal gDNA extraction—Invitrogen ChargeSwitch gDNA Normalized Buccal Cell Kit (CS) (Cat. No. CS11020, Life Technologies, Carlsbad, CA, USA).
2. Buccal swab—sterile, individually wrapped buccal swabs with ejectable swab head (Whatman Omni Swab WB100035 (Sigma-Aldrich)).

Cartridge Layout

(Shown with locations of initial loading highlighted)

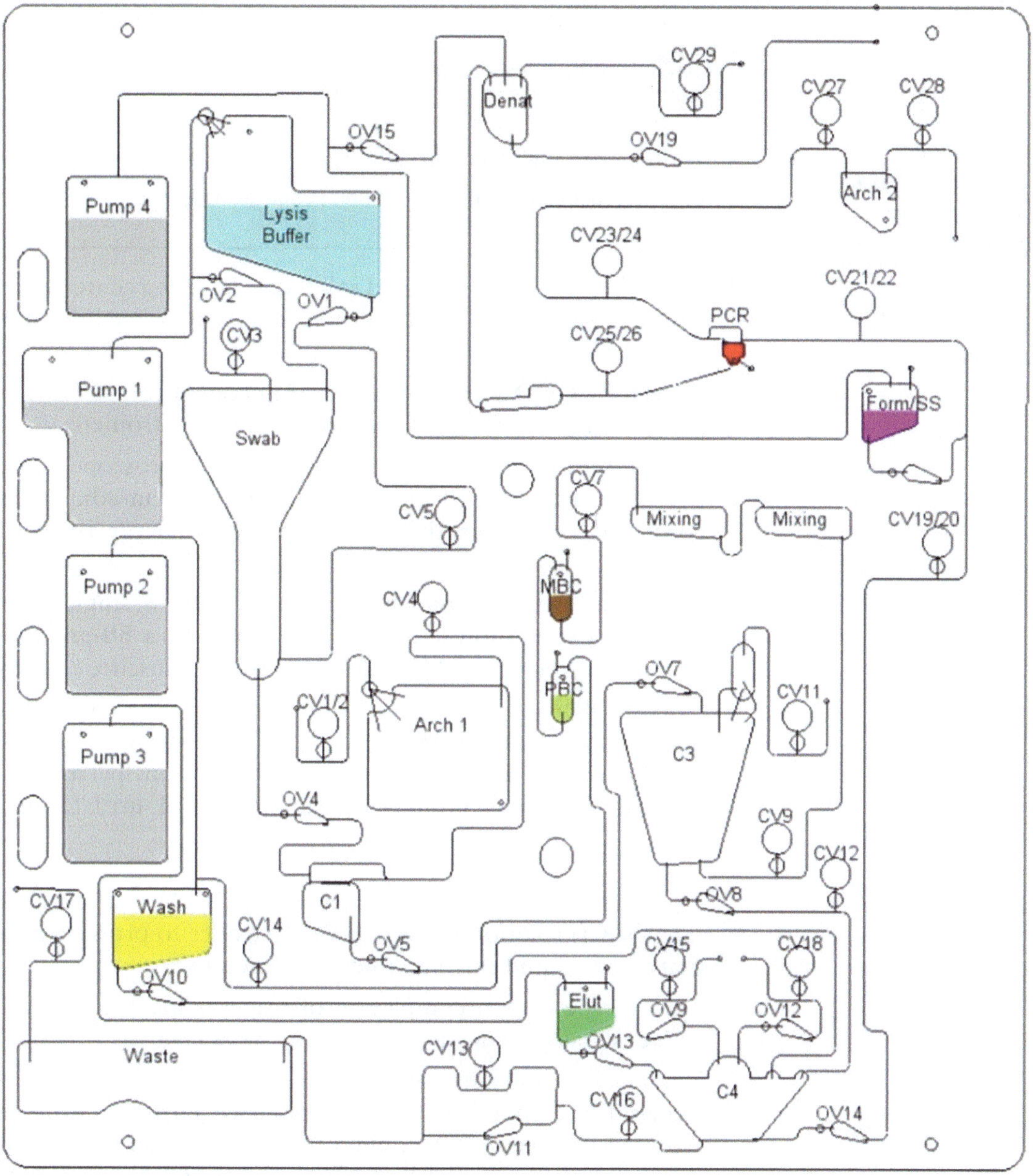

Fig. 3 Microfluidic cartridge layout with chamber labels and functional elements. Chambers loaded with reagents are indicated with colored dyes. The color indicators are pumps (×4), gray; lysis buffer, blue; purification buffer, light green; CS magnetic beads, brown; wash buffer, yellow; elution buffer, green; formamide plus size standard solution, purple; and PCR master mix, red

2.5 Multiplex Amplification Kit

Promega *PowerPlex ESI16* system constituents—per PCR reaction:

1. PowerPlex 16 10× primer pair mix	2.5 μL
2. PCR master mix (5×)	5.0 μL
3. Amplification grade water	2.5 μL
4. Metered purified DNA template	15 μL

2.6 Denaturation Solution and Size Standard (Per Cartridge, Mixing)

1. Hi-Di™ Formamide solution (Life Technologies, CA)	117.5 μL
2. WEN Internal Lane Standard (ILS500 Promega, Madison, WI)	7.5 μL

2.7 Capillary Electrophoresis System

1. Laser source—diode-pumped solid-state laser at a center wavelength of 491 nm (Cobolt Calypso, 50 mW max.) stabilized by a faraday isolator.
2. Detector—CCD sensor (Sony S7031) cooled to −5 °C coupled to a 120-mm focal diffraction grating spectrometer.
3. Optical setup—epifluorescence with a 40× microscope objective (Olympus LUCPLFLN, NA = 0.6) with an adjustment collar to correct for the microchip bottom layer thickness (0.7 mm).
4. Microchip device—borosilicate devices with a single cross injector, a separation length of 130 mm, and a 50-μm wide channel (Micronit Microfluidics B.V., Enschede, The Netherlands).
5. Separation matrix—3.5% w/v mix of hydroxyethyl cellulose (HEC, Sigma-Aldrich 308633) and polyvinylpyrrolidone (PVP, Polysciences 06067) containing 7.1 M urea (Sigma-Aldrich U6504).

2.8 Others

1. Electrochemical pump—electrolysis of 0.25 M sodium chloride (NaCl) solution between two electrodes to produce a gas that moves liquid through micro-channels [3].
2. Sealing tape—upon completion of reagents loading, the cartridge is sealed using laser-cut sealing tapes (Fig. 4).
3. Data analysis software—software to deconvolute fluorescence signals for each of the five fluorescence channels from all acquired CCD frames to obtain the multichannel electropherogram was customized by SoftGenetics LLC (NanoIdentity, version 1.2). A commercial sizing software (GeneMarker v 1.6, SoftGenetics LLC) is used to determine the size of each allele detected based on the calibration with the size standard.

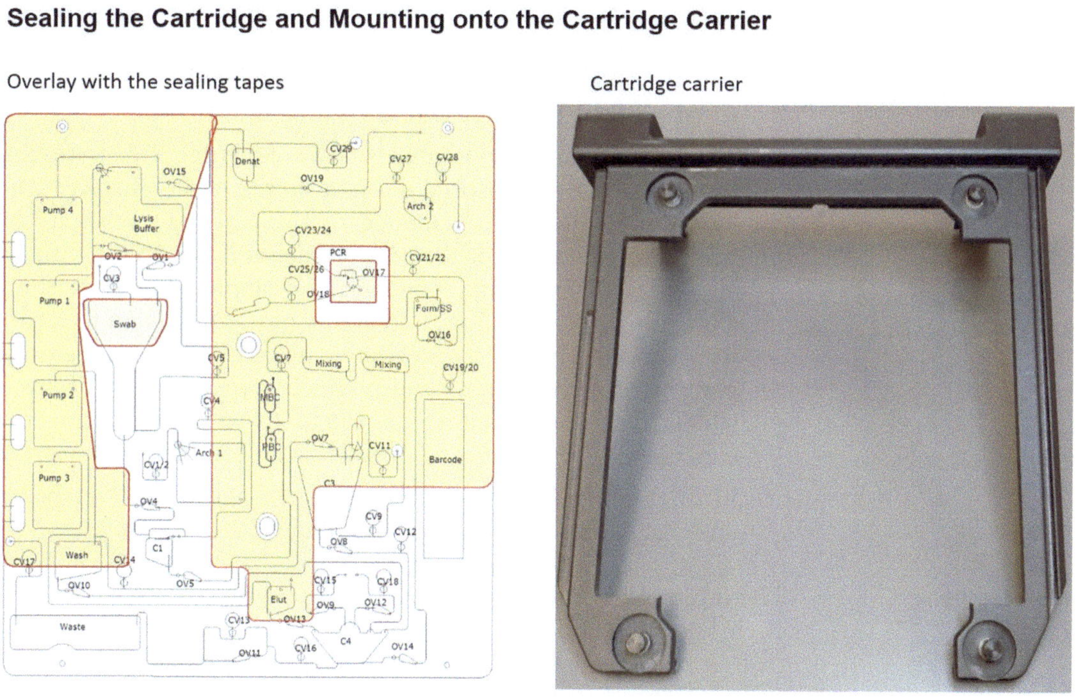

Fig. 4 Cartridge sealing and mounting onto the carrier frame. There are four precut pieces of sealing tape. The sealing tape areas are shown in the red outline. The sealed cartridge is mounted onto the carrier frame by screws at the four corners

3 Methods

3.1 Cartridge Fabrication

1. The cartridge is subtractively fabricated out of a 3-mm thick polycarbonate (PC) sheet. The 2D AutoCAD drawing is converted for CNC (computer numerical control) milling (Roland MDX-540). The 0.5-mm-thick polycarbonate capping layer is laser cut using a VersaLaser 30 W CO_2 laser system. All PSAs are laser cut to size using the 2D drawing file for reference. After machining, cartridges are deburred to remove debris.
2. The cartridges and capping layers are cleaned thoroughly using filtered water and ethanol mixtures (70% ethanol soak 5 min, 99% ethanol rinse) and then air-dried in a BSL/2 biosafety cabinet. UV sterilization is done for approximately 1 h before assembly.
3. Paraffin wax in a 20-mL glass vial is heated past melting point ($\geq$65 °C) on a hot plate. The liquid paraffin is then dispensed manually into each machined valve opening.
4. Install electrodes—Aluminum wire cut to 1.5 (top electrode) and 1.75 (bottom electrode) inches is inserted into the side

ports of the pump chambers via alignment channels and UV-glued to create a liquid seal. Additionally, the PEEK tubing cut to ~1″ length is pressed into the side-drilled pilot hole at the top of the cartridge and UV-glued into place.

5. The subassembly or joining of the cartridge substrate and capping layer is performed on a sterile table surface. The laser-cut 90106 PSA is applied to the cartridge substrate by peeling off one side of the release liner and the PSA is visually aligned to the cartridge substrate. The capping layer is carefully aligned with the cartridge and gently pressed by hand to secure its location. Then, the cartridge is placed in a hydraulic press plate (Carver Press) and subjected a compression of 178 lbs./sq. in. The compression is repeated twice after rotating the cartridge 90° each time, to ensure that air bubbles are removed.
6. Final cleaning—The cartridge is wiped down with 99% ethanol and placed in a ziplock bag for storage.

3.2 Electronic Control

The internal PC controls all aspects of the system. Custom printed circuit boards (PCBs) include embedded resistive heaters and control circuitry for managing heaters, Peltier controllers, and high-voltage controls for CE. The PC also controls the laser and CCD camera. The software provides the operational workflow protocol defining the sequences of operations for the sample preparation process, the PCR process, and the CE process. The PC gathers operational data from each run such as PCR cycling temperatures (interval, 1 s), the HVPS operational values (interval, 1 s), and the CCD acquisition frames (typical exposure, 140 ms).

3.3 General Operation

The instrument is self-contained with the exception of computer peripherals. It houses the electronic control systems and the laser-induced fluorescence detection system and holds the PCB to which the cartridge is precisely aligned for actuation of the pumps and valves. When the computer has booted, the initial start screen will be displayed. The unit is ready to use.

3.4 Run Preparation

The cartridge is individually packed in plastic bags, and gloves are required for handling initial setup (Fig. 3).

3.4.1 Initial Setup

The initial setup consists of sequentially loading the extraction/PCR/denaturation reagents into the cartridge chambers. Refer to Table 1 for details.

1. Eject the buccal swab head into the "key-shaped" swab chamber. Seal tightly with the "trapezoid-shape" PSA tape.
2. Wipe the loading areas with delicate wipes with low lint levels and low electrostatic discharge and install the precut sealing tape.

Table 1
Reagents and volumes required for preloading the microfluidic human ID cartridge

Chamber	Reagent	Volume required
Pump 1	0.25 M NaCl	1.2 mL
Pump 2–4	0.25 M NaCI	0.8 mL
Lysis Buffer	CS Lysis Buffer (L11)	1.0 mL
	Proteinase K	10 μL
PBC	Purification Buffer (N6)	30 μL
MBC	CS Magnetic Beads	15 μL
Wash	CS Wash Buffer (W12)	200 μL
Elut	CS Elution Buffer (E5)	100 μL
Form/SS	Hi-Di Formamide	117.5 μL
	IL500 CC5 Size Standard	7.5 μL
PCR	*PowerPlex ESI 16* Master Mix	10 μL

3. Make sure all the loading holes are sealed tightly (*see* **Note 2**).
4. Load the PCR reagent mix last. Avoid wiping the loading hole before placing the tape over it (*see* **Note 3**).
5. Mount the loaded cartridge onto the cartridge carrier (Fig. 4).

3.4.2 Installing into the Unit

Insert the cartridge carrier vertically from the top into the unit and align with the PCB using the alignment pins. Close the top lid. Run the automated DNA extraction program.

3.4.3 CE Chip Preparation

A general practice is to avoid air bubbles in the reservoir and spills onto the chip adapters. All solutions used for loading the CE microchip should be filtered with a syringe filter (average pore diameter 0.2 μm, hydrophilic polysulfone membrane, Pall Acrodisc 4496). The typical chip (Fig. 5) loading procedure is given below.

1. Fill reservoirs 1, 2, and 3 with 1 M HCl. Apply vacuum to reservoir 4 for 10 min.
2. Replace HCl in reservoirs 1, 2, and 3 with deionized filtered water (Millipore Milli-Q™) or molecular biology DNase-free water. Apply vacuum for 5 min.
3. Remove the buffer and fill the reservoirs with the separation matrix polymer. Apply vacuum to reservoir 4 for 25 min.

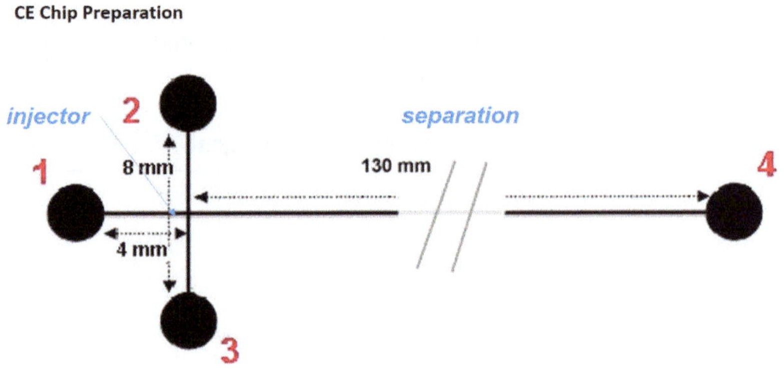

Fig. 5 CE microchip design, orientation, and nomenclature used for the preparation steps

4. Remove the excess polymer from the reservoirs. Add buffer to all reservoirs with only a few μL where sample will be transferred. Once loaded, the CE chip must be used within 30 min to prevent deterioration of the separation matrix.

3.5 Running the System

Once the initialization and hardware check are completed, log into the program with a user name and password from the database holding all authorized users (password and user level). The two levels of access are user and instrument manager.

3.6 Cartridge Run Operation

The user has the option to "Run System" which offers a complete, automated test (normal operation). The user can also run each module separately, review data, or test instrument functionalities (*see* **Note 4**). Insert the preloaded and sealed cartridge (*see* Subheading 3.4) into the unit and close the top lid. The instrument automatically checks the connections of the circuit board to the pump electrodes. The result of this test will be displayed in the connections indicator.

1. Enter the sample ID (*see* **Note 5**).
2. Select "Run System" and the desired protocols for each module (*see* **Note 6**).
3. Sample preparation operation (Fig. 6, the left panel)—The run sequence is Buccal Swab Lysis Module, DNA Extraction Module, and metering 15 μL of purified DNA solution in the PCR chamber (*see* **Note** 7). During the DNA extraction, a portion of the sample lysate (400 μL, metered in C1 chamber) passes through microfluidic channels and mixing chambers, picking up the CS beads and purification buffer. The excess lysate is detoured into Archive-1 (Arch 1) chamber. When the mixture arrives in the C3 chamber (Fig. 3), further mixing is achieved through a bubble-mixing mechanism, by keeping Pump-1 "ON" continuously for 3 min. The mixture of sample lysate

and beads is incubated in the C3 chamber for 3 min, allowing the bead-DNA complex formation. Upon flowing the sample mixture into the C4 chamber, the actuated magnet behind the C4 chamber captures the bead-DNA complexes. The unbound lysate constituents follow the microfluidic circuit into the waste chamber. Sequentially, after the washing step, the elution buffer is delivered to the C4 chamber. Incubation at 60 °C for 3 min completes the separation of DNA molecules from the beads into the elution buffer. Upon switching on the magnet, the beads are collected at the bottom of the C4 chamber, leaving the detached DNA in solution. The purified DNA solution is then delivered to the DNA archive chamber (Arch 2), while 15 μL of DNA template is left in the PCR chamber. The excess samples from both the swab lysate (Arch 1) and purified DNA sample (Arch 2) are temporarily stored in the archive chambers while the cartridge is processed and retrieved after cartridge run completion (*see* **Note 8**).

4. PCR module operation—After metering 15 μL of purified DNA sample in the PCR chamber, the valves surrounding the PCR chamber are closed (in the following order: CV19/20, CV21/22, CV23/24, and CV25/26). When CV27 and CV28 are closed, the microfluidic loop for sample preparation module is closed. A two-step PCR protocol was developed for a rapid STR analysis using the Promega *PowerPlex ESI 16 plus* kit. The short thermal cycling protocol is used: 95 °C for 2 min (initial denature step) and then 1 cycle of 95 °C for 10 s and 58 °C for 2 min, followed by 26 cycles of 95 °C 10 s and 58 °C 2 min. The final extension step is at 58 °C for 30 min (*see* **Note 9**). During the PCR (Fig. 6, the middle panel), the information available to the user includes a real-time graph of the process, the temperature set point and actual reading, the elapsed time, and the number of cycles that remain.

5. Transfer module operation—The transfer is done in two parts. First, the PCR products are transferred to the denature chamber, and then the denatured PCR solution is transferred to the CE reservoir. The whole transfer process is automated. The denature solution containing formamide and size standards is delivered to the denature chamber by Pump 4, passing through OV16, CV21/22, OV17, OV18, and picking up the PCR product (Fig. 3). During denaturation at 95 °C for 5 min, the vent of the denature chamber is closed by actuating CV29. The denatured PCR product is then delivered to the CE device through OV19 and a PEEK tube connecting the cartridge to the CE chip.

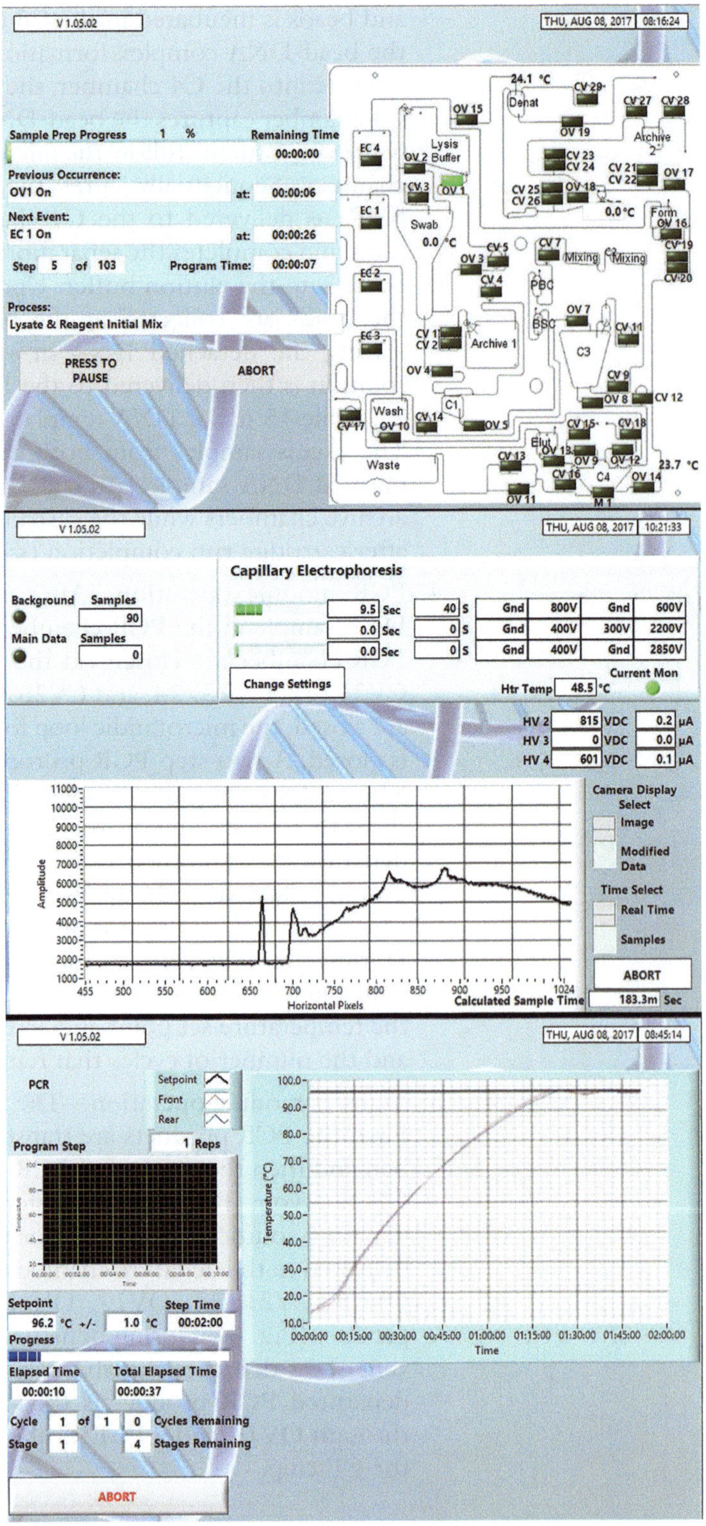

Fig. 6 On MiDAS images—sample preparation, PCR, and CE modules in operation. The left screen shows the live monitoring of DNA extraction in process. The light indicates when an element is in function within the sequence defined by the

3.7 Capillary Electrophoresis (CE) Module

The CE chip is installed after a software prompt upon PCR completion. The software will then call for aligning the CE chip with the incident light beam.

3.7.1 CE Chip Positioning

1. Place the CE microchip in its designed slot and secure it with the clips.
2. Connect the tubing from the sample preparation cartridge to the CE chip.

3.7.2 Alignment

1. Using the alignment software interface, maximize the chip autofluorescence peak at pixel #615 using the DC servo controls (transverse alignment, *y*-axis) in combination with the piezo positioner (focusing, *z*-axis).
2. Occasionally, maximize the overall signal from the background spectrum of a loaded glass chip with the focusing lens to enter the spectrometer slit. With the CE chip installed, the program will initiate the laser and show the alignment screen.

3.7.3 Voltage Parameters (See Note 10)

Typical migration voltages for a 140-mm glass chip using a gated injection:

1. Phase I (loading): E1 = 0 V; E2 = 800 V; E3 = 0 V; E4 = 600 V for 40 s.
2. Phase II (injection): E1 = 0 V; E2 = 400 V; E3 = 300 V; E4 = 2200 V for 2 s.
3. Phase III (migration): E1 = 0 V; E2 = 400 V; E3 = 0 V; E4 = 2850 V for 850 s.

3.7.4 Transfer to the CE Chip

Once the transfer is complete, the CE process begins. Figure 6 (the right panel) shows an example of the CE screen during sample loading. The CE process will execute the three-stage protocol, average 100 background acquisitions, record, and store each CCD acquisition frame, with the details of the voltage, current, and heater temperature.

Fig. 6 (continued) selected protocol. The middle screen shows the PCR in progress: set point temperature (black trace), front Peltier temperature (red trace), and rear Peltier temperature (blue trace). Both cycle numbers and time can be read out on the screen. The PCR progress is visualized with blue LED bar on MiDAS instrument. The right screen shows the live CE chip running screen. The screen displays the sequence of voltages performed and a live trace of the acquired CCD spectrum binned over all horizontal pixels. In this particular case, the system is still performing the initial loading of the denatured PCR product into the chip injector

3.8 Retrieval

1. Sample retrieval: When the cartridge run is completed, the archived samples are retrieved (*see* **Note 8**). The retrieved samples are stored in sterile tubes at −20 °C for further processing to serve as benchtop controls for on-cartridge DNA extraction (with archived sample lysate) and amplification (with archived isolated DNA). Therefore, comparison between conventional benchtop and on-cartridge performance is possible with the same buccal swab sample.
2. Data retrieval: Data files are stored under the name associated with the testing (*see* **Note 11**). The data files are in text format.

3.9 Data Review and Analysis

1. Run log data review: The user has the option to review the operational data (log) recorded during each test. The data available for review includes the operation temperature of the C4 and denature chambers as well as the Peltier during sample preparation and transfer, the operation temperature of the Peltiers during PCR, and the voltage, current, and temperature operations of the CE run.
2. Data deconvolution—*NanoIdentity*: The deconvolution is a two-step process. First, five horizontal binning areas (each defined by pixel, range, and relative ratio) are defined on the full spectrum acquired on the CCD. The binning areas are best obtained by running a matrix standard sample for each fluorescent probe used in the PCR kit. The ratios and the range are used to adjust the intra-color balance between fluorophores given the difference in emission intensity. Then, each fluorescent channel is corrected for cross talk using a 5 × 5 correction matrix subtracting the secondary contributions to each major peak. This is best obtained by running a full allelic ladder. The resulting processed data is saved into a format readable by DNA sizing software.
3. Profile analysis—peak calling with *GeneMarker software*: The size of each detected fragment is attributed by interpolation with a calibration curve obtained from the internal lane standard peaks mixed with the PCR product during the denaturation step (Fig. 7).

In summary, MiDAS employs a self-contained microfluidic cartridge, which contains four modular components: buccal swab lysis and DNA extraction, multiplexed PCR, PCR products transfer for analysis, and microchip CE separation. The user loads only the assay reagents thereby maintaining the flexibility to perform different assays, making the system a more flexible and versatile platform. The instrument provides all the elements required for completion of rapid DNA analysis, namely, a main power supply, controlled thermal cycling, on-cartridge pumping and valving, optical detection, process control, and computer subsystems. It can be termed a swab sample-in and DNA profile-out.

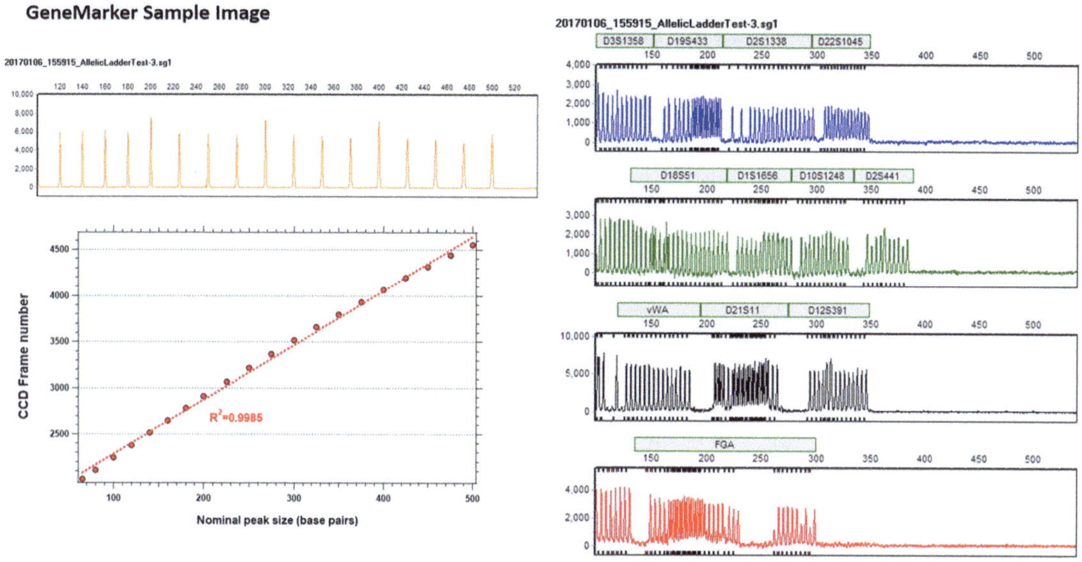

Fig. 7 Size calibration curve peaks and corresponding linear fit from an allelic ladder run on a microchip (left). The sized electropherogram peaks for each fluorescence channel are shown (right), but peak labels are omitted for clarity. GeneMarker v1.6 sizing software was used to generate the graphs

4 Notes

1. The architecture of the fluidic system is designed such that outlets from one chamber to the next are on the bottom side of the chamber, taking advantage of gravity and thus removing bubbles.
2. There are total of three pieces of precut cartridge sealing tape—left side, right side, and swab chamber (trapezoid shape). The PCR chamber sealing tape (small square shaped) is supplied separately. In Fig. 4, the red contour outlines the tapes and cartridge overlay.
3. Slowly dispense the PCR master mix into the PCR chamber (at the bottom) and hold the pipette tightly while withdrawing it away from the loading hole to avoid spills. Do not wipe the PCR loading hole area before placing the sealing tape.
4. The operation runs pre-programmed text files to perform the various processes. The Run System button performs sample preparation followed by the PCR process. When the PCR protocol is complete, the user is prompted to install and align the CE chip. Then, the program resumes by performing the transfer to the CE chip where the electropherogram is then automatically acquired. Each module can be run separately. The Test Connections button performs a self-test of the

pumps on the cartridge to verify that the electrodes are properly connected.

5. Although there is an opportunity later, it is recommended to enter the sample ID in the main menu by keyboard, on-screen keyboard, or barcode scanner reading the barcode on the cartridge.
6. When Run System is pressed, the test menu screen is displayed. The start screens vary based on selections. The operation of each is fundamentally the same. The program loads the default protocol for sample preparation and the default profile for PCR. The Read New SP Protocol button for sample preparation or the Read New PCR Profile button for PCR allows running different versions of either protocol.
7. Figure 6 shows screenshots during sample preparation, PCR, and CE. The user can monitor progress, verify the latest and next step in the process sequence, and view a brief description of the current process. The user can also abort the process. If the sample preparation process is aborted, the system proceeds to the PCR.
8. When the cartridge run is completed, the archived sample can be retrieved. A sterile 18G 1½ needle (BD) is used to gently poke a hole through the sealing tape over the retrieving hole, located at the bottom of each archive chamber. The purified DNA sample (around 60 μL) is collected from the Arch 2 chamber bottom hole with a sterile long-tip pipette, and the residual lysate sample (around 600 μL) is collected from the Arch 1 lysate archive chamber. The retrieved samples are stored at -20 °C in sterile microfuge tubes.
9. To achieve the optimized internal temperature for PCR, the actual set point temperature on the Peltier assembly is 96 °C for 2 min (initial denature step), and then 1 cycle of 95 °C for 10 s and 58.2 °C for 2 min, followed by 26 cycles of 95.7 °C for 10 s and 58.2 °C for 2 min. The final extension step is at 59 °C for 30 min. Programming the PCR protocol is accomplished using the PCR programmer option. To read a different protocol, press READ. To exit, press DONE.
10. The voltage parameters for the CE test are stored in a configuration file. Changes can be made by editing this file.
11. Multiple files are produced with data for each process. One file records all the processes of the test including software version, detector type, user name, and test times and processes including the temperature readings for the PCR run. A second file is a record of the temperatures for the CE chip heater. The CE process creates five files in its run. One file has the time of each CCD reading. Another contains the raw data readings from the CCD. A third has background readings from the CCD. A

fourth stores voltage, current, and temperature readings. The last file provides all the voltage and time settings as well as the equipment (laser, CCD, etc.) used for the CE process.

Acknowledgments

This work was initially sponsored by grants from Forensic Science Services, Ltd. (UK) and the European Commission Seventh Framework Project (FP7)—MiDAS Consortium. The authors would like to thank Matthew Barrett and Baiju Thomas for the cartridge design and fabrication, Brett Duane for his technical support for the MiDAS instrument, and Carla Brooks for her experimental contribution, at the Center for Applied Nano-Bioscience and Medicine.

References

1. Manz A, Graber N, Widmer HM (1990) Miniaturized total chemical analysis systems: a novel concept for chemical sensing. Sens Actuators B Chem 1:244
2. Njoroge SK, Chen HW, Witek MA, Soper SA (2011) Integrated microfluidic systems for DNA analysis. Top Curr Chem 304:203–260
3. Liu RH, Yang J, Lenigk R, Bonnano J, Grodzinski P (2004) Self-contained, fully integrated biochip for sample preparation, polymerase chain reaction amplification, and DNA microarray detection. Anal Chem 76:1824–1831
4. Lagally ET, Scherer JR, Blazej RG, Toriello NM, Diep BA, Ramchandani M, Sensabaugh GF, Riley LW, Mathies RA (2004) Integrated portable genetic analysis microsystem for pathogen/infectious disease detection. Anal Chem 76:3162–3170
5. Easley CJ, Karlinsey JM, Bienvenue JM, Legendre LA, Roper MG, Feldman SH, Hughes MA, Hewlett EL, Merkel TJ, Ferrance JP, Landers JP (2006) A fully integrated microfluidic genetic analysis system with sample-in-answer-out capability. Proc Natl Acad Sci U S A 103:19272–19277
6. Lounsbury JA, Bienvenue JM, Landers JP (2014) Chapter 15: Sample-to-result STR genotyping systems potential and status. In: Shewale JG, Liu RH (eds) Forensic DNA analysis: current practice and emerging technologies. CRC Press Taylor and Francis Group, Boca Raton, FL, pp 350–370
7. Horsman KM, Bienvenue JM, Blasier KR, Landers JP (2007) Forensic DNA analysis on microfluidic devices: a review. J Forensic Sci 52:784–799
8. Bruijins B, Asten AV, Tiggelaar R, Gardeniers H (2016) Microfluidic devices for forensic DNA analysis: a review. Biosensors 6:41. https://doi.org/10.3390/bios6030041
9. Roux DL, Root BE, Hickey JA, Scott ON, Tsuei A, Li J, Saul DJ, Chassagne L, Landers JP, Mazancourt P (2014) An integrated sample-in-answer-out microfluidic chip for rapid human identification by STR analysis. Lab Chip 14:4415–4425
10. Moreno LI, Brown AL, Callaghan TF (2017) Internal validation of the DNAscan/ANDE™ rapid DNA analysis™ platform and its associated PowerPlex-16 High Content DNA Biochip Cassette for use as an Expert System with reference buccal swabs. Forensic Sci Int Genet 29:100–108
11. Tan E (2011) Rapid DNA analysis in the ANDE system: fully integrated, fully PT automated generation of STR profiles from buccal swabs. In: Presented at 22nd international symposium on human identification. National Harbor, MD
12. Manna AD, Nye JV, Carney C, Hammons JS, Mann M, Shamali FA, Vallone PM, Romsos EL, Marne BA, Tan E, Turingan RS, Hogan C, Selden RF, French JL (2016) Developmental validation of the DNAscan™ Rapid DNA Analysis™ instrument and Expert System for reference sample processing. Forensic Sci Int Genet 25:145–156
13. Jovanovich S, Bogdan G, Belcinski R, Buscaino J, Burgi D, Butts ELR, Chear K, Ciopyk B, Eberhart D, El-Sissi O et al (2015) Developmental validation of a fully integrated sample-to-profile rapid human identification

system for processing single-source reference buccal samples. Forensic Sci Int Genet 16:181–194

14. Salceda S, Barican A, Buscaino J, Goldman B, Klevenberg J, Kuhn M, Lehto D, Lin F, Nguyen P, Park C, Pearson F, Pittaro R, Salodkar S, Schueren R, Smith C, Troup C, Tsou D, Vangbo M, Wunderle J, King D (2017) Validation of a rapid DNA process with the RapidHIT ID System using GlobalFiler Express chemistry, a platform optimized for decentralized testing environments. Forensic Sci Int Genet 28:21–34

15. Hopwood AJ, Hurth C, Yang J, Cai Z, Moran N, Lee-Edghill JG, Nordquist A, Estes MD, Haley JP, McAlister CR, Chen X, Brooks C, Smith S, Elliott K, Koumi P, Zenhausern F, Tully G (2010) Integrated microfluidic system for rapid forensic DNA analysis: sample collection to DNA profile. Anal Chem 82:6991–6999

16. Yang J, Brooks C, Estes MD, Hurth CM, Zenhausern F (2014) An integratable microfluidic cartridge for forensic swab sample lysis. Forensic Sci Int Genet 8:147–158

17. Estes MD, Yang J, Duane B, Smith S, Brooks C, Nordquist A, Zenhausern F (2012) Optimization of multiplexed PCR on an integrated microfluidic forensic platform for rapid DNA analysis. Analyst 137:5510–5519

18. Hurth C, Yang J, Barrett M, Brooks C, Nordquist A, Smith S, Zenhausern F (2014) A miniature quantitative PCR device for directly monitoring a sample processing on a microfluidic rapid DNA system. Biomed Microdevices. https://doi.org/10.1007/s10544-014-9895-8

19. Hurth C, Smith S, Nordquist A, Lenigk R, Duane B, Nguyen D, Surve A, Hopwood AJ, Estes MD, Yang J, Cai Z, Chen XJ, Lee-Edghill JG, Moran N, Elliott K, Tully G, Zenhausern F (2010) An automated instrument for human STR Identification: design, characterization, and experimental validation. Electrophoresis 31:1–8

20. Jung W, Yang J, Barrett M, Duane B, Brooks C, Hurth C, Nordquist A, Smith S, Zenhausern F (2014) Recent improvement in miniaturization and integration of a DNA analysis system for rapid forensic analysis (MiDAS). J Forensic Investig 2(1):7

Chapter 15

Achieving Stable Electrospray Ionization Mass Spectrometry Detection from Microfluidic Chips

Iulia M. Lazar

Abstract

The past two decades have witnessed remarkable advances in the development of microfluidic devices as bioanalytical platforms for the analysis of biological molecules. The implementation of mass spectrometry (MS) detection systems on these devices has become inevitable, and various chip-MS ionization interfaces have been developed. As electrospray ionization (ESI) is particularly relevant for the analysis of large biological molecules such as proteins or peptides, efforts have focused on advancing interfaces that meet the demands of nano-separation techniques that are typically used prior to MS detection. Achieving stable ESI conditions that enable sensitive MS detection is, however, not trivial, especially when the spray is generated from a microfabricated platform. This chapter is aimed at providing a step-by-step protocol for producing stable and efficient electrospray sample ionization from microfluidic chips that are used for capillary electrophoresis (CE) separations.

Key words Microfluidics, Mass spectrometry, Electrospray ionization, Signal stability

1 Introduction

To date, a variety of microchip-ESI/MS interfaces [1–19], relying mostly on nano-ESI microfabricated nebulizers, on liquid sheath or liquid junction approaches, and recently on surface acoustic wave nebulizers and electrostatic-spray ionization [20–22], have been developed. In the simplest form, electrospray can be generated directly from the chip edge [1–3], from inserted capillaries [4–9], or from chip-integrated microfabricated emitters [10–12, 15–19]. Interfacing CE separations to mass spectrometry detection is particularly challenging due to the fact that optimal CE separation conditions that involve high concentrations of nonvolatile buffer systems or additives (100–200 mM) are not tolerated by the electrospray ionization process. Approaches relying on the use of a liquid sheath or liquid junction that dilutes the CE separation buffer prior to ESI have proven to be successful. Nevertheless, the inability to use nonvolatile buffer systems, potentially the best for a

Debashis Dutta (ed.), *Microfluidic Electrophoresis: Methods and Protocols*, Methods in Molecular Biology, vol. 1906,
https://doi.org/10.1007/978-1-4939-8964-5_15,

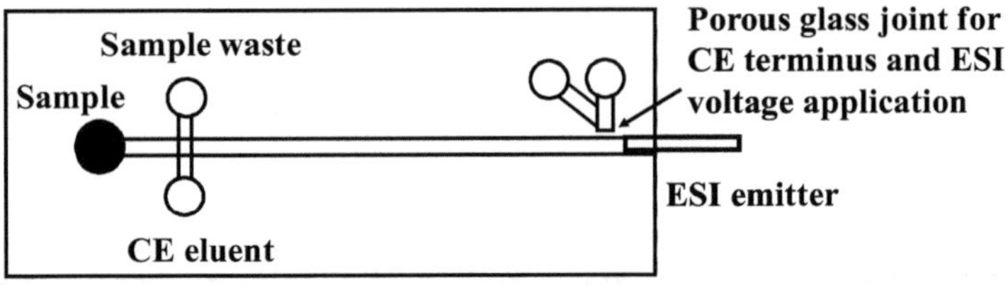

Fig. 1 Schematic diagram of a CE microfluidic chip with an ESI emitter inserted in the CE separation channel. The ESI voltage is applied through a porous glass junction placed at the end terminus of the CE separation channel

given separation, continues to be a challenge. Channel coatings that bleed and generate intense background ions add to the list of concerns. Moreover, additional challenges relate to the chip-MS interface design and operation, i.e.:

(a) The contribution to peak broadening should be minimal; therefore, the dead volumes associated with the interface should be very small, at the pL level.

(b) The chip must produce sufficient EOF to sustain a stable spray; therefore, separations must be conducted in high pH buffers.

(c) Organic solvent modifiers (CH_3OH, CH_3CN) must be added to the CE buffer to enable a stable and efficient ESI process.

(d) The electrical field strengths on the chip must be set properly to maintain balanced electroosmotic flows (EOFs) throughout all channels that do not alter the flat EOF profiles and separation efficiencies [23, 24].

It is beyond the purpose of this work to present in detail the various ESI-MS interfaces that have been developed for CE separations on a chip. Therefore, for the protocol that is described in this chapter, we will consider the case of a simple cross-channel glass chip configuration that has been equipped with a fused silica capillary emitter for the generation of an electrospray, and that has the CE terminus/ESI voltage applied through a porous glass junction [9–25]—a design that interferes minimally with the CE separation performance (*see* Fig. 1). Glass chips are preferred for such applications due to the native ability to generate EOF (300–400 nL/min) at pH ~ 7–8, a flow that is an ideal match for nano-ESI sources that operate at 50–300 nL/min for achieving sensitive, amol-fmol detection. For microfluidic chips that comprise various other implementations of an ESI-MS interface, the contribution of the interface to peak broadening and spray stability and sensitivity should be explored by experiments that are specific to that particular interface. The principles that are described in here, however, are applicable for chip preparation and testing the ESI signal quality.

2 Materials

2.1 Microfluidic Chip Preparation for Analysis

Use only high-purity reagents and HPLC-grade organic solvents and water.

1. Microfluidic chips prepared from glass.
2. Stand for mounting the microfluidic chips.
3. Binocular benchtop microscope.
4. Microanalytical balance (e.g., Mettler Toledo).
5. Ultrasonic bath (e.g., Branson 1/2 gallon).
6. Syringe infusion pump (e.g., Harvard Pump 11 Elite).
7. Syringe: 250 μL (e.g., Hamilton 1725 RN).
8. Vacuum system.
9. Disposable hypodermic syringe: 1 mL, gauge 26, 1/2 in. needle.
10. Conical polypropylene tubes: 15 mL.
11. Fused silica capillary tubing: 20 μm i.d. × 90 μm o.d. (Polymicro Technologies).
12. Fused silica capillary tubing: (50–100) μm i.d. × 360 μm o.d. (Polymicro Technologies).
13. Sodium hydroxide solution: 0.1 M NaOH. Weigh 40 mg NaOH with the analytical microbalance, and dissolve in 10 mL water in a 15 mL conical tube. Disperse by sonication and filter the solution, if necessary, by using a 0.2 μm filter. Store the stock solution at room temperature.
14. Epoxy glue (e.g., Epo-Tek, two-component glue, Epoxy Technology).
15. E6000 glue (Eclectic Products).
16. Methanol.

2.2 Fluidic Flow Testing and Optimization

1. Nikon Eclipse TE-2000U inverted epi-fluorescence microscope with 4×, 10×, or 20× objectives and appropriate fluorescence filter cubes. Other fluorescence microscope systems of similar performance can be used. CCD camera and image processing software are optional.
2. LabSmith microfluidics high-voltage power supply. Other microfluidics power supplies of similar performance can be used.
3. Picoammeter (e.g., Keithley 6487).
4. Digital multimeter (e.g., Fluke 179).
5. High-voltage probe (e.g., Fluke 80K-40).
6. Tabletop microcentrifuge (e.g., Eppendorf 5424).

7. Amber glass sample vials: 4 mL.
8. Polypropylene vials: 1.5–2 mL.
9. Fused silica capillary tubing: 200 μm i.d. × 360 μm o.d. (Polymicro Technologies).
10. CE buffer solution: 10 mM NH_4HCO_3. Weigh 8 mg NH_4HCO_3 with the analytical microbalance, and dissolve in 10 mL aqueous/organic solvent mixture (e.g., H_2O/CH_3OH 70:30 v/v), in a 15 mL conical tube. Disperse by sonication and filter the solution, if necessary, by using a 0.2 μm filter. Store the stock solution at room temperature (*see* **Notes 1** and **2**).
11. Rhodamine 610 stock solution: 2.6 mM. Weigh 5 mg Rhodamine 610 with the analytical microbalance, and dissolve in 4 mL water, in an amber glass vial. Disperse by sonication and filter, if necessary, by using a 0.2 μm filter.
12. Acetic acid, glacial.

2.3 ESI-MS from the Chip

1. Mass spectrometer equipped with nano-ESI source, e.g., LTQ-XL (Thermofisher Scientific). Other MS systems of similar performance can be used.
2. XYZ stage.
3. Thermoshaker (e.g., Eppendorf ThermoMixer® C).
4. Ammonium bicarbonate solution: 50 mM NH_4HCO_3, pH ~ 7.8. Weigh 40 mg NH_4HCO_3 with the analytical microbalance, and dissolve in 10 mL water in a 15 mL conical tube. Disperse by sonication and filter, if necessary, by using a 0.2 μm filter (*see* **Note 2**).
5. Bovine serum albumin (BSA) stock solution: 1.25 mg/mL BSA. Weigh 5 mg BSA with the analytical microbalance, and dissolve in 4 mL ammonium bicarbonate solution (50 mM) in an amber glass vial. Disperse by sonication and filter, if necessary, by using a 0.2 μm filter (*see* **Note 3**).
6. Peptide standard stock solutions: 1 mg/mL. Weigh 4 mg of one or more standard peptides (e.g., angiotensin II, bradykinin, leucine-enkephalin, or gramicidin S) with an analytical microbalance, and dissolve in 4 mL water in an amber glass vial. Disperse by sonication and filter, if necessary, by using a 0.2 μm filter (*see* **Note 3**).
7. Trypsin, mass spectrometry grade: 0.1 μg/μL. Prepare the trypsin solution by dissolving 20 μg sequencing-grade trypsin in 200 μL ammonium bicarbonate solution (50 mM). Disperse briefly by sonication (*see* **Note 4**).

2.4 Data Analysis

1. Computer with operating system compatible with the MS data processing software.
2. Software package for analyzing MS raw files: Proteome Discoverer (Thermofisher Scientific). Other software packages that handle the raw data files generated by the mass spectrometer can be used.
3. Statistical analysis software. Any software package that handles basic statistical data analysis can be used.

3 Methods

3.1 Microfluidic Chip Preparation for Analysis

1. Prepare microfluidic chips in-house [26, 27] or purchase from various manufacturers, e.g., Micronit, Micralyne, Agilent, etc. We note that specialized designs, with electrospray ionization capability, must be fabricated. It is beyond the purpose of this chapter to detail the fabrication of microfluidic devices; however, for describing the procedure, we will consider the use of (a) glass chips that enable either pressure or electrically driven fluid flows (*see* Fig. 1), (b) capillary emitters inserted in a microfluidic channel for generating an electrospray (*see* Fig. 2), and (c) a nanoporous glass junction for applying the ESI high voltage (*see* Fig. 1).
2. Attach reservoirs to the microfluidic chips. The reservoirs can be prepared from glass tubing (e.g., 2–5 mm i.d. × 4–7 mm o.d.), cut in 7–8 mm length, and attached with a two-component epoxy glue. Cure at 90–110 °C for 1 h. Alternatively, commercial polymeric unions can be used that enable the connection of a syringe pump to the chip (*see* **Note 5**).
3. Insert a piece of a fused silica capillary tubing (20 μm i.d. × 90 μm o.d. × 1 cm long) in the end of the microfluidic channel that will deliver the fluid to the ESI-MS interface, seal with E6000 glue, and cure at room temperature overnight (*see* **Note 6**). This will be the ESI emitter.
4. Secure the chip in a stand. For in-house made chips, the stands are also in-house prepared. Such stands can be fabricated from PEEK sheets and braces that help fasten the chips. The stand can also accommodate fittings for fluid delivery [28].
5. Clean/rehydrate the microfluidic channels with NaOH solution (0.1 M). Dispense the solution with the aid of a hypodermic syringe in one reservoir, and allow the channels to fill by capillary action. Fill the chip from only one reservoir to avoid trapping gas bubbles in the microchannels. Apply a slight vacuum from the other reservoirs to help the NaOH solution flow through the channels. Continue the process for 30 min (*see* **Note 7**). Alternatively, if unions are bonded to the chip, the

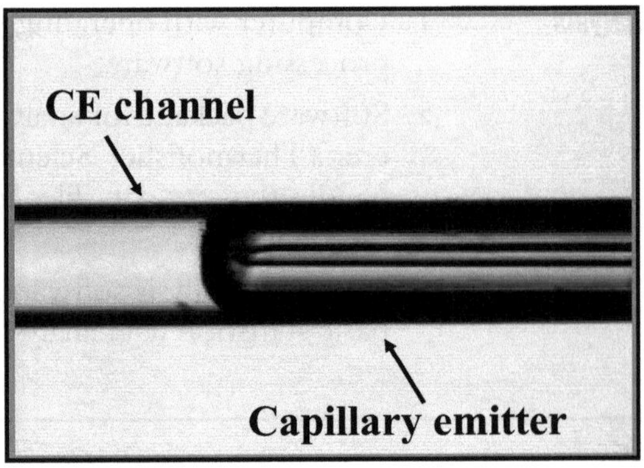

Fig. 2 Picture of a fused silica capillary (20 μm i.d. × 90 μm o.d.) inserted in the CE separation channel

channels can be flushed by using a 250 μL syringe connected via a (50–100) μm i.d. × 360 μm o.d. fused silica capillary. The syringe can be actuated manually or by a syringe pump.

6. Remove the solution of NaOH from all reservoirs and channels through vacuum suction, and rinse first the chip reservoirs with copious amounts of water. Next, fill the channels with water, as described above, and rinse sufficiently to completely remove the NaOH solution. Check with pH paper the pH in the chip reservoirs.
7. Remove the water from all reservoirs, and fill the reservoirs and channels with methanol. Remove the methanol from all microfluidic channels through vacuum suction, and allow the chip to dry.

3.2 Fluidic Flow Testing and Optimization

1. Fill the chip with CE buffer solution, as described in Subheading 3.1, **step 5**.
2. Visually check the chip under a benchtop microscope to ensure that air bubbles are not trapped in the microfluidic channels or reservoirs.
3. Secure the chip in a stand, and introduce Pt electrodes in all reservoirs.
4. Connect the high-voltage power supply, and test for electrical continuity by applying a low electric field strength, e.g., 100–250 V/cm, between various chip reservoirs. Measure the electrical current through the microfluidic channels. This can be accomplished by applying the voltage to one reservoir and grounding the opposite reservoir through the picoammeter. For channel sizes of ~(10–50) μm depth × (50–100) μm wide, the electrical currents should be ~10–100 μA (*see* **Note 8**).
5. If the current measurements are satisfactory and reproducible, increase progressively the field strength to the maximum value

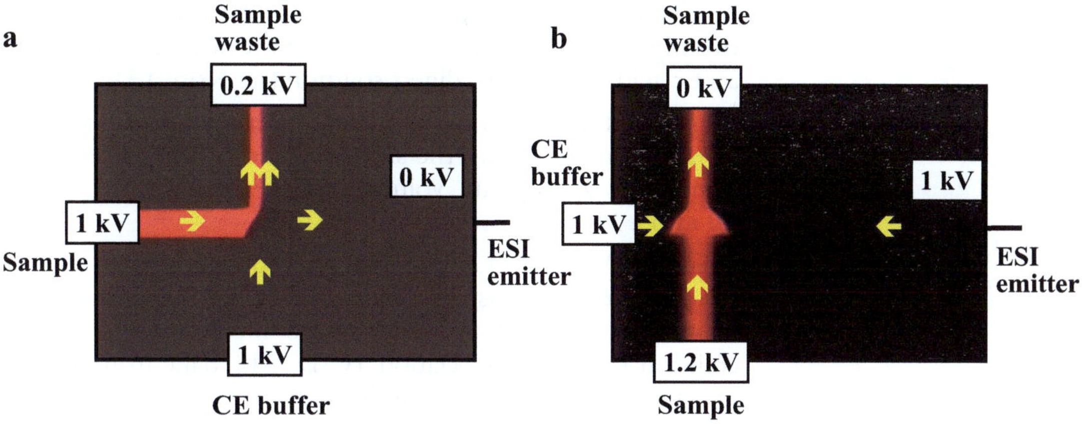

Fig. 3 Pictures of Rhodamine 610 injection in a microfluidic CE channel. The shown voltages enable either a (**a**) gated injection or (**b**) pinched injection

that will be used during analysis (e.g., 400–500 V/cm), repeat the above current measurements, and check for reproducibility (*see* **Note 9**).

6. Measure the fluid flow (EOF) that is delivered through the ESI emitter. This can be accomplished with a simple experimental setup that involves sliding a 200 μm i.d. capillary (~10 cm long) over the 90 μm o.d. ESI emitter and monitoring the displacement of the liquid meniscus vs. time. For stable ESI generation, the flow rate should be at least ~100 nL/min for capillary emitters of 20 μm i.d. × 90 μm o.d. If the chip is equipped with a fine silicon spraying nozzle, e.g., ~(5–10) μm i. d. × 20 μm o.d., lower flow rates (~50 nL/min) will be tolerated (*see* **Note 10**).
7. Prepare a Rhodamine 610 solution of 10–20 μm by diluting an appropriate volume of the stock solution with CE buffer solution (*see* **Note 11**).
8. Place the chip on the fluorescence microscope, and fill the sample reservoir with Rhodamine 610 solution (20 μm). Perform either a gated or pinched CE injection [29] of Rhodamine 610 by applying the appropriate high voltages to the corresponding reservoirs (*see* Fig. 3) (*see* **Notes 12** and **13**).
9. Optimize the sample injection/analysis voltages, and check again the electrical currents between different reservoirs. Monitor the exit EOF through the ESI emitter, as described above, and adjust all voltages proportionally to ensure the presence of sufficient EOF for stable ESI-MS operation (*see* **Note 14**).

3.3 ESI-MS from the Chip

1. Prepare a BSA digest solution by mixing 800 μL of the BSA stock solution (1.25 mg/mL) with the 200 μL trypsin solution in a 1.5 mL Eppendorf polypropylene vial. Mix by vortexing or

brief sonication, and place the vial in the thermomixer for overnight digestion, at 37 °C. Quench the next day with glacial acetic acid, 10 μL/mL digest solution (*see* **Note 15**).

2. Prepare dilute solutions of standard peptides or BSA digest (1–10 μm) by diluting the stock peptide or BSA digest solutions with appropriate volumes of CE buffer (NH_4HCO_3 10 mM). Mix by vortexing.
3. Create an MS and a tandem MS data acquisition method. The MS method will be used for optimizing the ESI baseline signal from the chip, and the tandem MS method for collecting collision-induced dissociation (CID)-MS data from sample injections and separations on the chip.
4. Burn off with a lighter the polyimide coating from the chip ESI emitter for a length of 1–2 mm (*see* **Note 16**).
5. Fill the microfluidic device with fresh CE buffer, place the device in the stand, and secure the stand on an XYZ stage positioned in front of the mass spectrometer. Adjust the position of the chip such that the tip of the ESI emitter is ~2 mm away from the mass spectrometer inlet capillary (*see* **Note 17**).
6. Apply the optimized voltages (*see* Subheading 3.2, **step 9**) to the chip reservoirs, and double-check the presence of flow at the tip of the ESI emitter.
7. Next, raise all voltages, proportionally, until the level of the ESI voltage is sufficient for generating a stable Taylor cone and signal from the ESI emitter. For the emitter used in this study (20 μm i.d. × 90 μm o.d.), the optimal ESI voltage is ~1800–2000 V. If the ESI source is equipped with a CCD camera, the onset of the spray can be monitored visually.
8. Allow for the spray to stabilize for 4–5 min, and assess thereafter the stability of the signal by acquiring data with the MS acquisition method. Process the results and display the total ion chromatogram (TIC). For a stable ion trace, the RSD is within ±10% of the signal average. If signal spikes are present and frequent, the flow rate, the position of the ESI emitter in the ion source, or the ESI voltage should be adjusted (*see* Fig. 4a). Averaging 3–10 mass spectra improves the stability of the signal (*see* **Notes 18–20**).
9. Perform an injection, initially of a single peptide or a mixture of 3–5 standard peptides and then of a more complex sample such as the BSA digest. Use the tandem MS method for data acquisition.
10. Process the raw MS files with the software package at hand. Identify the peptide sequences and assess separation efficiency.

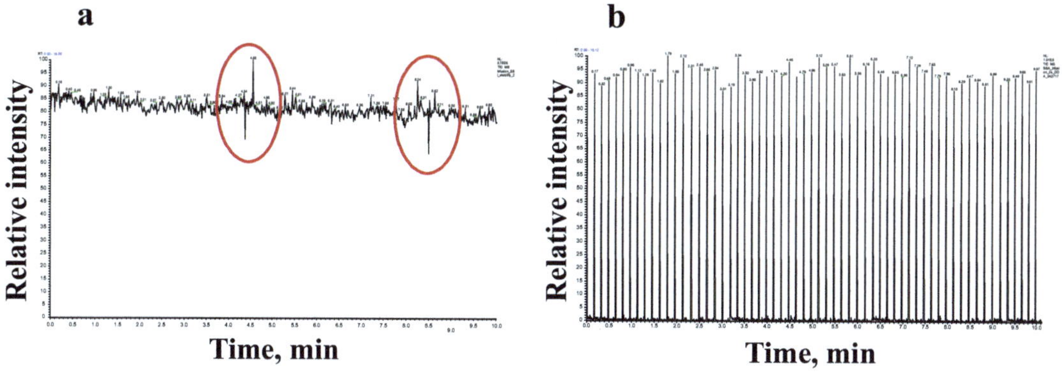

Fig. 4 Chromatograms displaying the stability of the ESI signal during an infusion experiment. (**a**) TIC generated with an MS acquisition method, the spikes represent instances of spray instability; (**b**) TIC generated with a data-dependent acquisition method, the peaks represent instances of full mass spectra, while the valleys represent instances of tandem mass spectra

11. Perform multiple injections, and evaluate sensitivity, sample migration reproducibility, and chip reusability. Low fmol range detection of peptides from simple mixtures should be a routine. If the results are satisfactory, the overall microfluidic method can be used for further separations.

4 Notes

1. The CE buffer solutions must be optimized for a given separation. Nevertheless, when CE separations are interfaced to ESI-MS detection, only volatile buffer systems can be used for separation, such as NH_4HCO_3, CH_3COOH, or HCOOH. Nonvolatile salts, such as phosphate buffers, deposit on the MS inlet ion optics elements, deteriorate the electrical fields, and very rapidly suppress the ESI signal. Such salts can also contribute to irreversible corroding of the MS ion optics components. In addition, the volatile buffers must be dissolved in an aqueous/organic solvent mixture that has at least 20% organic solvent (e.g., CH_3OH), preferably more (30–50%), to ensure stable ESI operation. In case that the chip has a design that enables the delivery of an organic solution that is mixed with the CE buffer just prior to ESI, then the CE buffer solution can be prepared with water alone. The pH of the CE buffer can be adjusted to a range of 6–8; however, lower pH values should be avoided as the EOF will be too small to sustain stable ESI.
2. All solutions that are electrosprayed from the chip and that contain volatile components must be prepared fresh, daily, prior to analysis.

3. The solubility of protein and peptide stock solutions in water (or other solution of choice) should be checked prior to preparation. The stock solutions can be preserved in a solid-frozen state at −20 °C. In case that the solutions contains organic solvents, freeze at −80 °C.
4. The trypsin solution should be prepared fresh, right before performing the digestion of the proteins, to avoid losing its enzymatic activity.
5. Any epoxy glue that is used in the fabrication of microfluidic chips should be evaluated for possible bleeding of components that generate background noise in the ESI mass spectra of the aqueous/organic eluents that are used during analysis.
6. The E6000 glue continues to be soft after curing. This enables the removal of the glue and the ESI emitter, in case of damage and replacement. If this procedure is attempted, the glue should be peeled off carefully such that its removal extracts the emitter.
7. Do not allow the chip to dry out while filled with NaOH solution, as this may result in channel clogging. Replenish the reservoirs continuously with solution while the chip is treated with NaOH, and/or preserve the chip in a water-saturated chamber.
8. The presence of organic solvents reduces the magnitude of the electrical current.
9. Avoid the use of high electric field strengths that produce high current, Joule heating, bubble formation, and possible catastrophic failure of the experiment.
10. Very fine ESI emitters, of a few micrometer i.d., are capable of pulling their own flow in the presence of a strong electrical field. No pumping or EOF is necessary. Clogging of such fine emitters is, however, frequent.
11. The concentration of the Rhodamine 610 solution that will be injected in the chip should be sufficient to enable visualization with the available fluorescence microscope.
12. Rhodamine 610 is a fluorescent dye that at neutral and high pH values has a zwitterionic structure. It is neutral and can be used as a tracer dye to monitor the direction of the fluidic flows. At low pH, Rhodamine 610 is (+) charged; therefore, in addition to its electroosmotic mobility, it will also display electrophoretic mobility.
13. When the chip is interfaced to an MS detector and flow is necessary for the generation of an electrospray, gated injections are preferred as such injections do not interrupt the flow toward the ESI emitter and do not destabilize the spray. Pinched injections can be used if can be performed fast. Small

perturbations during injection, due to voltage switching, are expectable, however, in either case.

14. If the experimental setup in the lab includes a laser point detection system from the chip, the movement of a fluorescent plug of Rhodamine 610 through the microfluidic separation channel can be observed and used for the calculation of the flow rate. Rhodamine 610 can be used for this purpose only at neutral or basic pH, when it does not possess electrophoretic mobility.
15. Typically, to improve the enzymatic digestion process, the proteins are first denatured in a urea solution (6–8 μm), the disulfide bonds are reduced in the presence of dithiothreitol (5 mM), and the free cysteine thiol groups are alkylated with iodoacetamide to block the random regeneration of disulfide linkages. In such cases, the protein digest solutions should be first cleaned with C18 desalting cartridges before attempting injection on the chip. For the purpose of this demonstration, reduction and alkylation were ignored, as these steps did not affect the outcome of the optimization process.
16. The polyimide coating should be removed from the ESI emitter, as during operation it can peel off and destabilize the electrospray ionization process.
17. In most cases the ESI-MS interface of the mass spectrometer must be modified to accommodate the microfluidic chip. An XYZ stage represents a simple solution. The chip and the stand should be enclosed in a transparent plastic chamber to prevent accidental exposure to high voltages.
18. Averaging mass spectra is useful for improving signal stability; however, the number of mass spectra that can be averaged is limited by the data acquisition speed of the particular MS platform that is used in the experiment. The time needed to generate the averaged mass spectrum should be kept short, e.g., <1 s, to avoid contributions to peak broadening and loss of separation resolution.
19. It is useful to have "practice" chips that encompass only a sample infusion channel that can be connected through a polymeric union to a syringe pump. With such a setup, a known, stable flow rate can be generated, and the sample solution can be continuously infused for optimizing the ESI signal and the MS data acquisition parameters.
20. Whether using a regular CE or an infusion chip, EOF, or syringe pumping for fluid delivery, the performance of the chip ESI-MS platform can be also evaluated with the data-dependent acquisition method while infusing a peptide mixture (*see* Fig. 4b). A stable signal will result in equally spaced "signal spikes" of approximately the same height. The signal

spikes in Fig. 4b represent occurrences of full mass spectra when a TIC is recorded, while the valleys represent the period of time that the instrument spends on acquiring the tandem mass spectra and when a TIC is not recorded.

Acknowledgment

This work was supported by grant NSF/DBI-1255991 to I.M.L.

References

1. Xue Q, Foret F, Dunayevskiy YM, Zavracky PM, McGruer NE, Karger BL (1997) Multichannel microchip electrospray mass spectrometry. Anal Chem 69:426–430
2. Ramsey RS, Ramsey JM (1997) Generating electrospray from microchip devices using electroosmotic pumping. Anal Chem 69:1174–1178
3. Xue Q, Dunayevskiy YM, Foret F, Karger BL (1997) Integrated multichannel microchip electrospray ionization mass spectrometry: analysis of peptides from on-chip tryptic digestion of mellitin. Rapid Commun Mass Spectrom 11:1253–1256
4. Figeys D, Ning Y, Aebersold R (1997) A microfabricated device for rapid protein identification by microelectrospray ion trap mass spectrometry. Anal Chem 69:3153–3160
5. Figeys D, Gygi SP, McKinnon G, Aebersold R (1998) An integrated microfluidics-tandem mass spectrometry system for automated protein analysis. Anal Chem 70:3728–3734
6. Xu N, Lin Y, Hofstadler SA, Matson D, Call CJ, Smith RD (1998) A microfabricated dialysis device for sample cleanup in electrospray ionization mass spectrometry. Anal Chem 70:3553–3556
7. Li J, Thibault P, Bings NH, Skinner CD, Wang C, Colyer C, Harrison J (1999) Integration of microfabricated devices to capillary electrophoresis-electrospray mass spectrometry using low dead volume connection: application to rapid analyses of proteolytic digests. Anal Chem 71:3036–3045
8. Lazar IM, Sundberg SA, Ramsey RS, Ramsey JM (1999) Subattomole-sensitivity microchip nanoelectrospray source with time-of-flight mass spectrometry detection. Anal Chem 71:3627–3631
9. Lazar IM, Ramsey RS, Jacobson SC, Foote RS, Ramsey JM (2000) Novel microfabricated device for electrokinetically induced pressure flow and electrospray ionization mass spectrometry. J Chromatogr A 892:195–201
10. Schultz GA, Corso TN, Prosser SJ, Zhang S (2000) A fully integrated monolithic microchip electrospray device for mass spectrometry. Anal Chem 72:4058–4063
11. Kameoka J, Orth R, Ilic B, Czaplewski D, Wachs T, Craighead HG (2002) An electrospray ionization source for integration with microfluidics. Anal Chem 74:5897–5901
12. Sjödahl J, Melin J, Griss P, Emmer A, Stemme G, Roeraade J (2003) Characterization of micromachined hollow tips for two-dimensional nanoelectrospray mass spectrometry. Rapid Commun Mass Spectrom 17:337–341
13. Sen AK, Darabi J, Knapp DR (2009) Design, fabrication and test of a microfluidic nebulizer chip for desorption electrospray ionization mass spectrometry. Sensors Actuators B Chem 137:789–796
14. Deng Y, Henion J, Li J, Thibault P, Wang C, Harrison DJ (2001) Chip-based capillary electrophoresis/mass spectrometry determination off carnitines in human urine. Anal Chem 73:639–646
15. Chambers AG, Mellors JS, Henley WH, Ramsey JM (2011) Monolithic integration of two-dimensional liquid chromatography-capillary electrophoresis and electrospray ionization on a microfluidic device. Anal Chem 83:842–849
16. Chambers AG, Ramsey JM (2012) Microfluidic dual emitter electrospray ionization source for accurate mass measurements. Anal Chem 84:1446–1451
17. Sainiemi L, Nissila T, Kostiainen R, Franssila S, Ketola RA (2012) A microfabricated micropillar liquid chromatography chip monolithically integrated with an electrospray ionization tip. Lab Chip 12:325–332

18. Sainiemi L, Sikanen T, Kostiainen R (2012) Integration of fully microfabricated, three-dimensional sharp electrospray ionization tips with microfluidic glass chips. Anal Chem 84:8973–8979
19. Mao P, Gomez-Sjoberg R, Wang D (2013) Multinozzle emitter array chips for small-volume proteomics. Anal Chem 85:816–819
20. Heron SR, Wilson R, Shaffer SA, Goodlett DR, Cooper JM (2010) Surface acoustic wave nebulization of peptides as a microfluidic interface for mass spectrometry. Anal Chem 82:3985–3989
21. Ho J, Tan MK, Go DB, Yeo LY, Friend JR, Chang HC (2011) Paper-based microfluidic surface acoustic wave sample delivery and ionization source for rapid and sensitive ambient mass spectrometry. Anal Chem 83:3260–3266
22. Gasilova N, Yu Q, Qiao L, Girault HH (2014) On-chip spyhole mass spectrometry for droplet-based microfluidics. Angew Chem 126:4497–4501
23. Lazar IM, Rockwood AL, Lee ED, Sin JCH, Lee ML (1999) High-speed TOFMS detection for capillary electrophoresis. Anal Chem 71:2578–2581
24. Lazar IM (2007) Microfluidic devices with mass spectrometry detection. In: Landers JP (ed) Handbook of capillary and microchip electrophoresis and associated microtechniques, 3rd edn. CRC Press, Boca Raton, FL, pp 1459–1506
25. Khandurina J, Jacobson SC, Waters LC, Foote RS, Ramsey JM (1999) Microfabricated porous membrane structure for sample concentration and electrophoretic analysis. Anal Chem 71:1815–1819
26. Armenta JM, Dawoud AA, Lazar IM (2009) Microfluidic chips for protein differential expression profiling. Electrophoresis 30:1145–1156
27. Iliescu C, Taylor H, Avram M, Miao J, Franssila S (2012) A practical guide for the fabrication of microfluidic devices using glass and silicon. Biomicrofluidics 6:016505
28. Lazar IM, Kabulski JL (2013) Microfluidic LC device with orthogonal sample extraction for on-chip MALDI-MS detection. Lab Chip 13 (11):2055–2065
29. Ermakov SV, Jacobson SC, Ramsey JM (2000) Computer simulations of electrokinetic injection techniques in microfluidic devices. Anal Chem 72:3512–3517

Chapter 16

Microchip-Based Electrophoretic Separations with a Pressure-Driven Backflow

Ling Xia and Debashis Dutta

Abstract

It is well known that the resolving power of capillary zone electrophoretic separations may be improved with an increase in the applied electric field strength and separation time. While large electric fields may be realized in short analysis channels commonly employed in microfluidic systems, this experimental design is not suitable for achieving long separation times. In this chapter, we describe the use of a steady and/or a periodic pressure-driven backflow to increase the separation time in short microchannels thereby enabling the analysis of closely related species on microchip devices. The reported backflow was realized in our assays using an on-chip pressure-generation capability that relied on the partial blockage of electroosmotic flow around a junction of two glass channel segments having different depths. Although the noted strategy led to additional band broadening in the system, the resolving power of our device was observed to substantially improve upon introduction of the reported steady/periodic pressure-driven backflow for analysis channels shallower than 5 μm.

Key words Band broadening, Capillary zone electrophoresis, Electroosmosis, Microfluidic pump, Pressure-driven backflow

1 Introduction

Capillary zone electrophoresis (CZE) is an established liquid phase analysis technique that allows the separation of closely related chemical and biological species under relatively gentle operating conditions [1]. With the emergence of microfluidic systems, further improvements in the separation performance of CZE have been possible primarily through miniaturization of the analysis column. In particular, the shorter cross-sectional dimensions of microfluidic channels have allowed the injection of narrower plugs as well as diminished mass transfer limitations and Joule heating effects in these systems, thereby minimizing the broadening of analyte bands. In addition, the ability to readily integrate other sample preparation and analysis procedures to CZE on microchip devices has enabled the realization of complicated assays with high efficiencies and

Debashis Dutta (ed.), *Microfluidic Electrophoresis: Methods and Protocols*, Methods in Molecular Biology, vol. 1906, https://doi.org/10.1007/978-1-4939-8964-5_16,

using small sample sizes. Nevertheless, the inability to accommodate long separation columns in microfluidic systems has limited the applicability of microchip-based separation methods to relatively simple samples. In this situation, there is a clear need for developing strategies that would allow an improvement in the peak capacity of microfluidic separations enabling them to analyze complex mixtures of practical significance.

Over the years, a variety of approaches have been proposed to shorten the separation length of CZE assays. A key objective in these approaches has been to slow down the migration of the analyte molecules through the assay column without compromising their electrophoretic transport rates such that the sample is subjected to the electric field for the maximum possible time. To this end, the use of static and dynamic coatings to reduce electroosmotic flow (EOF) in CZE systems is well known. Such a flow is usually undesirable in electrophoretic assays as it advects all molecules with the same speed and therefore does not contribute to the separation process. Notably, dynamic coatings are easier to apply and can withstand high pH values, but they have been often reported to be less stable than their static counterpart. Moreover, these coatings usually included as additives in the separation buffer can alter the electrophoretic transport of the analyte species besides interfering with the direct integration of the CZE technique to several detection and/or downstream analysis methods. Static coatings on the other hand are particularly unsuitable for miniaturization of the CZE method due to increasing difficulties in applying them to the channel walls as the size of the analysis column is reduced to the micro-/sub-micrometer length scale. Recently, the application of external voltages at the walls of microfluidic channels has been shown to allow local control of zeta potential, and therefore EOF, in a dynamic fashion [2]. Although this approach overcomes some of the disadvantages of static/dynamic coatings, it offers limited control over the range of electroosmotic transport rates that may be realized in the system and tends to complicate the instrumentation aspects of the CZE method. The use of a steady pressure-driven flow to counteract electroosmosis in CZE analysis presents an alternative approach to reducing the separation length of this technique [3]. The ability to periodically vary the strength of this flow field moreover can allow the realization difficult charge-based separations in short microfluidic ducts based on the flow counterbalanced capillary electrophoretic (FCCE) method as described by Culbertson and Jorgenson [4, 5]. The FCCE technique relies on the use of a pressure-driven backflow to periodically reverse the direction of sample migration without inverting the electric field in it. Consequently, sample molecules can be kept in the separation field much longer than otherwise possible under normal CZE conditions, significantly increasing the effective separation length and voltage for the assay.

Unfortunately, the application of a steady or a periodic pressure-driven backflow introduces hydrodynamic dispersion of analyte bands [3–5] often compromising the resolving power of the CZE method. However, because such sample dispersion scales directly with the narrowest cross-sectional dimension of the analysis channel [6], it is possible to diminish this process by either packing CZE columns with small particles or miniaturizing their lateral dimensions. Previous studies have shown that the use of packed capillaries for CZE analyses with a pressure-driven backflow to be ineffective in reducing hydrodynamic dispersion likely due to non-uniformities in the particle packing density [7]. On the other hand, our limited ability to dynamically control pressure-driven transport within micro-/nanofabricated structures presents a significant challenge in the miniaturization of CZE methods with counteracting pressure gradients. It must be pointed out that while pressure-driven flows may be generated in micrometer-sized channels using an external syringe pump, the dynamic control over this flow velocity is often poor in such an arrangement as a result of the large dead volumes involved at the microchip-pump interconnect. Moreover, the hydrodynamic resistance to liquid transport increases sharply with a reduction in the lateral dimensions of a microfluidic duct which compromises our ability to realize a wide range of pressure-driven flow velocities primarily due to leakage issues at the microchip-pump junction. In this chapter, we describe the use of a simple on-chip pumping unit that allows precise introduction of a steady or a periodic pressure-driven backflow into a microfluidic separation channel enhancing the resolution of CZE assays performed in it [3, 5, 8]. The backflow in the reported devices was produced by fabricating a shallow segment (0.5 μm deep) downstream of the analysis column (5 μm deep) and applying an electric field across it. A mismatch in the electroosmotic transport rate at the interface of this segment was shown to then yield a pressure gradient that counteracted the electrokinetic migration of the analyte bands either slowing them down or reversing their flow direction altogether without significantly compromising the separation field in the system. Although such a pressure gradient also led to additional band broadening, the overall separation resolution of our device was observed to improve upon introduction of the reported pressure-driven backflow owing to the increased separation time in the assay.

2 Materials

Use only high purity reagents and deionized water for preparation of buffers and samples.

2.1 Microfluidic Chip Preparation for Analysis

1. Microfluidic chips made from glass.
2. Cylindrical glass tubes about 4 mm i.d. and 8 mm tall.
3. A stereo microscope (e.g., Nikon SMZ).
4. Microanalytical balance (e.g., Mettler Toledo).
5. Ultrasonic bath (e.g., Branson 1/2 gallon).
6. Disposable hypodermic syringe: 1 mL, gauge 26, 1/2 in. needle.
7. Vacuum system.
8. Sodium hydroxide solution: 0.1 M NaOH. Weigh 40 mg NaOH with the analytical microbalance and dissolve in 10 mL water in a 15 mL conical tube. Disperse by sonication and filter the solution, if necessary, by using a 0.2 μm filter. Store the stock solution at room temperature.
9. Ultraviolet curable glue (e.g., Norland Products Inc.).
10. Methanol.

2.2 Fluidic Flow Testing and Visualization

1. Nikon Eclipse TE-2000U inverted epi-fluorescence microscope with 4×, 10× or 20× objectives and appropriate fluorescence filter cubes. Other fluorescence microscope systems of similar performance can be used. An attached CCD camera and image processing software is also needed for accurate flow visualization in the microchannels.
2. LabSmith microfluidics high voltage power supply. Other microfluidics power supplies of similar performance can be used.
3. Picoammeter (e.g., Keithley 6487).
4. Digital multimeter (e.g., Fluke 179).
5. Amber glass sample vials: 20 mL.
6. Buffer solution: 1 mM sodium tetraborate. Weigh 3.81 mg of $Na_2B_4O_7{\cdot}10H_2O$ with the analytical microbalance and dissolve in 10 mL deionized water in a 15 mL conical tube. Disperse by sonication and filter the solution, if necessary, by using a 0.2 μm filter. Store the stock solution at room temperature (*see* **Note 1**).
7. Rhodamine B solution: 10 μM. Weigh 47.9 mg Rhodamine B with the analytical microbalance and dissolve in 10 mL of the 1 mM sodium tetraborate solution, in an amber glass vial. Disperse by sonication and filter, if necessary, by using a 0.2 μm filter. Prepare fresh before use.

2.3 Electrophoretic Separation

1. LASER, e.g., 125 mW Argon ion tunable laser, Melles Griot with tunable emission wavelengths between 454 and 514 nm.
2. A single-point laser-induced fluorescence (LIF) detection setup with appropriate optical and mechanical components [9].

3. FITC-labeled arginine solution: 30 μM. Prepare by adding 156.6 mg of arginine and 116.7 mg of fluorescein isothiocyanate (FITC) to a 10 mL solution of the 1 mM sodium tetraborate buffer and allow the reagents to react at 37 °C for 1 h. Disperse by sonication and filter, if necessary, by using a 0.2 μm filter. Prepare fresh before use.
4. FITC-labeled glutamic acid solution: 30 μM. Prepare by adding 132.3 mg of glutamic acid and 116.7 mg of FITC to a 10 mL solution of the 1 mM sodium tetraborate buffer and allow the reagents to react at 37 °C for 1 h. Disperse by sonication and filter, if necessary, by using a 0.2 μm filter. Prepare fresh before use.
5. FITC-labeled glutamine solution: 30 μM. Prepare by adding 131.4 mg of glutamic acid and 116.7 mg of FITC to a 10 mL solution of the 1 mM sodium tetraborate buffer and allow the reagents to react at 37 °C for 1 h. Disperse by sonication and filter, if necessary, by using a 0.2 μm filter. Prepare fresh before use.
6. Amino acid sample 1: Mix the FITC-labeled arginine, glutamic acid, and glutamine solutions using a Thermoshaker (e.g., Eppendorf ThermoMixer® C) to prepare a 30 mL solution of the amino acid sample 1.
7. Amino acid sample 2: Mix the FITC-labeled glutamic acid and glutamine solutions using a Thermoshaker (e.g., Eppendorf ThermoMixer® C), and add 10 mL of the 1 mM sodium tetraborate buffer to the mixture to prepare a 30 mL solution of the amino acid sample 2.

2.4 Data Analysis

1. MATLAB with Curve Fitting toolbox for fitting Gaussian curves to the analyte peaks recorded on the electropherograms and subsequently estimating their migration times as well as standard deviations.

3 Methods

3.1 Microfluidic Chip Preparation

1. Prepare microfluidic chips in-house [3, 5] or purchase from various manufacturers, e.g., Micronit, Micralyne, Agilent, etc. We note that specialized microchip designs that allow on-chip pressure generation must be fabricated, e.g., with shallow segments at appropriate locations in the microfluidic network (*see* Figs. 1 and 2).
2. Attach the cylindrical glass tubes at the channel ports to serve as liquid reservoirs. Apply the ultraviolet curable glue to the bottom rim of the glass tube and then place it around the channel port. Place the microchip with the glass tubes under

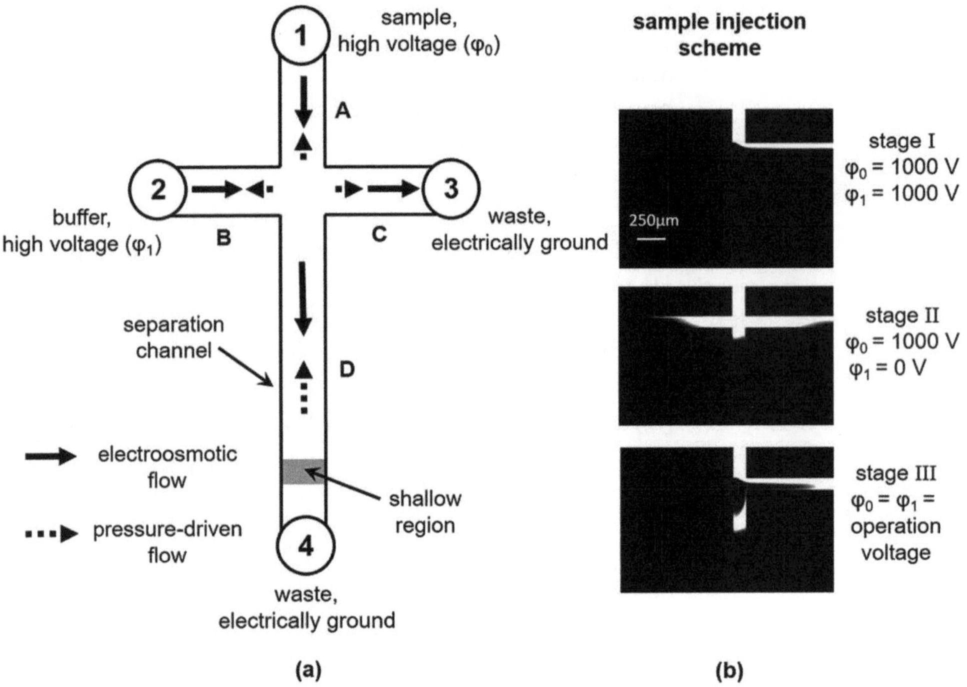

Fig. 1 (**a**) Schematic of the microchip device used for performing CZE analysis with a steady pressure-driven backflow. Notice that the magnitude of the pressure-driven flow in segments A, B, and C is expected to be about a third of that in segment D in this design. (**b**) Gated injection scheme employed for introducing sample plugs into the separation channel of the microchip device shown in (**a**). The fluorescence images presented above were obtained using a 10 μM Rhodamine B sample prepared in a 1 mM sodium tetraborate buffer applying $\varphi_0 = \varphi_1 = 1$ kV in stage III of the injection process (Reproduced with permission from ref. 3, Copyright©2012, American Chemical Society)

ultraviolet lamp (e.g., Blak-Ray UV lamp from UVP, LLC) for 30 min (*see* **Note 2**).

3. Clean/re-hydrate the microfluidic channels with the NaOH solution. Dispense the solution with the aid of a hypodermic syringe in one reservoir, and allow the channels to fill by capillary action. Fill the chip from only one reservoir to avoid trapping gas bubbles in the microchannels. Apply a slight vacuum from the other reservoirs to help the NaOH solution flow through the channels. Leave the NaOH solution in the channel for 15 min (*see* **Notes 3** and **4**).
4. Remove the NaOH solution from all reservoirs and channels through vacuum suction, and rinse the chip with copious amounts of deionized water. Next, fill the channels with water, as described above, and rinse sufficiently to completely remove the NaOH solution. Check with the pH in the chip reservoirs with pH paper.

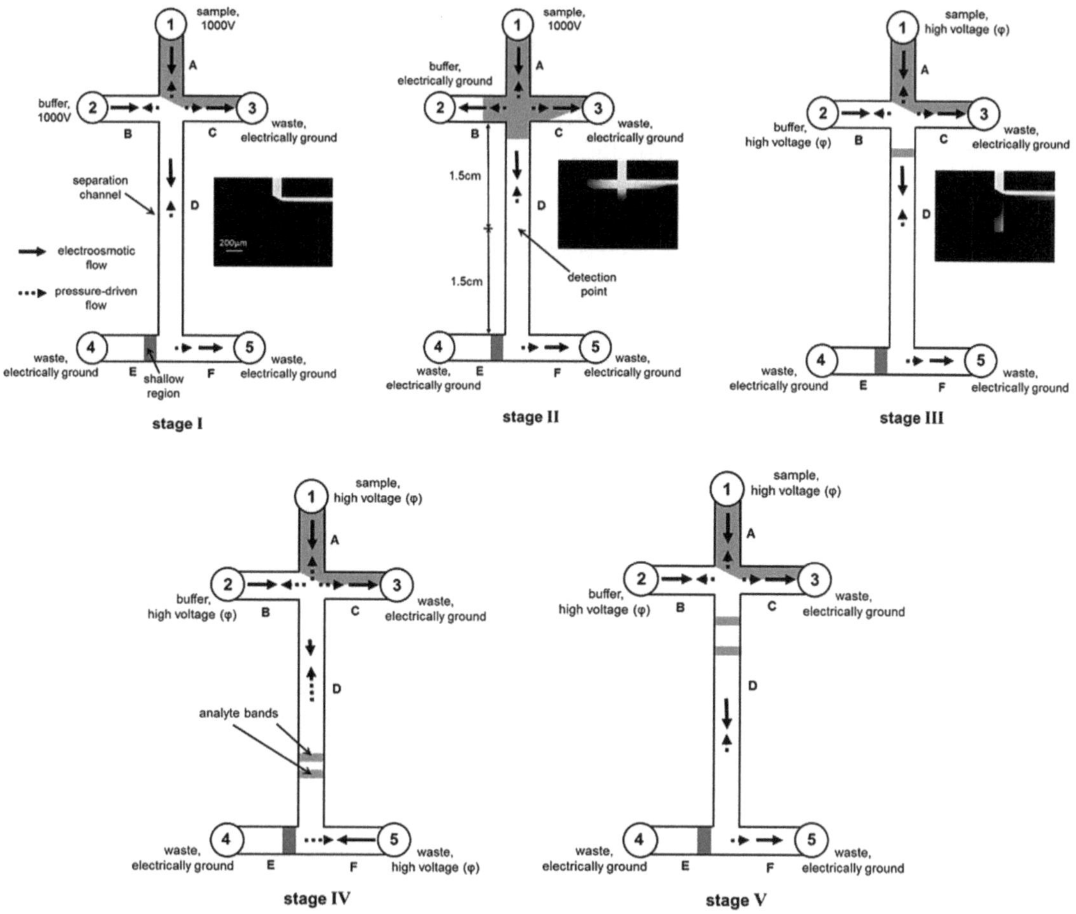

Fig. 2 Schematic of the microchip device used for performing flow counterbalanced capillary electrophoresis with a periodic pressure-driven backflow. Notice that the magnitude of pressure-driven backflow in segment D was expected to be about 23% of that in segment F in this design. The fluorescence images presented above were obtained using a 10 μM Rhodamine B sample prepared in a 1 mM sodium tetraborate buffer applying $\varphi = 1$ kV in stage III of the injection process (Reproduced from ref. 5 with permission from The Royal Society of Chemistry)

5. Remove the water from all reservoirs and fill the reservoirs and channels with methanol. Remove the methanol from all microfluidic channels through vacuum suction and allow the chip to dry.

3.2 Fluidic Flow Testing and Optimization

1. Fill the chip with the buffer solution, as described in Subheading 3.1, **step 3**.
2. Visually check the chip under a benchtop microscope to ensure that air bubbles are not trapped in the microfluidic channels or reservoirs.

3. Secure the chip on the epi-fluorescence microscope, for example, using paper tapes, and introduce Pt electrodes connected to the terminals of a high-voltage power supply in all reservoirs.
4. Test for electrical continuity by applying a low electric field strength, e.g., 100–250 V/cm, between various chip reservoirs and measure the electrical current through the microfluidic channels. This can be accomplished by applying the voltage to one reservoir and grounding the opposite reservoir through the picoammeter.
5. If the current measurements are satisfactory and reproducible, increase progressively the field strength to the maximum value that will be used during analysis (e.g., ~1000 V/cm), repeat the above current measurements, and check reproducibility (*see* **Notes 5** and **6**).
6. Fill the sample reservoir shown in Figs. 1 and 2 with the 10 μM Rhodamine B solution (*see* **Note** 7).
7. Perform a gated injection [3, 5] of Rhodamine B by applying the appropriate high voltages to the corresponding reservoirs (*see* Figs. 1 and 2) (*see* **Note 8**).
8. Optimize the sample injection/analysis voltages and check again the electrical currents between different reservoirs.

3.3 Separation Process

1. Fill the appropriate amino acid sample in the sample reservoir (*see* **Note 9**).
2. Place the microchip on the XYZ stage in the LIF detection setup, and secure with a paper tape. Adjust the position of the stage and the optical components to maximize signal recorded by the detector, e.g., photomultiplier tube (Hamamatsu Photonics) (*see* **Notes 10** and **11**).
3. Apply the optimized voltages (*see* Subheading 3.2, **step 8**) to the chip reservoirs and record the separation peaks on an electropherogram.
4. Perform multiple separation runs for a chosen operating voltage to estimate the reproducibility in the assay.
5. Next, raise all voltages, proportionally, and record all corresponding electropherograms to optimize the separation voltage (*see* Figs. 3 and 4).
6. Extract the electropherograms from the LIF detection setup as an ASCII file (e.g., .txt file), and analyze the data using MATLAB.
7. Estimate the peak variance and resolution to assess separation performance.

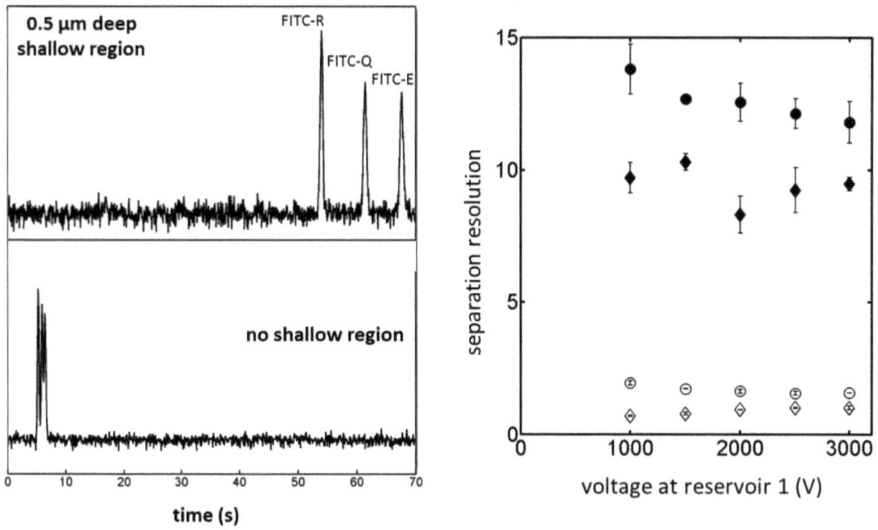

Fig. 3 (**a**) Enhancement in the CZE analysis of a FITC-labeled amino acid sample using the reported microfluidic device. The symbols R, Q, and E in the electropherograms presented above refer to the amino acids arginine, glutamine, and glutamic acid, respectively. (**b**) Measured separation resolution for the amino acid peaks shown in (**a**). The symbols ● and ◆ here refer to the separation resolutions between the FITC-R, FITC-Q, and FITC-Q, FITC-E peaks, respectively. The open and filled symbols here correspond to measurements made in separation channels without a shallow region and with a 0.5 μm deep one, respectively. The error bars included in figure (**b**) are based on five measurements (Reproduced with permission from ref. 3, Copyright©2012, American Chemical Society)

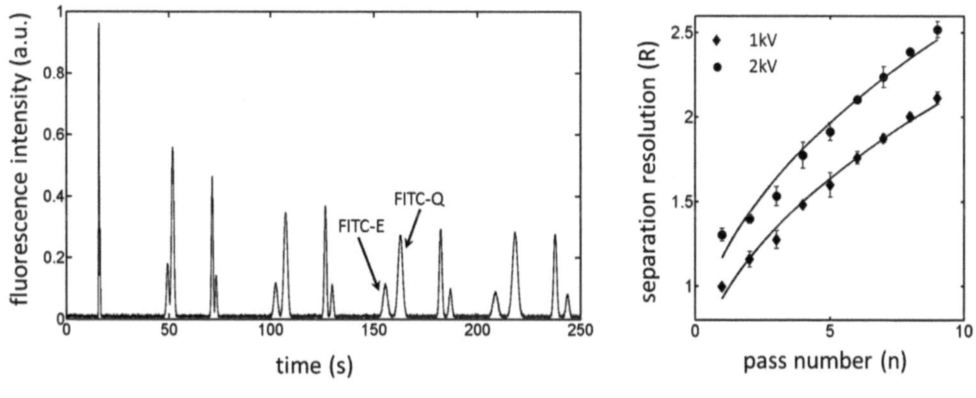

Fig. 4 Separation of a FITC-labeled amino acid sample in the reported FCCE device. The symbols Q and E here refer to glutamine and glutamic acid, respectively. The solid lines in the separation versus pass number graph are the best-fitted curves to the experimental data and can be described by the equations $R = 0.349 + 0.616\sqrt{n}$ (1 kV case) and $R = 0.522 + 0.596\sqrt{n}$ (2 kV case). The error bars for the separation resolution here represent the standard deviation in this quantity based on five measurements (Reproduced from ref. 5 with permission from The Royal Society of Chemistry)

4 Notes

1. The choice for the buffer concentration offers a balance the magnitude of the pressure-driven backflow and separation resolution. For a given buffer pH and applied voltage, while the pressure-driven flow velocity is observed to increase with a reduction in the buffer concentration, stronger adherence of the analyte molecules to the channel walls are seen under these conditions which contributes to band broadening and asymmetry compromising the separation resolution. For buffers with ion concentrations less than 1 mM, the pH also tends to drift toward that of deionized water, i.e., 7.
2. Any ultraviolet curable glue used for attaching the cylindrical glass tubes to the channel ports should be evaluated for possible bleeding of components that can alter electroosmotic flow rates, increase adherence of the analyte molecules to the channel walls, and/or block liquid/analyte transport in the fluidic ducts.
3. Do not allow the chip to dry out while filled with the NaOH solution, as this may result in channel clogging. Replenish the reservoirs continuously with solution while the chip is treated with NaOH, and/or preserve the chip in a water-saturated chamber.
4. Alternatively, if unions are bonded to the chip, the channels can be flushed by using a 250 μL syringe connected via a (50–100) μm i.d. × 360 μm o.d. fused silica capillary. The syringe can be actuated manually or by a syringe pump.
5. For channels that are ~5 μm deep and 100 μm wide, the electrical current should be ~0.5–5 μA for a 10 mM sodium tetraborate buffer.
6. Avoid the use of high electric field strengths that produce high current, Joule heating, bubble formation, and possible catastrophic failure of the experiment.
7. The concentration of the Rhodamine B solution that will be injected in the chip should be sufficient to enable visualization with the available fluorescence microscope.
8. Rhodamine B is a fluorescent dye that at neutral and high pH values has a zwitterionic structure. When neutral, it can be used as a tracer dye to determine the direction and magnitude of fluid flow. At low pH however, Rhodamine B can acquire a positive charge and, therefore, in addition to its electroosmotic transport can also display electrophoretic migration.
9. Wash the chip sequentially with methanol and the buffer solution between working with the Rhodamine B and amino acid sample in the sample reservoir.

10. It may be convenient to do a preliminary optimization of the position of the XYZ stage and optical components using a similar microchip with its channels filled with the Rhodamine B solution prior to working with the actual separation device.
11. It is recommended to visualize the separation process using the epi-fluorescence microscope first prior to operating the microchip with the single-point LIF detection setup.

References

1. Jorgenson JW, Lukacs KD (1981) Capillary zone electrophoresis. Science 222:266–272
2. Polson NA, Hayes MA (2000) Electroosmotic flow control of fluids on a capillary electrophoresis microdevice using an applied external voltage. Anal Chem 72:1088–1092
3. Xia L, Dutta D (2012) A microchip device for enhancing capillary zone electrophoresis using pressure-driven backflow. Anal Chem 84:10058–10063
4. Culbertson CT, Jorgenson JW (1994) Flow counterbalanced capillary electrophoresis. Anal Chem 66(7):955–962
5. Xia L, Dutta D (2013) Microfluidic flow counterbalanced capillary electrophoresis. Analyst 138:2126–2133
6. Taylor GI (1953) Dispersion of soluble matter in solvent flowing slowly through a tube. Proc R Soc (London) 219:186–203
7. Henley TH, Wilburn RT, Crouch AM, Jorgenson JW (2005) Flow counterbalanced capillary electrophoresis using packed capillary columns: resolution of enantiomers and isotopomers. Anal Chem 77:7024–7031
8. Yanagisawa N, Dutta D (2010) Pressure generation at the junction of two microchannels with different depths. Electrophoresis 31:2080–2088
9. Fu JL, Fang Q, Zhang T, Jin XH, Fang ZL (2006) Laser-induced fluorescence detection system for microfluidic chips based on an orthogonal optical arrangement. Anal Chem 78:3827–3834

INDEX

A

Ablation
  rate 48, 49
  threshold fluence 48
Accelerated Nuclear DNA Equipment (ANDE) 208
Advection 146, 147, 151, 154, 157, 169, 172, 187
Advection-diffusion equation 146, 151, 157, 169
Affinity 87, 90, 96, 145
Ampholyte 114, 117, 118, 121
Analyte 55, 67, 87, 100, 114, 126, 133, 144, 167, 197, 239
Anolyte 114, 118, 120

B

Band broadening ....v, 184, 241, 248
Boundary conditions 157, 170, 172, 180, 188
Bovine serum albumin (BSA) 135, 136, 138–141, 228, 231, 232
Buccal
  gDNA extraction 210
  swab 208–210, 214, 216, 220
Buffer
  borate 58, 60, 72, 75, 200
  Britton-Robinson 114
  high-conductivity 72, 127, 128
  4-(2-hydroxyethyl)-1-piperazineethanesulfonic acid (HEPES) 72, 75, 76
  ion 100, 103, 108
  leading 100, 102, 103, 108, 145
  low-conductivity 127, 128, 151, 158
  pH 114, 118, 128, 168, 226, 248
  phosphate 72, 75, 127, 200, 232
  run 58, 59, 200
  strength 149
  terminating 100, 102, 103, 108
  tris(hydroxymethyl)amino-methane (Tris) 72, 88–93, 127, 200
Buffered oxide etchant 4–6, 9
Burgers' equation 153

C

Capacitively coupled contactless conductivity detector (C^4D) 134
Capillary
  electrochromatography (CEC) 145, 146, 161
  electrophoresis (CE) 1, 55, 66, 79, 87, 89, 91, 94, 95, 197, 202, 207, 210, 212, 219
  isotachophoresis (CIETP) 145
  molding 27, 31, 43, 44, 51
Catholyte 114, 118
Cetyltrimethylammonium bromide (CTAB) 76
Charge coupled device (CCD) 59, 71, 114, 131, 210, 212, 214, 219, 220, 222, 227, 232, 242
Charge-to-size ratio 87
Chrome mask
  Etchant 4, 6, 7
Collision induced dissociation (CID) 232
Computer numerical control (CNC) micromachining 21, 22, 34, 38–40
Concentration 16, 45, 67, 72–76, 80, 88, 94, 99–102, 106, 109, 118, 123, 128, 131, 133, 135, 139, 140, 144–146, 149, 151, 153, 155, 157, 161, 162, 169, 179, 181, 184, 194, 201, 202, 234, 248
Conductivity
  contactless 134
  detection 135, 136, 138
Creatinine 135, 136, 138–140
Crystallite melting point 16
Cyclic olefin copolymers (COC) 1, 84

D

Debye
  Hückel 67, 75, 158
  layer 143, 158, 185
  length 144, 146, 156, 158, 159
Deep reactive-ion etching (DRIE) 126, 128
Detection
  absorbance 95, 145
  amperometric 79, 204
  electrochemical 79–84, 198, 199, 203
  fluorescence 57, 59, 95, 134, 214
  mass spectrometry 79, 225

Debashis Dutta (ed.), *Microfluidic Electrophoresis: Methods and Protocols*, Methods in Molecular Biology, vol. 1906, https://doi.org/10.1007/978-1-4939-8964-5,

Diffusion
coefficient 169, 170, 172, 175, 179
length 170, 203
time 151, 157, 184
Dirichlet problem 159
Dispersion
anomalous 147, 151, 156, 158, 160, 161
electrodynamic 172, 176–183
electromigration 151
hydrodynamic 170, 172–175, 177, 184, 191, 192, 241
Taylor-Aris 156, 158, 160, 161, 172, 176, 177, 191
DNAscan 208

E

Electric
charge 143
current 67, 75, 83, 147
double layer 144
field 62, 67, 73, 74, 87, 100, 120, 125, 143–146, 149, 151, 159, 160, 164, 167–169, 173, 177, 179, 185, 188, 189, 193, 198, 202, 203, 230, 234, 240, 241, 246, 248
permittivity 144, 146
voltage 50, 56–59, 61, 62, 66, 67, 71, 73, 76, 79, 80, 83, 84, 95, 103, 108, 115, 120, 121, 128, 133, 135, 136, 138, 199, 202, 204, 210, 214, 227, 229–231, 235, 242, 246
Electrode
pencil 134–137
platinum (Pt) 71, 73–76, 126, 138
reference 81–83
working 82, 83
Electrolyte
background (BGE) 80, 88, 136, 198
Electroosmotic flow (EOF) 57–62, 65, 67, 69, 72–75, 77, 79, 87, 89, 96, 146, 152, 153, 160, 172, 178, 194, 198, 203, 226, 231, 232, 234, 235, 240
Electropherogram 89, 90, 139, 140, 212, 221, 243, 246, 247
Electrophoresis
capillary 1, 55, 66, 79, 87, 89, 91, 94, 95, 114, 143, 197, 202, 207, 210, 212, 219, 245
gel 133
Electro-spray ionization (ESI)
emitter 226, 229, 231, 232, 234, 235
source 226, 228, 232
voltage 226, 232
Elution 87, 211, 217
End-labeled free Solution electrophoresis (ELFSE) 144
Explosives 90
External calibration method (ECAL) 101, 109

F

Faceted design 152
Field-amplified sample stacking (FASS) 65–68, 72–76
Flow Counterbalanced Capillary Electrophoresis (FCCE) 240, 247
Fluorescence quantum yield 77
Free flow
electrophoresis (FFE) 113, 167
isoelectric focusing (FFIEF) 113, 176
zone electrophoresis (FFZE) 170, 172, 176–179, 187
Function
bessel 144, 158
complementary error 154
dirac delta 153
Haarhoff-Van der Linde (HVL) 156

G

Gated injection 65–67, 73–75, 83, 219, 231, 244, 246
Gaussian
curve 243
distribution 49
fit 243
peak 60
shape 49
voltage 130
GeneMarker software 220
Glass
borosilicate 2, 7, 82, 84, 116
soda-lime 2
Glass transition temperature 15, 16, 38, 46
Gradient
conductivity 127, 128, 130
electrophoretic 125
focusing 128, 130
pH 113, 114, 123, 127, 128, 145, 176, 194
pressure 159, 172, 178, 183, 241
voltage 130

H

Helmholtz-Smoluchowski 146
High salt stacking 91, 95
Hot embossing 16, 19, 34, 38, 46
Hydrodynamic voltammograms 204, 205
Hydrophilic 215

Hydrophilize ... 72
Hydrophobic ... 69, 89, 96

I

Injection molding ... 16, 20, 34, 39
Instrumentation ... v, 56, 82, 94, 125, 133, 135, 198, 199, 240
Internal standard method (ISTD) ... 101–103, 105, 106, 108–110
Ion beam etching ... 25, 29, 30, 36, 43, 50
Isoelectric
  Focusing (IEF) ... 113
  point (pI) ... 113
Isotachopherograms ... 107, 109
Isotachophoresis (ITP) ... 99–103, 105, 106, 109, 110

J

Joule heating ... v, 72, 75, 95, 147, 158, 168, 169, 179, 184–193, 234, 239, 248

K

Kirchhoff's rules ... 62

L

Lab on a chip (LOC) ... 197
Langmuir model ... 162
Laplacian operator ... 159
Large-volume sample stacking with electroosmotic flow pump (LVSEP) ... 65, 67, 69, 71–74, 76, 77
Laser
  argon ion ... 59, 60
  CO_2 ... 21, 48, 136, 141, 213
  excimer ... 48
  Nd:YAG ... 48
  photoablation ... 21, 34, 40, 48
Laser-induced fluorescence (LIF) ... 70, 71, 198, 199, 242, 246, 249
Luer connectors ... 122

M

Mass
  spectra ... 232–236
  spectrometry (MS) ... 79, 121, 198, 199, 225–236
Metal deposition system ... 80
Method-of-moments ... 167, 169, 173, 174, 182, 186, 189, 193
Micellar electrokinetic chromatography (MEKC) ... 69, 87–90, 94, 145
Micelle ... 69, 73, 76, 77, 87, 89, 96, 145
Microchip
  bonding ... 7, 9
  electrophoresis (ME) ... 55, 59, 65, 80, 81, 83, 99, 135, 137, 197–205
  isotachophoresis (μITP) ... 99
Micro contact printing ... 14, 46
Microfluidic ... v, vi, 1–11, 20, 21, 23–25, 56, 60, 61, 65, 113–123, 149, 151, 156, 173, 185, 187, 197, 199, 203, 207–223, 225–236, 239–244, 247
Microfluidic-based Integrated rapid DNA Analysis System (MiDAS) ... 207–223
Microfluidic paper-based analytical devices (μPADs) ... 134
MicroITP software ... 106
Micro stereolithography ... 30, 32, 36, 44, 51, 52
Micro total analysis systems (μTAS) ... 55, 79
Micro transfer molding ... 15, 34, 37
Mobility ... 60, 67, 75, 89, 100, 102, 103, 105, 143–145, 147, 154, 156, 170, 179, 181, 198, 200, 234, 235

N

Nanochannel ... 29, 31, 125–131
Nanofluidic ... 126, 131
Nanoporous glass junction ... 229
Nitroaromatic ... 90–93
Number of theoretical plates ... 147, 160

O

Off-stoichiometric thiol-ene polymers (OSTE) ... 84
One-variable-at-a-time (OVAT) ... 203
Online
  pre-concentration ... 203
Orientation factor ... 172, 175, 177, 192

P

Paper ... 14, 47, 133–141, 149, 230, 246
Peak
  asymmetry ... 154
  height ... 74–77, 92, 93
  tailing ... 163
  width ... 92, 93, 147, 202
Péclet numbers ... 151, 154, 170, 172, 180, 181, 187, 192
Photoinitiator ... 116, 117
Photolithography ... 14, 27, 41, 58, 84, 127–129, 199
Photomask ... 2–6, 9, 35, 40, 116, 117, 122, 126, 127

Photoresist
developer ... 4
spinner ... 3, 5, 9
Picoammeter ... 227, 230, 242, 246
Pinched injection (PI) ... 57, 58, 74, 231, 234
Plasma etching ... 25, 27, 29, 35, 36, 42, 43, 128
Plateau mode ITP ... 100, 101
Plate height ... 147, 149
Pleated channels ... 152
Point-of-analysis (POA) ... 207
Polyamide (PA) ... 16
Polycarbonate (PC) ... 46, 84, 209, 213
Poly(dimethylsiloxane) (PDMS) ... 1, 15, 25, 29, 34–38, 42–45, 50, 51, 60, 71, 72, 75, 80–82, 84, 198
Polyelectrolyte ... 144
Polyether ether ketone (PEEK) ... 25, 29, 35, 36, 42, 50, 210, 214, 217, 229
Polyimide coating ... 232, 235
Polymerase chain reaction (PCR) ... 207–209, 211, 212, 214–217, 219–222
Polymethyl methacrylate (PMMA) ... 1, 16, 20, 22, 23, 25, 26, 28, 31, 34–36, 39, 41, 42, 50, 84, 198
Polyoxymethylene (POM) ... 16, 25, 34
Polystyrene (PS) ... 1, 34, 39, 103, 107
Polysulfone (PSU) ... 16, 34, 215
Poly(vinyl alcohol) (PVA) ... 72, 75, 77
Pressure-driven flow ... 167–169, 172, 177, 182, 184, 186, 187, 191, 192, 240, 241, 244
Printed circuit board (PCB) ... 214, 215
Proteome Discoverer ... 229
Pseudostationary phase ... 69, 76

R

Race-track effect ... 149, 151
Radius of curvature ... 151
RapidHIT ... 208
Relative standard deviation (RSD) ... 110, 232
Relative step height (RSH) ... 100, 101, 107, 109
Replica molding (REM) ... 14, 37, 45
Resolution ... 5, 22, 30, 33, 52, 76, 77, 114, 116, 125, 133, 147, 158, 168, 202–204, 235, 241, 246–248
Response Surface Methodology (RSM) ... 203
Retention factor ... 76

S

Sample
concentration ... 74, 151, 169, 201
detection ... vi, 70, 101, 233
injection ... v, 55, 65, 68, 92, 103, 134, 157, 231, 232, 246
loading ... 154, 219
matrix ... 67, 72, 76, 77, 87–93, 95, 96
pH ... 67
pre-concentration ... 203, 207
variance ... 176
Selectivity ... 79, 92
Self-assembled monolayers (SAM) ... 14, 16, 37
Sensitivity ... 65, 68, 77, 79, 87, 92, 96, 114, 121, 125, 133–135, 151, 198–200, 202, 203, 226, 233
Separation
buffer ... 94, 168, 200, 225, 240
resolution ... 235, 241, 247, 248
time ... 89, 95, 204, 241
voltage ... 67, 89–93, 95, 204, 205, 246
Sheath flow ... 114, 118, 121
Short tandem repeat (STR) ... 208, 209, 217
Signal spikes ... 232, 235
Signal to noise (S/N) ... 204, 205
Silicon ... 1, 15, 31, 34, 35, 37, 38, 40, 46, 80, 84, 126–130, 231
Sodium dodecyl sulfonate (SDS) . 69, 72, 73, 75–77, 96, 145, 205
Solvent-assisted micromolding (SAMIM) ... 15
Stationary phase ... 145, 146
Stokes-Einstein relation ... 156, 186
Stream broadening ... 167, 169, 173, 174, 182, 186, 189, 193
Sudan III ... 89, 90, 96
Surfactant ... 73, 76, 87, 92–94, 96, 205
Sutherland's formula ... 147, 149
Sweeping ... 65, 69, 70, 73, 75–77, 91, 95

T

Taylor cone ... 232
Theory
band broadening ... 143, 241
lubrication ... 158, 161, 164
Taylor-Aris ... 157
Thermoelectric cooler (TEC) ... 210
Total ion chromatogram (TIC) ... 232, 233, 236
Turn angle ... 151

U

UV
absorbance ... 95, 145
light ... 9, 22, 23, 35, 36, 42, 43, 52, 81, 82, 117
patterning ... 23, 25, 26, 35, 41, 42

V

Van Deemter equation 174
Viscosity 27, 44, 50, 52, 56, 144, 146, 147, 149, 179, 186, 189, 191

W

Wall adsorption kinetics 161

X

X-ray lithography (XRL) 22

Z

Zeta potential 75, 144, 146, 153, 158–162, 164, 240
Zwitterion 113, 234, 248

MIX
Papier aus verantwortungsvollen Quellen
Paper from responsible sources
FSC® C105338

If you have any concerns about our products,
you can contact us on
ProductSafety@springernature.com

In case Publisher is established outside the EU,
the EU authorized representative is:
Springer Nature Customer Service Center GmbH
Europaplatz 3, 69115 Heidelberg, Germany

Printed by Libri Plureos GmbH
in Hamburg, Germany